SKY AND SYMBOL

SKY AND SYMBOL

The Proceedings of the Ninth
Annual Conference of the Sophia Centre
for the Study of Cosmology in Culture,
University of Wales, Trinity Saint David,
4–5 June 2011

Edited by Nicholas Campion and Liz Greene

SOPHIA CENTRE PRESS

First published in 2013.

Sophia Centre Press
University of Wales, Trinity Saint David,
Ceredigion, Wales SA48 7ED, United Kingdom.
www.sophiacentrepress.com

Publisher's Cataloging-in-Publication
(Provided by Cassidy Cataloguing Services, Inc.).

Names: Sophia Centre for the Study of Cosmology in Culture. Annual Conference (9th : 2011 : Bath, England), author. | Campion, Nicholas, editor. | Greene, Liz, editor.

Title: Sky and symbol : the proceedings of the Ninth Annual Conference of the Sophia Centre for the Study of Cosmology in Culture, University of Wales, Trinity Saint David, 4-5 June 2011 / edited by Nicholas Campion and Liz Greene.

Description: Ceredigion, Wales, United Kingdom : Sophia Centre Press, 2013. | Series: Studies in cultural astronomy and astrology ; vol. 4 | Includes bibliographical references and index.

Identifiers: ISBN: 978-1-907767-03-6 (paperback) | 978-1-907767-53-1 (ebook)

Subjects: LCSH: Archaeoastronomy--Congresses. | Signs and symbols--Religious aspects--Congresses. | Cosmology--Congresses. | Sky--Congresses.

Classification: LCC: GN799.A8 S66 2013 | DDC: 520.9--dc23

Printed worldwide by Ingram Spark.

Contents

IMAGES

Acknowledgements

We would like to thank the Sophia Trust for its generous funding of the Sophia Centre, and Jennifer Zahrt for her exemplary editing skills.

SKY AND SYMBOL

INTRODUCTION

Nicholas Campion and Liz Greene

The papers included in this volume are all concerned, in one way or another, with various perceptions of the world of celestial phenomena as symbols. The papers range from art history and analytical psychology to critiques and accounts of astrology, and roam from western to non-western cultures, and from ancient to modern. 'Symbol', however, is one of those words whose definition and nature are subject to an ongoing debate. The argument has been running, sometimes quite heatedly, for many centuries, for it rests on one of the most fundamental questions: how we perceive and define reality. The debate is reflected today in a number of recent scholarly works, often deeply opposed to each other in methodology, but all devoted to explaining the nature of symbols in religion, art, and society. The ongoing debate also continues among practitioners of psychology and the arts, as these fields are particularly rich in what many would understand as symbolism.

The word 'symbol' is derived from the Greek *sumbolon,* derived from *sumballein,* which means 'to join together'. But what is it that is 'joined together' in order to generate a symbol? Different social connotations, different objects, or different levels of reality? Is the symbol an exclusively cultural product, shaped entirely by social factors and therefore exhibiting changeable meanings according to cultural context? Is it the image of an archetypal pattern or inherent cognitive 'representation' that transcends cultural differences and retains a consistent meaning despite its adaptability to various cultural milieux? Or is it a combination of these in varying degrees? And whatever a symbol is, are all astrologies, for example, ultimately 'symbolic systems'? Or are some literal and some symbolic? Historical, sociological, religious, artistic, and psychological approaches often arrive at conclusions that may be wildly contradictory; so what is the appropriate methodology to explore the nature of celestial symbols? It is these kinds of questions that hover in the background of the material covered in the papers that follow.

The papers, most of which were given at the University of Wales Trinity Saint David Sophia Centre Conference in Bath, UK, on 4–5 June, 2011, have been grouped into three sections.[1] The first section, containing two presentations and entitled 'The Nature of Symbols', is concerned specifically with the issue of what is meant by a symbol and how this relates to the portrayal and interpretation of celestial phenomena. The first paper, by Nicholas Campion, explores the issue of what astrologers mean—or think they mean—when they use the phrase 'symbolic language' to describe what they do, and whether some astrologies might be more 'literal' or 'veridical' than others. Dr. Campion traces current approaches to astrological symbolism back to the legacies of both Plato in the fourth century BCE and the Theosophical movement of the late nineteenth century, raising important questions about the ways in which astrological perceptions are coloured by particular cosmological and philosophical currents even when the practitioner is unaware of the heritage on which their work is based. The second paper, by Gary Wells, examines older representations of celestial phenomena in art as modes of expressing the ineffable and the literal simultaneously, and the striking contrast between this more inclusive understanding of the role and meaning of the visual image—as both pedagogic and inspirational, scientific and imaginal—and the post-Enlightenment imposition of apparently clear boundaries between what is 'symbolic' and what is 'real', resulting in profound changes in the collective understanding of both science and art.

Part Two, entitled 'Ancient, Medieval, and Early Modern Expressions', contains five papers, which examine celestial symbolism in very different cultural contexts that span a time-frame from the Mesopotamian world of the fourteenth century BCE to the alchemical-astrological explorations and experiments of sixteenth-century Western Europe. Ola Wikander focuses on mythological narratives as well as omen texts from Ugarit (in what is now modern Syria) in the second millennium BCE, considering the ambiguous role of

[1] Two of the papers included in this volume—those by Professor Elliot R. Wolfson and Dr. Peter Forshaw—were given at a conference entitled 'Imagining Astrology', which took place at the University of Bristol on 10–11 July, 2010.

the Sun and its presiding deity, in both benevolent and destructive forms, as a prototype of later symbolic understandings of the Sun in astrological texts. Andrea Lobel examines Jewish astrological symbolism and its derivations from Babylonian and Hellenistic cultures, considering the ways in which these older elements, previously considered 'foreign', were integrated and developed within a specifically Jewish religious and cultural framework. Darrelyn Gunzburg explores a particular medieval artifact from central Italy, the Fontana Maggiore at Perugia, which displays not only religious but also celestial imagery, demonstrating the ways in which the artistic expression of astrological images can serve as a bridge between human life and the heavens. Elliot Wolfson's paper concerns one of the most complex themes in esoteric literature: the 'astral body' and its role in the works of the Jewish Pietists and Spanish Kabbalists of the twelfth to thirteenth centuries. Professor Wolfson focuses specifically on the medieval Jewish idea of renouncing the body in order to cultivate a celestial vision resulting in a form of mystical transformation, raising the question of whether symbols in themselves have the power to transform consciousness. The last paper in this section, by Peter Forshaw, considers early modern alchemy with its attendant proliferation of astrological symbols in both texts and images, examining the complex relationship between these two discrete yet intimately related spheres of expression in the works of important historical figures such as Paracelsus, whose explorations into alchemy as a healing art continue to influence both complementary medical practice and depth psychology in today's world.

Part Three is entitled 'Celestial Symbols in the Contemporary World', and includes four papers examining the use of celestial symbolism in particular cultures in the twentieth and twenty-first centuries. Anthony Thorley explores the mysterious landscape phenomenon known as the 'Glastonbury Zodiac', discussing its understanding as both a literal and a symbolic expression and the influence on its interpretation that has arisen from the Theosophical doctrines of H. P. Blavatsky. Christel Mattheeuws discusses the perceptions of the landscape, including the sky, expressed by the indigenous people of Zanadroandrena Land in central east Madagascar, comparing their approach to the ways in which a horoscope is read as both a symbol of individual birth and an image of the origin of the world, linking personal, communal, and cosmological narratives in a single symbolic thread. Jennifer Zahrt examines celestial symbolism in the German silent cinema of the 1920s, revealing the filmmakers' ideas of the acquisition of magical power through understanding the heavens symbolically, and the ways in which these films attempted to express esoteric knowledge through the modern vehicle of the cinematic experience. The last paper, by Liz Greene, is focused on C. G. Jung's controversial private diary, published in 2009 and known as *The Red Book*. This paper explores Jung's profound involvement in astrology and its symbolism, his incorporation of a dynamic, psychological form of astrology in the textual and visual expressions of *The Red Book*, and the extent to which the importance he assigned to astrological symbols influenced and perhaps even definitively shaped many of his later psychological theories.

None of the papers in this volume attempts to assert an inarguable definition of what a symbol is. All the papers raise as many questions as they attempt to answer, and reveal, through historical, sociological, and philosophical investigation into a number of different cultures and periods of history, the manifold ways in which the symbols of the heavens have been understood and incorporated, visually, ritually, and through narrative, into virtually every aspect of human life.

PAPERS FROM THE 'SKY AND SYMBOL' CONFERENCE NOT INCLUDED IN THIS VOLUME:

One paper from the conference was accepted for publication in *Culture and Cosmos*, Vol. 15, No. 2 (Autumn/Winter 2011): Pam Armstrong, 'Ritual Ornamentation: From the Secular to the Religious'. This paper explores the results of a survey into whether the ritualised use of domestic ornaments suggests that the owner invests those objects with religious significance.

Four conference papers were not submitted for publication:

1. 'THE LOTUS AND THE SUNFLOWER: THE ROLE OF SYMBOL AND SOUL IN LATE ANTIQUE THEURGIC RITUAL':
Crystal Addey (Cardiff University and University of Wales, Trinity St. David)

It is indeed, in imitation of it [the light of the gods] that the whole heaven and cosmos performs its circular revolutions, is united with itself, and leads the elements round in their cyclic dance,...causes lasts to be joined to firsts, as for example earth to heaven, and produces a single continuity and harmony of all with all.

—Iamblichus, *De Mysteriis*, 1.9

This talk explores the role and uses of 'symbols' in theurgic ritual. Theurgy, which literally means 'god-working', was a type of ritual widely respected and extensively used in the late antique period (from the third century CE onwards) by Neoplatonist philosophers. Symbols (*symbola*) form the central nexus of theurgic ritual—yet these 'symbols' were not considered to have merely a metaphorical, representational power (a common meaning of 'symbol' in the contemporary world), but an inherent, ontological connection with divine realms. They were considered to be a direct and ineffable link with divine truth, operating on the level of a talisman. The paper investigates theurgic symbols, focusing particularly on the lotus and the sunflower, and explores their inherent connections with celestial phenomena such as the planets.

2. 'PLUTARCH'S COSMOLOGICAL DREAMS AND CELESTIAL SYMBOLISM':
Dorian Greenbaum (University of Wales Trinity St David)

Plutarch of Chaeronea may be best known for his *Lives*, but his essays (*Moralia*) are rich and creative forays into philosophy and religion. Within several of these essays are embedded visions of cosmological, astrological, and religious significance, giving insight into Plutarch's views on the nature of the soul, the mind, daimons, and God. This paper explores two of these visions, that of Thespesius in *On the Delays of Divine Vengeance*, and the Myth of Timarchus in *On the* Daimonion *of Socrates*. Both visions utilise the sky as their setting, and the paper examines why Plutarch chose the sky for this role, and explores the celestial metaphors imagined by Plutarch as he weaves his own visions of soul, mind and spirit.

3. 'INTRODUCING "THE SAXL PROJECT": A NEW DIGITAL RESEARCH RESOURCE':
Kristen Lippincott (Founding Member, The Exhibitions Team; former deputy Director National Maritime Museum and Director Royal Observatory Greenwich)

The Saxl Project is an on-line collaboration between Kristen Lippincott and the Warburg Institute (University of London), which will update and complete Fritz Saxl's landmark publications on illustrated astrological manuscripts from the Middle Ages and Renaissance (1915–66). The project was scheduled to go 'live' in January 2011 and the website is intended to provide full descriptions and illustrations from over 250 manuscripts.

4. 'THE SYMBOLISM OF THE SKY IN MEDIEVAL HEBREW TEXTS':
Ilana Wartenberg (University College, London)

This paper discusses the scientific as well as symbolic role of the heavens as portrayed in medieval Hebrew literature. Examples taken from scientific as well as non-scientific sources are provided, and their role in the larger context of medieval Jewish life is analysed, with a focus on poetic, astrological and calendrical treatises.

5. 'HEAVEN AS THE COINCIDENCE OF FIRE AND WATER: THE SYMBOL OF THE SKY IN MEDIEVAL JEWISH MYSTICISM':
Elliot R. Wolfson (New York University):

This paper explores the symbol of the sky in kabbalistic texts from Provence and Spain in the thirteenth-fourteenth centuries, focusing on the mystical appropriation of the mythical idea expressed in rabbinic literature concerning the constitution of heaven (*shamayim*) as a mixture of fire (*esh*) and water (*mayim*). From the perspective of the kabbalists, the heavens symbolize the coincidence of opposites that is expressive of the unity of God. All that takes place on earth is a mirror reflection of the heavens, which are a mirror reflection of the divine. Astrological knowledge thus provides the means to access the space of the double mirroring wherein we envision the image that is real.

PART ONE:

The Nature of Symbols

IS ASTROLOGY A SYMBOLIC LANGUAGE?

Nicholas Campion

ABSTRACT: It is common for western astrologers to rationalise astrology as a symbolic language. However, it is also widely believed that astrological claims are true; that astrology can, for example, make accurate forecasts. This paper questions what astrologers actually mean by the phrase 'symbolic language'. There are apparent contradictions in the astrological position: do astrologers really think symbolically? Do some, but not all uses of astrology rely on the interpretation of symbols? Does some astrology rely on the assumption of absolute, literal, truth, rather than the reading of symbols? And, are these approaches compatible? This study draws on previous work by Liz Greene and explores the understanding of symbolism through the western astrology of the English-speaking world, focusing on the legacy of Platonism and theosophy.

INTRODUCTION: SYMBOL AND CONFUSION

Central to an understanding of astrology's diversity in the western tradition is the distinction between 'Natural' and 'Judicial' Astrology, which originates in classical discourse in the 1st century BCE.[1] Natural astrology requires no more than the observation of seasonal phenomena and natural influences supposedly deriving from celestial motions. Judicial astrology, by contrast, depends on the astrologer's interpretative judgment, making complex deductions from horoscopes, which are schematic maps of the sky cast for a precise time, place and date. While the distinction between the two forms of astrology, Natural and Judicial, is not absolute, in this paper, by astrology I mean Judicial astrology in the western tradition.

Western astrology is widely defined as a symbolic language. The claim that this is so is propagated via writings

[1] Nicholas Campion, *Astrology and Cosmology in the World's Religions* (New York: New York University Press, 2012), p. 16.

by modern astrologers across the world-wide-web.[2] It is also supported by some of the leading astrologers of the twentieth century: in 1982 John Addey, former President of the Astrological Association, stated that astrology 'is a system for symbolism of a high order which can be most valuable'.[3] 'When we enter the realm of symbolism', he added, 'we open the field to the higher human faculties of reason and intuition and to science and philosophy in their true and integral sense'.[4]

The idea that astrology is symbolic also occurs in some academic quarters: Aldo Mazzucchelli introduced his argument on the matter thus: 'In the second part of this paper, after having characterised the astrological chart as a symbol, we shall develop the idea presented before: its interpretation may only be formulated through the use of metaphors'.[5] Dean and Mather, arguing from a sceptical perspective, agree, suggesting that, 'The great popularity of astrology is due largely to its symbolic and non-scientific nature, for this makes it instantly helpful in understanding, i.e. in areas of belief...Astrology is symbolism based on astronomy'.[6] John Hayes provides an example, speaking as an astrologer:

> **Key to Astrological Symbolism**—Axial Rotation and Orbital Revolution
>
> (1) Orbital Revolution—the Earth circles around the Sun and (2) Axial Rotation—the Earth rotates around its axis. Orbital

---

[2] Clare Goodwin, 'The Language of Astrology', 2000, http://www.abgoodwin.com/therapy-and-readings/language-of-astrology.html, accessed 4 November 2012.

[3] John M. Addey, 'Astrology as Divination', *Astrology* 56, no. 2 (1982): p. 41.

[4] Ibid.

[5] Aldo Mazzucchelli, 'Celestial Weathercock, Diagrams & Metaphorical Web: Some Semiotic Considerations on Western Astrology', *European Journal for Semiotic Studies* 12, no. 4 (2000), reproduced as 'Astrology, Hermeneutics and Metaphorical Web', http://cura.free.fr/xv/12mazz-e.html, (accessed 15 November 2012).

[6] Geoffrey Dean and Arthur Mather, *Recent Advances in Natal Astrology: A Critical Review 1900–1976* (Subiaco, Australia: Analogic, 1977), p. 3.

> Revolution and Axial Rotation give us a dualism of life-direction, a dualism of individual and collective.[7]

Hayes regards the astrological symbol as embodied in mathematical astronomy, and embedded in the fabric of the cosmos, whereas for Dean and Mather, it is representational and hence arbitrary, a difference of opinion which, I will argue, runs to the heart of the debate not just between astrologers and sceptics, but amongst astrologers themselves.

The most authoritative study of symbolism to date is that by Peter Struck, who identified the emergence of the notion of the symbol in classical Greece and its dual function both as representative of the thing symbolised, and as an embodiment of it.[8] The latter definition is significant for, if the symbol is an embodiment of the thing symbolised, both symbol and symbolised share a real, substantial, essential, connection, whether physically, psychically or both. Struck influenced Liz Greene, the author of the only academic inquiry into symbolism in astrology, from which this paper has developed.[9] Greene made three contributions to the debate, in relation to astrology, and which are relevant to this paper:

1. The existence of the notion of an astrological symbol can be identified in Hellenistic and late-Classical Neoplatonic literature
2. The emphasis on astrology-as-symbolism in modern literature is particularly due to the influence of theosophy.
3. The word 'symbol' possesses dual meanings as either (a) a representation of an object (the thing symbolised), with only an arbitrary connection to the object or (b) an embodiment of an object, in which the symbol and the object share an essential, real, connection, being embedded in the fabric of the cosmos. These two

---

[7] John Hayes, 'Introduction to the Symbolic Language of Astrology', http://www.johnhayes.biz/Articles/symbolicLanguageAstrology.htm, (accessed 4 November 2012).

[8] Peter Struck, *The Birth of the Symbol* (Princeton: Princeton University Press, 2004).

[9] Liz Greene, 'Signs, Signatures, and Symbols: the Languages of Heaven', in *Astrologies: Plurality and Diversity*, ed. Nicholas Campion and Liz Greene (Lampeter: Sophia Centre Press, 2010), pp. 17–45.

> definitions of 'symbol', as either representational or embodied, pervade modern astrological literature, although the distinction is usually unacknowledged. Greene pointed out that astrological writings tend to move between the two definitions but that the latter, the symbol as embodiment, has the dominant philosophical lineage, being evident in late-classical Neoplatonism. There is some ambiguity, though, in that a glyph, the shorthand figure used to indicate a planet, may be both representational and embodied.

The notion of symbol as glyph, that is, as visual shorthand, occurs throughout astrological literature. For example, Ivy Goldstein-Jacobsen, who was highly respected in post-war America, referred to, 'Mercury, whose symbol is the caduceus or wand of the physician'.[10] Notions of glyph as symbol were developed by Alan Leo in the 1890s and 1900s. He saw the structure of the planetary glyphs as symbols of different forces; the circle represents Spirit, the semi-circle symbolises the polarity of spirit-matter, or 'Self-Not-self', Night-Day, or any form of duality.[11]

For a Platonic astrologer symbols were necessarily embedded in the cosmos, simply because everything was interconnected and interdependent. This much was argued by Plotinus (204/5–270).[12] For Iamblichus (c. 245–c. 325) a symbol might be an image or an action and, in each case, it both represented the thing it symbolised, and shared the same essence with the thing it symbolised [13] However, while astrology may indeed be a symbolic language, astrologers rarely, as Greene observed, discuss the consequences of this statement, or what is meant by the term 'symbol'. Rare examples include discussions by Dane Rudhyar, who was deeply Platonic and theosophical, and Michael Harding, who

---

[10] Ivy Goldstein-Jacobsen, *Here and There in Astrology* (no place: no publisher, 1961), p. 36.

[11] Alan Leo, *Esoteric Astrology* (London: N. Fowler, 1925), pp. 2–3.

[12] Plotinus, 'On Whether the Stars are Causes', *Enneads*, Vol. 2, trans., A. H. Armstrong (Cambridge, MA: Harvard University Press, 1929), 8. 5–10; 13.11–13; 13.30–37; 15.10–15. .

[13] Iamblichus, *On the Mysteries of the Egyptians, Chaldeans, and Assyrians*, trans. Thomas Taylor (1821; repr., Frome: Prometheus Trust 1999), p. 301, notes 7–9.

rejected Platonism and theosophy as the rationale for astrology as a language.[14]

There is a suspicion amongst sceptics that the definition of astrology as symbolic is favoured by astrologers because it diverts critics of astrology away from the claim that astrology lacks either a credible mechanistic rationale or convincing scientific verification of its claims. For example, this statement is on several sceptical websites, copied from Phil Plait's 'Bad Astronomy' website: 'To avoid making testable claims about the driving forces of astrology, astrologers rely on mumbo-jumbo to bamboozle the public'.[15] Curry, citing Geoffrey Cornelius, has argued that social scientists typically practice what Cornelius calls 'a common avoidance strategy', which is, 'to avoid allowing the material to touch the observer as truth for the observer'.[16] This may well be a weakness of traditional social science's failure to interpret the world, but just because something has truth for the observer, does not mean that it has either any necessary value, or that it possesses truth for anyone else. Indeed for other people, the perceived truth may be false. The sceptical stance is therefore that astrologers' simultaneous reliance on the subjective truth of personal experience on the one hand, and the objective truth of astronomical measurement on the other, is itself an avoidance strategy, deflecting any further discussion of astrology's nature. This is a widely held view amongst anti-astrology sceptics and has occurred in the angry exchanges that pervade discussions on the 'Astrology' page on Wikipedia. The following was posted by a pseudonymous editor on 18 July 2012:

---

[14] Dane Rudhyar, *The Planetarisation of Consciousness* (New York: Aurora Press, 1977); Michel Harding, *Hymns to the Ancient Gods* (London: Arkana, 1992).

[15] http://www.rulit.net/books/bad-astronomy-misconceptions-and-misuses-revealed-from-astrology-to-the-moon-landing-hoax-read-233351-53.html, (accessed 15 November 2012).

[16] Patrick Curry, 'The Historiography of Astrology: A Diagnosis and a Prescription', in *Horoscopes and Public Spheres: Essays on the History of Astrology*, ed. Kocku von Stuckrad, G. Oestmann and H. Darrel Rutkin (Berlin: Walter de Gruyter, 2005), pp. 261–74, here p. 270; Geoffrey Cornelius, 'Verity and the Question of Primary and Secondary Scholarship in Astrology', in *Astrology and the Academy*, ed. Nicholas Campion, Patrick Curry, and Michael York (Bristol: Cinnabar Books 2004), pp. 103–13.

> The whole 'astrology isn't scientific' claim is an excuse, a case of special pleading so that people who hold ridiculous beliefs can avoid having their beliefs scrutinized. Astrology tried to be a science, and it failed without being left a single leg upon which to stand, so people who are alive today who either lack competence/intelligence and/or are emotionally attached to astrology claim that astrology isn't scientific and therefore isn't amenable to testing.[17]

Plait himself states that astrology is not a science, in spite of what he implies (mistakenly) are claims to the contrary by astrologers, but is based on analogies which are themselves false because they are not based on physical causes.[18] The astronomer Philippe Zarka agrees that, if astrology is a symbolic language, it is, of necessity, false.[19] The examples of analogy given by Plait, for example, that Mars is red and therefore angry and dangerous, are what might be called symbolic associations elsewhere. Plait's argument, however, is not quite as clear as he would have it, for modern science can also argue from analogy, or from symbol if the definitions of the two words overlap. [20] Argument from analogy or symbolic association is, therefore, not the problem, but testability is: as Chris French argued, 'Although analogies may provide a fruitful means of generating new and interesting hypotheses, the final verdict must always depend upon the results of direct, empirical tests of those hypotheses'.[21] Therefore, according to French, arguments

---

[17] Sædontalk 06:53, 18 July 2012 (UTC), http://en.wikipedia.org/w/index.php?title=Talk%3AAstrology&diff=prev&oldid=502919299

[18] Philip Plait, *Bad Astronomy: Misconceptions and Misuses Revealed, from Astrology to the Moon Landing 'Hoax'* (New York: Wiley, 2002), p. 215.

[19] Philippe Zarka, 'Astrology and Astronomy', in *The Role of Astronomy in Society and Culture*, International Astronomical Union Symposium 260, UNESCO, Paris, 19–23 January 2008, ed. David Valls-Gabaud and Alec Boksenberg (Cambridge: Cambridge University Press, 2008), pp. 420–25.

[20] Mary Hesse, *Models and Analogies in Science* (Notre Dame: University of Notre Dame Press, 1966); Paul Bartha, *By Parallel Reasoning: The Construction and Evaluation of Analogical Arguments* (Oxford: Oxford University Press, 2010).

[21] Chris French, 'Astrologers and other inhabitants of parallel universes', *The Guardian*, 7 February 2012.

from analogy should be tested scientifically, and either proven or disproven. A number of astrologers, such as Jeff Mayo—who will be discussed later—agree. There are subtle differences between Plait and French: whereas French argues that analogies should be empirically testable, Plait regards astrology as necessarily false because it is based on self-evidently erroneous premises. Plait's discussion, though, considers astrology independently of the Platonism which continues to exert a profound effect on its philosophy.[22] This may not be a problem if astrology's efficacy is at stake, but it is a weakness if astrology's historical context and cultural identity are to be understood. As John Robb pointed out, decontextualising the symbol is a means of making it seem ridiculous: 'By severing the use of symbols from their context of meanings, the informational view makes belief irrational and hence merely disadvantageous in followers and cynically optional in elites'.[23]

The widespread notion of astrology as a means of communication is evident in attempts to find similes for it. Roger Beck, who reminded his readers that, 'we should never forget that for the ancients the stars spoke', proposed that the word astrology be replaced by 'Star Talk', while Edgar Laird suggested 'Star Study'.[24] Ed Krupp's term, 'Sky Tales' (referring to constellation myths rather than astrology) may also be useful.[25] All three suggestions assume that astrology is a matter of narrative, of an account of the way the universe communicates, or the manner in which its truths are communicated, rather than of any fundamental 'laws' which underpin the stars' relationships with humanity.

---

http://www.guardian.co.uk/science/2012/feb/07/astrologers-parallel-universes?fb=native&CMP=FBCNETTXT9038

22 Nicholas Campion, *A History of Western Astrology*, Vol. 2. The Medieval and Modern Worlds (London: Continuum, 2009), pp. 252–53.

23 John E. Robb, 'The Archaeology of Symbols', *Annual Review of Anthropology* 27 (1998): pp. 329–46, here p. 334.

24 Roger Beck, *The Religion of the Mithras Cult in the Roman Empire: Mysteries of the Unconquered Sun* (Oxford: Oxford University Press, 2006), p. 153 and chap. 8; Edgar Laird, 'Christine de Pizan and Controversy Concerning Star Study in the Court of Charles V', *Culture and Cosmos* 1, no. 2 (Autumn/Winter 1997): pp. 35–48.

25 E. C. Krupp, 'Sky Tales And Why We Tell Them', in *Astronomy Across Cultures: the History of Non-Western Astronomy*, ed. Helaine Selin (Dordrecht: Kluwer Academic Publishers, 2000), pp. 1–30.

Another option is Star Stories, which can include the stories stars tell about humanity as much as the ones which people tell about the stars.[26] If any question of an external objective reality—the measurement of planetary cycles or celestial influences—is discounted, astrology can be seen as fundamentally a narrative, a discourse. Astrology may have its believers in the literal truth, and in the absolute objectivity of its truth-claims, but it may still function as a conversation, its participants being people—astrologers and their clients—and the cosmos—time, eternity, pattern, rhythm, and fate in its many forms. Even believers in astrology's literal truth must engage in language and narrative discourses. There is, then no necessary incompatibility between the notions of astrology-as-literal-truth and astrology-as-symbolic-language. Astrology tells stories and provides everyone with a celestial biography.[27] This may go some way to explaining its popularity.

It is therefore astrology's narrative quality, its ability to personalise what Brian Swimme and Thomas Berry have called 'The Universe Story', which is responsible for its popularity; according to Ivan Kelly, although its claims are false, 'many people appear to be attracted to astrology because it seems simple and speaks to their lives'.[28] Such is astrology's appeal that between 25% and 75% of the adult population of the UK and USA 'believes' in it.[29] Given that between 80% and 93% of astrologers in the UK and USA define astrology as a 'language', and many describe it as a symbolic language, astrology may, then, be considered an instance of a symbolic system to which a majority of adults subscribe.[30] If astrologers constitute what may be called a 'symbol-using community', one which assumes that meaning

---

[26] Campion, *Astrology and Cosmology in the World's Religions*, p. 17.

[27] Kocku von Stuckrad, 'From Astronomy to Naturphilosophie, From Matter to Spirit: Astrology in German Romanticism', paper presented at 'Astrologies', the eighth annual Sophia Centre Conference, Bath, 24–25 July 2010.

[28] Ivan Kelly, 'Why Astrology Doesn't Work', *Psychological Reports* 82 (1998): pp. 527–46, here p. 542; Brian Swimme and Thomas Berry, *The Universe Story* (San Francisco: Harper, 1992).

[29] Nicholas Campion, *Astrology and Popular Religion in the Modern West: Prophecy, Cosmology and the New Age Movement* (Abingdon: Ashgate, 2012), p. 153.

[30] Ibid., p. 181.

can be derived, the past understood, the present managed and the future predicted, all from the study of symbols, then the larger part of British and American adults subscribe to a common symbolic means of understanding their place in the cosmos.

The definition of astrology as language has a respectable lineage, being rooted in the word's Greek origin: astro-*logos*, translated variously as the 'word', 'logic' or 'reason' of the stars, implies that the main function of astrology is to communicate. This supposition is consistent with the view in first millennium BCE Mesopotamia, in which the movement of the stars was understood as *šitir šamê*, the 'writing of heaven'.[31] Divinity supplied information about the world to humanity and the world, it was thought, could in turn be 'read' and understood in a logical manner. The notion of the world as 'text' which is supposedly a characteristic of Postmodernism, has therefore been a feature of western astrology since its earliest recorded history.[32]

The question for astrologers is whether they think that it is the stars which are speaking, or whether astrology is a constructed language. The notion of astrology as a constructed language implies that it is a matter of meaning-making, of the construction of meaning, as opposed to the identification of meaning which is already actually embodied or embedded in the cosmos. That is, the difference may be between the proposition that astrology makes people in its own image, on the one hand, and the assumption that people make astrology in their image, on the other.

As a language, astrology relies on metonymy, using one word to mean another, so that when the modern western astrologer utters the word 'Mars', her, or his, colleagues hear the words 'anger', 'danger' and 'energy'.[33] When astrology

---

[31] Erica Reiner, *Astral Magic in Babylonia* (Philadelphia: American Philosophical Society 1995), p. 9. See also C. J. Gadd, *Ideas of Divine Rule in the Ancient East*, Schweich Lectures of the British Academy (1945; repr., Munich: Kraus Reprint 1980), p. 57. Jean Bottéro, *Mesopotamia: Writing, Reasoning and the Gods* (Chicago: University of Chicago Press, 1992), p. 133.

[32] Nicholas Campion, 'Astrology's Place in Historical Periodisation: Modern, Premodern or Postmodern?', in Campion and Greene, *Astrologies: Plurality and Diversity*, pp. 217–54.

[33] Margaret Hone, *The Modern Textbook of Astrology*, 4th ed. (1951; repr., London: L.N. Fowler, 1973), pp. 159–62.

says 'Venus' it is code for love, peace and desire or, in Aztec and Maya culture, war and violence.[34] Michael Harding, who avoided the word, 'symbol', regards such associations as metaphorical, rather than metonymic.[35] The point, though, is not whether astrology uses metaphor or metonymy, but that it uses one word to mean another. Some of astrology's modern adherents claim its language is universal, discussing such terms as the 'archetypal' feminine; in relation to the moon, archetypal here being a synonym for universal. Archetype in the sense used here is a simile for Plato's notion that there are universal Ideas, or Forms, which underpin all phenomenal existence.[36] Yet, while it may be a matter of faith that archetypes are universal, their manifestation need not be. The ambiguity is justified by C.G. Jung (1875–1961), the founder of analytical psychology, who wrote that,

> Archetypes are like river beds which run dry when the water deserts them, but which it can find again at any time. An archetype is like an old water course along which the water of life has flowed for centuries digging a deep channel for itself. The longer it has flowed in this channel the more likely it is that sooner or later it will return to its own bed.[37]

Much of the literature, consistent with Jung, expressly rejects the notion that the manifestation of an archetype should be universal, a point Nor Hall made in relation to the Moon and the feminine.[38] At the very least, archetypes are sufficiently divergent in their manifestation to suggest that they are not universal. For example, while the Greeks had a Moon goddess, in Babylon, the Moon-god, Sin, was a man, as was the Egyptian Thoth. Astrological symbols are, though, like

---

[34] Ibid., pp. 156–59; John B. Carlson, 'Venus-Regulated Warfare and Ritual Sacrifice in Mesoamerica', in *Astronomies and Cultures*, ed. Clive Ruggles and Nicholas Saunders (Niwot CA: University of Colorado Press, 1993), pp. 202–52.

[35] Harding, *Hymns to the Ancient Gods* (London: Penguin, 1992), p. 154.

[36] Plato, *Timaeus*, trans. R. G. Bury (Cambridge, MA: Harvard University Press, 1931).

[37] C. G. Jung, 'Civilisation in Transition', *Collected Works*, Vol. 10, trans. R. F. C. Hull (London: Routledge and Kegan Paul, 1964), ¶395.

[38] Nor Hall, *The Moon and the Virgin*, (London: The Women's Press, 1980).

any other symbols, polysemic: they have multiple meanings and require interpretation. From a symbolic perspective, then, the logic which leads a western astrologer to interpret Venus as peace, and an Aztec to look at it as war, does not deny the existence of a universal symbol. However, if the same archetype can be a peace-loving woman in one culture and an aggressive man in another, the value of symbols as a reliable representation of archetypes for purposes of astrological interpretation is called into question. The counter-argument, made by Hall and justified by Jung, is that the archetype is universal even if its manifestation is variable. Such a notion is also justifiable within Platonic cosmology; in the third century, Plotinus had argued that while some things do happen as a result of the movement of the heavens, others do not, a claim which for him did not contradict the concept of universal archetypes.[39] What we might retrospectively call the Platonic 'Uncertainty Principle' feeds directly on that form of classical scepticism which doubts that any statement of truth can be made. At the very least, Platonic uncertainty argues that statements of truth can be made at some times but not others. The consequences for astrology are problematic, for it becomes difficult to make statements based on astronomical motions with any certainty: the astrologer never knows whether the archetype associated with a planet will manifest in the phenomenal world and, therefore, whether a symbol can be read. This much is recognised by some forms of modern astrology influenced by theosophy and depth psychology.[40] It is also, paradoxically, codified in medieval astrology in the 'considerations before judgement', the rules which apply astrological indicators in order to decide whether the astrologer's reading can be successful.[41] The paradox of astrological uncertainty being dependent on the certain reading of astrological symbols is typical of the problems associated with the understanding of symbol within astrology.

Astrologers' confusion concerning the modern definition of symbol is reflected in the wider uncertainty in the English

---

[39] Plotinus, 'On Whether the Stars are Causes', 13.1–4.
[40] Campion, *Astrology and Popular Religion*, pp. 51–68.
[41] Olivia Barclay, *Horary Astrology Rediscovered* (West Chester, PA: Whitford Press 1990), pp.123–26.

language, and the slippage between the definition of symbol and those of similar words. Even in discussions which appear to share a Neoplatonic approach, there can be a retreat from the notion of the symbol as embedded. When Martin Ling, writing from a theological perspective, claimed that the sun and the moon symbolise the Spirit and the Heart respectively, it is unclear whether he thought that the relationship is in any way 'essential', or is simply representational.[42]

Any on-line dictionary makes the point about fluid and unclear definitions: an analogy is defined as 'similarity in some respects between things that are otherwise dissimilar'; an allegory can be a 'symbolic representation'; a symbol is 'something that represents something else by association', a 'printed or written sign' or an 'object or image that an individual unconsciously uses to represent repressed thoughts feelings, or impulses'. A sign 'suggests the presence of a fact, condition or quality', it may be dynamic, as an 'act or gesture', and is 'a division of the zodiac, represented by a symbol'.[43] The confusion extends to reference works. For example, *The Dictionary of Art* describes symbols as 'certain types of sign that are designed to extend the realm of representation, particularly so as to incorporate abstract ideas [and are] broadly less sophisticated in operation and meaning than allegories'.[44] A symbol therefore seems to be more than a sign, but less than an allegory. What Plait and French call analogy could therefore easily be called symbol elsewhere. Analogy, significantly, assumes that there are real similarities between the analogous phenomena, which is probably not what Plait intended, but does move closer to Greene's identification of the Neoplatonic assumption that there is a real connection between the symbol and the thing symbolised.

The confusion between the symbol as embodied within an ordered universe on the one hand, or arbitrary, representational and autonomous on the other can be traced to Plato. Plato's statement of the existence of the

---

42 Martin Ling, *Symbol and Archetype: A Study of the Meaning of Existence* (Louisville, KY: Fons Vitae, 2005), p. 3.

43 'Analogy', 'allegory', 'symbol', 'sign', in The Free Dictionary, http://thefreedictionary.com

44 Jane Turner, ed., *The Dictionary of Art* (London: Macmillan, 1996), Grove, Vol. 30, p. 163.

mathematically ordered cosmos underpins much western cosmology:

> Wherefore, as a consequence of this reasoning and design on the part of God, with a view to the generation of Time, the sun and moon and five other stars, which bear the appellation of 'planets' [i.e., 'wanderers'], came into existence for the determining and preserving of the numbers of Time.[45]

In this sense Platonic symbols move according to the same ordered cyclical motions as do the planets and, at the same time, they reveal the unfolding Ideas, or Archetypes, in the mind of god. If the symbol represents the archetype, then it must be mathematically regulated according to the same principles as the planets. For Rick Tarnas, arguing from this Platonic perspective, astrology, 'posits a systematic symbolic correspondence between planetary positions and the events of human existence'.[46] Jung emphasised an understanding of symbol, one also derived from Plato, in which symbols are necessarily uncertain. Jung wrote that symbols 'are tendencies whose goal is as yet unknown' and 'are the best possible formulation of an idea whose referent is not clearly known'.[47] In one version of Jungian cosmology, astrology then becomes a matter of human construction. 'We can see this most clearly', Jung wrote, 'if we look at the heavenly constellations, whose originally chaotic forms were organized through the projection of images. This explains the influence of the stars as asserted by astrologers'.[48] He added '"The starry vault of heaven" is in truth the open book of cosmic projection, in which are reflected the mythologems, i.e., the archetypes. In this vision astrology and alchemy, the two classical functionaries of the psychology of the collective unconscious, join hands'.[49] The concept of the archetype itself then becomes a kind of explanatory model, in which its

---

[45] Plato, *Timaeus*, 38C.

[46] Richard Tarnas, *Cosmos and Psyche* (New York: Penguin, 2006), p. 63.

[47] C. G. Jung, 'Mysterium Coniuntionis', *Collected Works*, Vol. 14, trans. R. F. C. Hull (London: Routledge and Kegan Paul, 1963), ¶668 and n. 54.

[48] C. G. Jung, 'The Structure of the Psyche', *Collected Works*, Vol. 8, trans. R. F. C. Hull (Princeton, NJ: Princeton, 1960), ¶325.

[49] C. G. Jung, 'On the Nature of the Psyche', ¶392.

manifestation through symbol can only be understood through the astrologer's act of interpretation.

The symbolic argument can therefore lead in two main directions. The first is into a Platonic cosmos in which symbolic readings of planetary meanings are located in an understanding that time moves in qualitatively fluctuating patterns which are themselves measured by astronomical motions. In this sense the object on which the symbol is based (that is, the star or planet), and the ordered movements of the cosmos which govern the motions of the planets, are primary. The second, which may be broadly characterised as Jungian, poses the symbol, and the astrologer reading it, as paramount, and any astronomical consideration as secondary. Although the Platonic/Jungian dichotomy is a simplification, for Jung himself was a Platonist, it is possible to identify two varieties of symbolic astrology: the essentialist or, to use Collingwood's term, the substantialist, on the one hand, and the representational on the other.[50] In the first the symbol is embodied in and embedded in the essence, or substance, in which the material world exists, while in the second, a symbol is no more than a matter of convenience. The slippage between the two is paralleled in the development of phenomenology from Edmund Husserl's Platonic argument that the study of any phenomenon on its own terms enables the perception of an underlying reality, to Merleau-Ponty's opinion that no underlying reality is necessary.[51] In Husserl's cosmology the act of contemplating an object allows the observer to eventually penetrate its deeper reality. For Merleau-Ponty there is no deeper reality and all the observer can do is describe the phenomenon. Drawing on Heidegger rather than Merleau-Ponty, but making a similar point, Michael Harding then argued for a purely descriptive phenomenological astrology, defined as a 'method for exploring the nature of or experience, and for providing a context in which these experiences may become more comprehensible'.[52] Harding described astrology as a

---

[50] R. G. Collingwood, *The Idea of History* (Oxford: Clarendon Press, 1946).

[51] Edmund Husserl, *Ideas: General Introduction to Pure Phenomenology* (1913; repr., London: Collier-MacMillan, 1972 [Eng. trns.1931]); Maurice Merleau-Ponty, *The Phenomenology of Perception* (London: Routledge, Kegan and Paul, 1962).

[52] Harding, *Hymns to the Ancient Gods*, p. 3.

language, but argued against the existence of any deeper, essential, reality. There is, then, no need to invoke archetypes as an explanation for astrology.[53] However, there is a third option, which is identifiable within astrological literature: in Platonism the identification and interpretation of the symbol can be an imaginative and creative act, and the symbol can therefore be at once both an imaginative construction and embodied in the real fabric of the cosmos.

SYMBOL IN ASTROLOGICAL LITERATURE: THEORY

Astrology, as a discipline, is largely devoid of theoretical discussion. Its practitioners are largely ignorant of the philosophical conversations concerning astrology over the last two millennia. As Geoffrey Cornelius put it,

> Most disciplines develop a set of descriptive categories above the how-to-do-it level of practical instruction, but in this respect astrology does not compare at all well. Despite its vast quantities of interpretative texts, its *theory* is little discussed and usually taken for granted. What a symbol means is one thing (interpretation); how it comes to mean that is quite another (theory).[54]

The problem Cornelius identified was that, although many astrologers are happy to describe astrology as a symbolic language, most do not know what they mean by the word 'symbol' and, with rare exceptions, are unable or unwilling to consider the consequences of the statement. As Greene argued, the notion of astrology as symbolic has been largely promoted in recent literature by theosophists. For example, Ronald Davison (1914–1985), a devout theosophist and President of the Astrological Lodge of the Theosophical Society through the 1970s wrote:

> The study of astrology is largely a study of symbology. There is a special significance attaching to every horoscopical factor. Each planet, sign and aspect, each house and quadrant of the horoscope has a particular root meaning

---

[53] Ibid., p. 88

[54] Geoffrey Cornelius, *The Moment of Astrology: Origins in Divination* (London: Arkana 1994), p. 236.

> and represents a natural principle, the working out of which can be traced in the manifold activities of everyday life.[55]

One widely quoted text defining astrology as symbolic was written by Isabel Hickey (1903-1980), an influential teacher of astrology in the USA in the 1960s and 70s and, like Davison, a theosophist. Hickey claimed that:

> Astrology deals with Symbols. The signs of the Zodiac are Symbols of great and potent forces. The physical planets are but the outer forms through which soul energies manifest. It is these that affect us, not the physical planets. When you meet a person you are not affected by his physical body. Yet you respond to his character and personality to your benefit or detriment. So it is with planetary energies.[56]

Hickey did not define what she meant by symbol, even when she equated the word with 'sign'.[57] By inference, though, from her other statements, she may have meant either the inner form of a planet, from which its 'soul energies' manifest, or the shorthand glyph used to represent it, or both.[58] Neither did Hickey use the word 'language' in this passage; that word is introduced by those who cite her, such as Cathy Pagano in her 2008 blog, having been popularised by Dane Rudhyar (1895–1985). [59] Rudhyar was a convinced theosophist and a contemporary of Hickey, but was by far the more important of the two in his formulation of twentieth century theosophical astrology. [60] Encouraged by such writers as Hickey and Rudhyar, the simple statement that astrology is a symbolic language occurs elsewhere, usually

---

[55] R. C. Davison, *The Technique of Prediction* (1955; repr., Romford, Essex: L. N. Fowler & Co. Ltd., 1983), p. 14.

[56] Isabel Hickey, *Astrology, a Cosmic Science* (1970; repr., Sebastopol, CA., 1992), p. 5.

[57] Ibid., pp. 7–10.

[58] Ibid., pp. 6, 10.

[59] Cathy Pagano, 'Symbolic Language Is Our Mother Tongue', 23 October 2008, http://www.astrology.com/symbolic-language-our-mother-tongue/2-d-d-48089,

[60] For Rudhyar see Nicholas Campion, *Astrology and Popular Religion*, pp. 61–64; Denis Ertan, *Dane Rudhyar: His Music, Thought and Art* (Rochester: University of Rochester Press, 2009); Dennis Frank, 'Dane Rudhyar's Reformulation of Astrological Theory', http://cura.free.fr/xx/17frank4.html, accessed 11 November 2012.

without a discussion of what it means. An example is provided by John Hayes, writing on 'Traditional Astrology':

> **Astrology is a symbolic language.**
> Astrology is the study of the motions and relative positions of the planets, Sun and Moon, interpreted in terms of human characteristics and activities.
> Astrology attempts to reduce life experience into a series of ordered cycles.[61]

In this sense, as a symbolic language, astrology is reductionist, reducing the complexity of human experience to ordered cycles. The notion of order needs to be emphasised: the appearance of astrological symbols is tied to the mathematical measurement of the positions of celestial bodies and astrological symbols cannot, therefore, be treated as arbitrary.

The tendency to confuse sign and symbol is evident in modern translations of classical texts. For example one passage in Plotinus is translated as 'All things are filled full of signs' by Armstrong, but as 'All teems with symbol', by MacKenna and Page.[62] Plotinus actually uses the word 'semion', an indication of anything and close to the standard modern meaning of sign, rather than 'symbolon', which would imply something more specific related to a deity. Nevertheless, MacKenna and Page chose to translate semion imaginatively as symbol, conveying a richer image in English, perhaps, but departing from the Greek. Hickey's own blurring of distinctions between sign and symbol is repeated by others who cite her on the web:

> In Astrology, every Planet has a symbol, every Sign has a Symbol, every aspect has a Symbol. Every Symbol has its own meaning, too, and they are important. To give you just a brief idea, look at the symbol for the Sun, since it is the oldest known Astrolical [*sic*] Symbol. It is a circle with a dot in the center. The circle represents the eternal self, the Soul. In Alan Oaken's [*sic*] Complete Astrology (Copyright 1988 by Alan Oaken [*sic*]) he explains: 'The Sun's glyph signifies

---

[61] Hayes, 'Introduction to the Symbolic Language of Astrology'.

[62] Plotinus (250ACE) 'The Second Ennead, Third Tractate—Are the Stars Causes?', 7.3, trans. by Stephen MacKenna and B. S. Page http://classics.mit.edu/Plotinus/enneads.2.second.html, accessed 17 March 2013.

> the emanation of light, i.e., life giving energy from the unlimited resources of the Divinity. The circle is a symbol of infinity, as it is a perfect shape without beginning or end'.[63]

In this account a symbol is therefore something that represents something else, such as an astrological glyph. All these examples, though, including Hickey, Goodwin and Hayes, suggest that the symbol itself is embedded in the spiritual and physical reality of the planetary bodies. For Hickey, a symbol may be representational, but it also exists within the soul energies which themselves inhabit the planet's physical form. It is the soul energy, rather than the planet's physical body, Hickey claimed, which affects humanity, just as it is an individual's personality which attracts or repels, rather than the physical condition.[64] The symbol's dual function as representation on the one hand, and as embedded and embodied on the other, may be linked, but not necessarily so: if embodied the symbol will also be representational, but if representational it does not have to be embodied. Hickey herself seems to be uncertain whether the symbol has power in itself or is a representative of power.

This uncertainty exists within astrological literature as confusion, but is also deeply embedded within the Platonic paradox in which the symbol can be seen as both tied to astronomical order and the product of the creative imagination. The problem for western astrologers is that few of them have read Plato. Platonism is absorbed second or third hand via Blavatsky and those, such as Rudhyar, who have read Blavatsky. Unlike Hickey, and rarely for an astrologer, Dane Rudhyar was not satisfied merely to describe astrology as a symbol and leave it at that. His use of astrological symbolism, according to Ertan, 'reflects the kind of unity and integration he sought [in] a world culture that would constitute only a part of the large cosmic cycle of wholeness'.[65] He acknowledged the problems of definition, which were so evident in Hickey's account, writing that:

---

[63] Anon., 'Astrology As a Symbolic Language', http://www.democraticunderground.com/discuss/duboard.php?az=view_all&address=245x5288, (accessed 4 November 2012).

[64] Hickey, *Astrology*, p. 6.

[65] Ertan, *Dane Rudhyar*, p. 89

> There has probably been no time in the history of human civilization when the word symbol or its equivalent in any language has been as much used or given so many varied meanings as it is today. Some philosophers and psychologists have coined new words in an attempt to make these meanings more precise. The words 'symbol' and 'sign' have been differentiated, and the distinction is useful provided no strict boundaries are established between the two sets of meaning....A sign, if precise and accurate, is strictly factual. It is a conventional and socially understandable way of presenting facts.[66]

If a sign, for Rudhyar, represents a fact in the mundane world, such as a mathematical formula, a symbol is a very different matter:

> ...when one deals with a symbol one is in the presence of something that goes beyond the rational and the factual, something *that is more than it is,* because the symbol describes not only what it appears to be rationally and objectively, but also the relationship between a specific human need and the possibility of satisfying this need. A symbol is formulated when there is a human need for it. It may be a strictly personal need...The symbol is 'person-centered.'
>
> Similarly, for the humanistic astrologer, a birth chart is a person-centered symbol. That is to say, it carries a 'message'—the symbolic formulation of the individual's dharma. It *suggests* how he can best actualize the innate potentialities of his particular and unique selfhood. It is a symbol, a mandala, or logos, a word of power. Astrology, seen from this point of view, is a language of symbols. Because it is a language, it implies a process of unfoldment of an idea of feeling-response.[67]

Rudhyar's words are a powerful indication of his syncretic merger of the Indian terminology introduced into astrology by theosophy with the language of depth psychology. He

---

[66] Dane Rudhyar, 'Symbols and the Cyclic Character of Human Existence', http://mindfire.ca/The%20Sabian%20Symbols/An%20Astrological%20Mandala%20-%20Symbols%20&%20the%20Cyclic%20Character%20of%20Existence.htm, (accessed 4 November 2012); See also Rudhyar, *The Planetarisation of Consciousness*, esp. pp. 226–31.

[67] Rudhyar, 'Symbols and the Cyclic Character of Human Existence'.

continued that, not only is the horoscope itself a symbol, but it contains other symbols: each degree of the zodiac has its own symbol: symbols exist within other symbols.[68] A horoscope as a single symbol may, then, contain hundreds, or thousands of other symbols. This is consistent with the Platonic theory of Ideas, Forms, or archetypes. Writing of Platonic Forms, Fuller explained that:

> They are, he [Plato] tells us, eternal and immutable, present always and everywhere, self-identical, self-existent, absolute, separate, simple, without beginning or end. They are complete, perfect, existent in every respect. They are without taint of sense or imagery, invisible to the eye, accessible only to the mind....Forming as they now did the intelligible structure of the entire universe, their scope had to be correspondingly extended. Wherever two or three data of sense are gathered together under a common name, there a Form is present also. Hence there must be as many Forms as there are possibilities of grouping things under headings and applying to them a common term.[69]

The model in which larger symbols contained smaller symbols *ad infinitum*, was well-known in European astrology: in the 1550s Henry Cornelius Agrippa wrote that, 'in the Soul of the world there be as many Seminal Forms of things, as *Ideas* in the mind of God'.[70] However, the existence of a two-thousand year old philosophical lineage did not help Rudhyar define 'symbol' clearly, as he himself admits. A symbol, he argued, is at once the answer given by both individuals and groups to experiences which reveal existential needs, may consist of a series of images abstracted from facts, and may be a verbal description as much as a visual representation.[71] Here, again, he is located within a Neoplatonic lineage: in the third century CE Iamblichus had looked at ritual actions as symbols; putting one's right shoe on first could be a symbolic act.[72]

---

[68] Rudhyar, *The Planetarisation of Consciousness*, p. 226.
[69] Benjamin Fuller, *A History of Philosophy* (New York: H. Holt & Co., 1938), pp. 130–31.
[70] Henry Cornelius Agrippa, *Three Books of Occult Philosophy*, facsimile of the 1651 translation (London: Chthonius Books, 1986), Bk I, ch. xi.
[71] Rudhyar, *The Planetarisation of Consciousness*, pp. 226–27, 254–55.
[72] Iamblichus, *On the Mysteries of the Egyptians*, p. 301.

Rudhyar's functional notion of a symbol as a means of solving problems, like Davison's notion of root meanings, is reminiscent of Sherry Ortner's definition of 'root metaphors' as, 'one type of key symbol in the elaborating mode' providing 'vehicles for sorting out complex and undifferentiated feelings and ideas, making them comprehensible to oneself, communicate to others, and translatable into ordered action'.[73] A symbol may therefore be a call to action as much as to self-understanding, the Christian crucifix being a prime example. Again, Iamblichus provides a precedent.

Plato's use of metaphor and allegory positions him as a story-teller. His theories of Forms, Ideas and the individual's ability to change and grow closer to the divine, give the impression of his cosmology as fluid and open the way to the creative use of symbol in modern western art, even if via his influence on theosophical artists, such as Wassily Kandinsky, the founder of abstract painting, via that Neoplatonic populariser, Mme. Blavatsky.[74] Rudhyar's view of the symbol accords well with the Idealism of early–twentieth century theosophical artists. Indeed, Rudhyar considered himself primarily to be an artist, a composer, who happened to be also an astrologer.[75] However, for Rudhyar a symbol is not just a product of the imagination; it also structures culture. His thinking is similar to that of the anthropologist Clifford Geertz who argued that culture itself is, 'an historically transmitted pattern of meanings embodied in symbols, a system of inherited conceptions expressed in symbolic form'.[76] Such a view finds a wider expression amongst other cultural theorists. Marshall Sahlins, for example, considered that, 'in all its dimensions, including the social and the material, human existence is symbolically ordered'.[77] Culture

---

[73] Sherry Ortner, 'On key symbols', *American Anthropologist* 75, no. 6 (1973): pp 1338–46, here pp. 1340–41.
[74] See Wassily Kandinsky, *Concerning the Spiritual in Art*, trans. M. T. H. Sadler (New York: Dover Publications, 1977).
[75] Ertan, *Dane Rudhyar*, pp. xi–xiv.
[76] Clifford Geertz, 'Religion as a Cultural System', in *Anthropological Approaches to the Study of Religion*. ed. Michael P. Banton (London: Frederick A. Praeger Press, 1966), p. 3.
[77] Marshall Sahlins, 'Two or Three Things that I Know about Culture', *The Journal of the Royal Anthropological Institute* 5, no. 3 (Sept. 1999): pp. 400, 401, 403.

is therefore transmitted through symbols which, in turn, structure culture. For Rudhyar symbols are both produced by culture and shape culture; they are both passive and active, imaginary and real. The imaginary, in a Platonic sense is, of course not necessarily false, as Henri Corbin argued when he developed the notion of the Imaginal.[78] The concept of the Imaginal involves a circular logic in which, if the existence of the world soul is accepted as real, then its consequences are real. There is, no problem with circular logic as long as it is recognised for what it is, in this case an attempt to provide a framework for the experience that the products of the imagination may be experienced as 'true' rather than 'false', and therefore offers a useful means of organising one's personal affairs. In Rudhyar's view symbols also evolve as consciousness evolves and have a tenuous relationship with facts, which are themselves represented by signs, which, except when they are representations such as glyphs, can be free-floating and autonomous. A symbol is, then, not the same as a sign, as Hickey implied by her lack of discussion, but can represent a sign.

Rudhyar's notion of astrology as a symbolic language was much encouraged by his reading of Jung, who reported that,

> I noticed to my amazement that European and American men and women coming to me for psychological advice were producing in their dreams and fantasies symbols similar to, and often identical with, the symbols found in the mystery religions of antiquity, in mythology, folklore, fairytales, and the apparently meaningless formulations of such esoteric cults as alchemy. Experience showed, moreover, that these symbols brought with them new energy and new life to the people to whom they came'.[79]

Rudhyar borrowed the Jungian view that symbols arise when they are needed in order to provide a response to cultural tensions and inspire change, and he regarded the current period, the 1960s, as characterised by 'symbol wars' in which competing world views vied for supremacy.[80] He

---

[78] Henri Corbin, *Mundus Imaginalis, or the Imaginary and the Imaginal,* 1964, http://www.hermetic.com/bey/mundus_imaginalis.htm
[79] C. G. Jung, 'Psychology and Alchemy', *Collected Works,* trans. R. F. C. Hull (Princeton, N.J., Princeton University Press: 1983), p.v.
[80] Rudhyar, *The Planetarisation of Consciousness*, p. 255.

also discussed the nature of language which, he wrote, 'should be considered as a complex group of symbols inasmuch as, with its special words and forms of syntax, it answers to a basic need of humanity: the need for communication'.[81] For Jung, who so influenced Rudhyar's views on language, the concept of a symbolic language is a tautology for all language is symbolic: 'Thus', he wrote, 'language, in its origins and essence, is simply a system of signs and symbols that denote real occurrences of their echo in the human soul'.[82] Generally, in Jungian terminology, a sign represents something which is known, a symbol something which is unknown. In a passage which inspired Rudhyar, Jung wrote:

> A word or image is symbolic when it implies something more than its obvious and immediate meaning. It has a wider 'unconscious' aspect that is never precisely defined or fully explained. Nor can one hope to define or explain it. As the mind explores the symbol, it is led to ideas that lie beyond the grasp of reason.[83]

Such a perspective can be framed within Lévy-Bruhl's theory of 'participation mystique', the pre-local consciousness defined as 'participation between persons and objects which form part of a collective representation'.[84] The Jungian model was first overtly adopted by (and adapted to his theosophical beliefs) by Dane Rudhyar in his first book, *The Astrology of Personality*, which was published in 1936. In 1974, thirty-eight years after its publication, *The Astrology of Personality* was voted second in 'a survey of 100–200 opinions on "The 7 Best Books in Astrology"', an indication of Rudhyar's ability to shape twentieth-century astrological thought.[85] It is reasonable to argue that Rudhyar's work was a principal medium for the transmission of the Platonic cosmology into

---

[81] Ibid., p. 228.
[82] C. G. Jung, 'Symbols of Transformation', *Collected Works*, Vol. 5, trans. R. F. C. Hull (London: Routledge and Kegan Paul, 1956), ¶13.
[83] C. G. Jung, 'The Importance of Dreams', in *Man and his Symbols*, ed. C. G. Jung (New York: Dell Publishing, 1954), p. 4.
[84] Lucien Lévy-Bruhl, *How Natives Think*, (Princeton, NJ: Princeton University Press, 1985), p. 76.
[85] Dean and Mather, *Recent Advances*, p. 3.

twentieth century practical astrology. Rudhyar linked the notions of symbol and language in astrology:

> Every birth chart is as 'good' as any other, in the sense that it symbolizes what the person potentially is and what he is meant to achieve, if he follows the 'instructions' which, as it were, are 'coded' in the pattern of the sky, as seen from the place and at the exact moment of birth.
>
> Such an astrology is not an empirical science. It is a 'language' which can reveal the archetype of what the total person (body, mind, feelings, etc.) essentially is—the 'Form' of his or her individuality.[86]

This is the argument to which Plait referred, although without citing Rudhyar, and using such sceptical jargon as 'mumbo-jumbo', rather than 'symbolic language': defining astrology as symbolic removes it from the gaze of empirical science. Rudhyar's argument finds considerable popular support, as Neil Spencer made clear, in response to a claim by Geoffrey Dean and Ivan Kelly that astrology lacked scientific support:

> Astrology is not a science but a symbolic, allusive language, and its practitioners, from Pythagoras to Kepler to Jung to W. B. Yeats, are diviners, searchers for meaning. The last quality, of course, is anathema to statisticians and scientists—useful people if you're building bridges, granted, but perhaps they should leave astrology to those for whom it is intended; philosophers, poets and lovers.[87]

However, while for Rudhyar to remove astrological symbology from the domain of empirical science is a legitimate move, for Plait it is not. Zarka finds agreement with Rudhyar in one sense, writing that, 'with a symbolic and esoteric discourse, astrology has none of the attributes of

---

[86] Dane Rudhyar, *The Astrology of Personality* (1936; repr., Garden City, NY: Doubleday, 1970), p. xi.
[87] Neil Spencer, 'Just Whose "Consciousness" Are We Talking About, Anyway?'
Short Cuts, *The Guardian*, 19 August 2003; See Geoffrey Dean and Ivan W. Kelly, 'Is Astrology Relevant to Consciousness and Psi?', *Journal of Consciousness Studies* 10, no. 6–7 (June-July 2003): http://www.imprint.co.uk/pdf/Dean.pdf

a true science', and, he adds, it is therefore positively dangerous, rather than efficacious.[88]

Rudhyar had made two important points. First he had followed that variety of Platonic astrology, traced back to Plotinus via the Renaissance philosopher Marsilio Ficino, which rejected the division of the astrological cosmos into malefic and benefic poles. Secondly, as already stated, he had claimed that to define astrology as a symbolic language removes it from the domain of empirical science. Yet, if astrological symbolism is based somehow on astronomical order, the issue of empirical observation cannot be avoided: if planetary aspects repeat then surely the symbolic language must exhibit such repetition, if astrological interpretation is to have any consistency. This is an issue that Rudhyar preferred to ignore. Two years later, in the book's second edition, Rudhyar described the *Astrology of Personality* as:

> ...an attempt to reformulate traditional astrology in terms of the modern philosophical and psychological outlook....It established new foundations for a consistent system of symbolism, using astrological factors as its symbols. Its goal was the formulation of an 'algebra of life', using organic life-qualities as its primary elements, defining these qualities particularly at the psychological level in terms borrowed from C. G. Jung's analytical psychology.[89]

However, if symbols are culturally conditioned, astrology, with its justification in a real, phenomenal, astronomical world of universals, transcending cultural conditioning, is in difficulty. Rudhyar's solution was to remove astrology not just from the demands of scientific rigour but entirely from any necessary connection with essential meaning or astronomical order, a step with no precedent in Neoplatonic cosmology:

> Astrology of itself has no more meaning than algebra. It measures relationships between symbols whose concreteness is entirely a matter of convention, and does not really enter into the problems involved—just as the symbols of algebra, *x*, *y*, *n*, are mere conventions....In other words, the astrological realm of moving celestial bodies is like the realm

---

[88] Zarka, 'Astrology and Astronomy', p. 425.

[89] Dane Rudhyar, *New Mansions for New Men* (New York: Lucis Publishing Company, 1938), p. xii.

> of logical propositions. Neither one nor the other has any real content. Both are purely formal, symbolical, and conventional.[90]

Rudhyar's algebraic metaphor was commented on by Robert Hand:

> Dane Rudhyar called astrology the algebra of life but I suggest it's much more like a taxonomy of life than an algebra. A taxonomy is a classification system. It classifies animals as reptiles or mammals, or humans as Aries or Taurus. If we could actually convert astrology into a symbolic algebra, into an algebra of quality, we would really have made a major advance in thought.[91]

The algebraic argument takes Rudhyar, the theosophist, into agreement with the anti-essentialist, anti-Platonic line taken from within an astrological perspective by Michael Harding.[92] Rudhyar's radical move was to create a paradox in which astrology is simultaneously justified by the calculation of the horoscope on the basis of a mathematically-measurable astronomical order, and is employed to make definite statements, yet simultaneously claims to be independent of that order. His rhetoric specifically broke the link between the subjective interpretation of symbol and the symbol's embodiment in objective astronomical motion. The astrological symbol is still both embedded in the Platonic essence, and freed from it; it is simultaneously certain and arbitrary. And while astronomical measurement is employed in order to construct a horoscope, it is no longer necessary to a justification of the symbol's efficacy. The conclusion, that astrology can be justified by a universe in which there is no reality beyond logical propositions is quite justifiable within Platonic cosmology and occurs notably in the opening verses of John's Gospel (I.1); 'In the beginning was the Word'. However, in this case, what is asked of astrology? To work with logical propositions? Such a conclusion is denied by the tone of Rudhyar's own writing on astrology.

---

90 Rudhyar, *The Astrology of Personality*, p. 48.

91 Robert Hand, 'The Proper Relationship of Astrology and Science', *Astrological Journal* 31, no. 6 (Nov./Dec. 1989): pp. 307–16, here p. 316.

92 Harding, *Hymns to the Ancient Gods*, pp. 3, 88.

Nevertheless, the appeal to algebra, to the Pythagorean component of astrology's Platonic heritage, is an attempt to establish astrology on universal principles—the concept that the entire cosmos is mathematically constructed and measurable—and protect it from the destabilising impact of the culturally-dependent symbol. However, the dominant logic of the Rudhyarian perspective is that astrology is a culturally dependent symbolic language with no intrinsic meaning. This was expressed most strongly in the words of Rudhyar's leading student, Alexander Ruperti, who claimed that the notion that, 'an astrology with a capital A which exists somewhere, has always existed, and to which we should all submit ourselves or be faithful to...is merely the result of our imagination'.[93] Gregory Szanto agreed, arguing that the horoscope has no fixed value, and the constructs used to interpret it are fabrications of the human mind.[94] Again, astrological and sceptical rhetoric coincide. Szanto's problem, though, is that he accepts that the interpretative models constructed by astrologers do yield truth.

The difficulty of navigating Platonic mathematics in the modern world is not confined to astrologers. It is shared by mathematicians such as Roger Penrose, who wrote,

> How is it that mathematical ideas can be communicated in this way? I imagine that whenever the mind perceives a mathematical idea, it makes contact with Plato's world of mathematical concepts...when one 'sees' a mathematical truth, one's consciousness breaks through into this world of ideas, and makes direct contact with it ('accessible via the intellect').[95]

Creativity and mathematical order are therefore not necessarily incompatible. The difficulty with the taxonomic solution, though, is that any taxonomy still requires description, language and metaphor. And, if language is

---

[93] Alexander Ruperti, *Meaning of Humanistic Astrology* (2002), p. 3, http://www.stand.cz/astrologie/czech/texty/rez-ru-a/rez-ru-a.htm, (consulted 25 Jan. 2003). See also Dane Rudhyar, *The Lunation Cycle: A Key to the Understanding of Personality* (Santa Fe: Aurora Press 1967), pp. 9, 12, 13.

[94] Gregory Szanto, *The Marriage of Heaven and Earth: the Philosophy of Astrology* (London: Penguin-Arkana, 1985), pp. 113–15.

[95] Roger Penrose, *The Emperor's New Mind: Concerning Computers, Minds and the Laws of Physics* (London: Vintage, 1991), p. 554.

required, perhaps there is no escape from symbol. The Rudhyarian paradox, that astrology is simultaneously a part of astronomical order and independent of it, was addressed by Geoffrey Cornelius, a successor to Davison as President of the Astrological Lodge of the Theosophical Society, who argued that 'the horoscopes we work with are not astronomical records of an event in the physical world. They are symbols in a world of human significance'.[96] In this sense astrology is no longer symbolism based on astronomy in Dean's sense, but an astrology based on symbols which happen to share the same names as astronomical bodies, and where the primary referent is human significance rather than planetary motion, echoing Jung's statement that astrology is a matter of projection. Cornelius continued, 'the ground for the coming-to-pass of astrological effects or showings is not founded in a coincidence in objective time of heavens above and event below...(rather) we should look in the direction of significant presentation of the symbol to consciousness'.[97] This hypothesis then produces a further tension, in which the need to cast horoscopes based on measured astronomical positions may disappear, and horoscopes cast for 'wrong' data may 'work'.[98]

The only role for astronomical calculation is, then, to provide a ritual framework for the divinatory process. In this case the question is whether astronomical calculation is of any more value than the casting of lots: Cornelius himself compared astrology to the I Ching.[99] Again, though, the argument is contained within astrological tradition; the influential tenth-century work, the *Centoloquium*, which obtained its authority from its attribution to Claudius Ptolemy, stated unambiguously that, 'He that is desirous to study any Art, hath in his Nativity without doubt some Star of the same Nature very well fortified'.[100] That is, the

---

[96] Geoffrey Cornelius, *The Moment of Astrology: Origins in Divination*, 2nd edn., (Bournemouth: The Wessex Astrologer, 2003), p. 253.
[97] Ibid., p. 38.
[98] Nicholas Campion, 'Mythical Moments in the Rectification of History', in *Astrology Looks at History*, ed. Noel Tyl (St. Paul, MN: Lllewllyn, 1995), pp. 24–63.
[99] Cornelius, *The Moment of Astrology*, p. 118.
[100] Claudius Ptolemy, 'Centiloquium', in John Partridge, *Mikropanastron, or an Astrological Vade Mecum, briefly Teaching the whole Art of Astrology* - viz., *Questions, Nativities, with all its parts, and*

astrologer is completely psychically integrated within the astrological process. Nevertheless, in modern discourse, the symbolic argument, detached from Platonic correlations with the heavens, divorces astrology from astronomy. The paradoxical consequence is that the pro-astrology Rudhyar-Cornelius view appears remarkably similar to the anti-astrology Kelly-Dean hypothesis: the astrologer and the sceptic begin to occupy the same ground and, although the former inclines to the view that astrology is meaningful and the latter that it is meaningless, both rest their case in the functional value of astrology as a subjective, symbolic system, which does not require any external agency. The only difference between astrologer and sceptic is a matter of opinion: whether or not astrology works. For the sceptic it is imaginary and false; for the astrologer—at least, for the Platonic astrologer—Imaginal and real.

The question is, then, whether the symbols with which astrology deals can be free-floating in the same way, for example, as those of the surrealist painter or theosophical composer. On this point astrologers are not in agreement. One who set out to test astrology empirically was Jeff Mayo, who played a leading role in astrological education in the 1960s. He had this to say about the birth chart:

> The birth chart is composed of symbols, presented as a 'pattern' unique to the person upon whose birth data it is based. Each symbol, astronomical by origin, is correlated with psychological features from which human behaviour can be interpreted.[101]

Mayo attempted to demonstrate the measurable reality of astrological symbols in his research from 1969–78, latterly with the psychiatrist Hans Eysenck, and considered that the existence of astrological symbols might be proved statistically, a somewhat difficult task if, as Rudhyar believed, they evolve.[102] Mayo's statistical approach suggests

---

the whole Doctrine of Elections never so comprised nor compiled before, &c. (London: William Bromwich, 1679), para 3.

[101] Jeff Mayo, *Astrology: A Key to Personality* (London: Penguin, 1995), p. 2.

[102] H.B. Eysenck and D.K.B. Nias, *Astrology: Science or Superstition*, 2nd ed. (Harmondsworth: Penguin, 1984), pp. 49–60; Mayo, *Astrology: A Key to Personality*, pp. 350–52; Jeff Mayo and Christine

that a symbol can have a permanent, universal, unchanging and readily identifiable nature. Mayo also explicitly refuted Rudhyar's claim that astrology as symbolic language is freed from the scrutiny of empirical science. (Mayo had some status as an astrologer, first as Principal of the Faculty of Astrological Studies and then as the founder of his own school, the Mayo School of Astrology, as well as the author of a number of textbooks.) Neither is there is any necessary escape from the purview of empiricism in the Neoplatonic tradition: two of the strongest supporters of empirical research into astrology in the late twentieth century were John Addey and Charles Harvey, both of whom were Presidents of the UK-based Astrological Association and active members of the Neoplatonic 'Universal Order'.

SYMBOL IN ASTROLOGICAL LITERATURE: PRACTICE

Astrological theory is one thing, astrological practice another, and there appears to be a 'rhetorical gap' between the former and the latter. Both Rudhyar's and Ruperti's writings on horoscope interpretation assume the correlation between an objectively measured time and real personal qualities. We can take any number of examples from Rudhyar's work. In his *Person Centred Astrology*, for example, we read that

> ...the opposition aspect [two planets separated by 180^0] represents, symbolically, at least, the moment at which...an 'objective structure of totality' is most likely to appear...It is particularly important in the case of the soli-lunar relationship, because the two Lights refer to the bi-polar life force which sustains and feeds the whole organism.[103]

The caveat 'symbolically, at least', aside, Rudhyar's cosmos was biological; he was concerned with the 'life-force', and considered that a planetary aspect is particularly important, not 'might be, may be or could be particularly important', depending on the time, location and context. In spite of his rhetoric, Rudhyar was not able to make final the ontological break between astrology and astronomy. The objective

---

Ramsdale, *Teach Yourself Astrology* (London: Hodder Headline, 1996), pp. 166–73.

[103] Dane Rudhyar, *Person Centered Astrology* (New York: Aurora Press 1980), p. 169.

nature of astronomical measurement is essential for the authentication of astrological symbolism, in spite of the rhetoric that this is not so. Rudhyar can use phrases such as 'represent symbolically, at least', but cannot escape the statement that the symbolic language is tied to the mathematically ordered cycles of the planets. As Rudhyar wrote, borrowing from Plato, 'Cyclic patterns that refer to human life in its most universal aspect have...the character of archetypes'.[104]

The astronomical opposition between two planets, measured geocentrically, correlates objectively with a developmental moment in the individual's life. In this sense, what freedom does the astrologer have to interpret? Rudhyar had written that the symbol lies beyond reason, and can neither be defined nor explained.[105] Yet here he is doing precisely the opposite, defining a symbol and using reason. His inspirational pronouncements about astrological symbols did not match his attempts to interpret those symbols.

The extent to which most practical astrology is symbolic, in the sense that it can be interpreted creatively, and is only a product of cultural convenience, as Ruperti suggested, is therefore problematic. The comparison is between symbolic and literal thinking, based on the distinction between allegorical and literal thinking in Biblical interpretation: was the world created in six actual days, or is the Genesis cosmogony a potential source of meaning about life on earth and the relationship between humanity and the ultimate? Much practical astrology has more in common with the habits of Biblical literalists, and it can be argued that astrologers may have to believe the literal truth of astrological claims; otherwise they are unable to perform the interpretative tasks required of them.[106]

Any search through astrological literature reveals far more examples of certainty than uncertainty. Without ascribing any negative qualities or characteristics to the astrologers quoted below, the examples provide valuable evidence. Dane Rudhyar's instructional writing on horoscope interpretation was characteristically vague, but no less

---

104 Dane Rudhyar, *Occult Preparations for a New Age* (Wheaton, IL: The Theosophical Publishing House, 1975), p. 111.

105 Jung, 'The Importance of Dreams', p. 4.

106 Campion, 'Mythical Moments in the Rectification of History', pp. 24–63.

problematic in its implicit certainty. For example, in *The Astrology of Personality* he wrote,

> When the Moon has north latitude one should be able to discover an underlying tendency toward paying great attention to a successful adjustment to the environment; and this can mean at the social level ambition, the strong desire to control the environment, and perhaps a yearning for personal fame or at least prestige. The individual makes an issue of his personal position.[107]

Where does this knowledge come from? Is it based on evidence, or tradition, or is it revealed? What does it mean to adjust to the environment? Surely all living creatures have to do this? Do not all people need some kind of prestige? And why is the uncertainty of the words 'can mean' and 'perhaps' contradicted by the sense that there really is a 'strong desire to control the environment'?

Neither is there much ambiguity in Alexander Ruperti's writing. To be clear, his language allows for uncertainty in the detailed manifestation of any astrological symbol, but of the general principles there is no doubt. As a randomly picked example, for north node in the 5th house and south node in the 11th house, he wrote:

> At this time one should strive to attain mastery over both his personal resources and his everyday circumstances through partnership activities and the comingling of efforts. This may occur in business, group activity or occult work, or through one-to-one relationships.[108]

Aside from the obvious all-encompassing breadth of the statement that the promised future may occur in group activities or in one-to-one relationships, the key word here is 'should'. This is neither humanistic nor person-centred, both terms Rudhyar applied to his astrology. It is a prescriptive statement of what the individual must do in order to perform their spiritual duty. Isabel Hickey did not share Rudhyar's sense for probabilities, and was far more certain both in her theory and practice. Astrology, she wrote,

---

107 Rudhyar, *The Astrology of Personality*, p. 264.

108 Alexander Ruperti, *Cycles of Becoming* (Vancouver, WA: CRCS Publications, 1978), p. 64.

> ...is a spiritual science. It involves the relationship between the larger universe outside you and the personal universe within. The same energies that function in your personal universe function in the larger 'out there'. The blueprint we call the horoscope, or the birthchart, plots the energies that flow in your magnetic field. At the moment of birth you took into your body, with the first breath, the vibrations manifested on that day and time at that particular spot on earth. The basic pattern goes with you throughout life...the influences of that moment and that place in space show how your potentials can be actualised in the future.[109]

Hickey then went on to give readings for planetary positions within the horoscope houses and zodiac signs, and aspects between planets. Her work was based in certainty, making a combination of psychological and sociological statements about present and past lives. To open a page at random, Taurus on the cusp of the fourth house, 'is extremely conservative in domestic surroundings', Gemini on the fifth house cusp, 'is not the type of parent who is basically emotional where his children are concerned', and Cancer on the sixth house cusp, 'is much more involved emotionally with his work and career then he is where personal relationships are concerned'.[110] She continues: 'Venus in (Aquarius) is cool, calm, collected and detached...has a clear cut mind but icy and cold emotions. Often has a background of an early life where demonstrations of lovingness were missing'; if the sign is afflicted by Saturn, Uranus or Mars, 'there would be a debt from a past life where love is concerned'.[111] How Hickey could make such statements is not clear. Perhaps she had experience of a client, or a few clients, who reported feeling unloved as a child, and extrapolated a general law from this. The use of the word 'often' suggests a confidence founded on quantification. Hickey needed to mimic the language of quantified research in order to give her statements power. More problematic is the statement about a 'debt from a past life'. In terms of the history of ideas the statement is a fascinating example of the import of Indian ideas into western astrology. But it is nevertheless problematic: where did Hickey acquire this

---

[109] Hickey, *Astrology*, p. 5.
[110] Ibid., p. 133.
[111] Ibid., p. 161.

information? Given that it is absolutely impossible to obtain any verifiable data about past lives, the certainty with which Hickey speaks is not justified.

These are statements of fact which cannot, in both Mayo's view and French's opinion, be tested. In Rudhyar's terminology they are signs rather than symbols. There is no subtle teasing out of symbolic meaning and the text indicates that individual potentialities are not to be actualised in the future but are present in the here and now.

For Hickey a symbol was not something to interpret, but something that both really exists in a state of self-evident truth, and reveals correct factual information about the past, (including past lives), present and future. Such statements exist in an ontological context closer to, say, scriptural accounts of angelic visitations, or other theological statements of absolute truth. In this sense, such astrology exists in an epistemological realm which parallels that inhabited by Christian fundamentalism and has an equal claim either to the truth or lack of it.

Other examples of literal certainty abound. This is from Steven Forrest and Jeffrey Wolf Green, analysing a horoscope at a seminar in September 1999:

> And we see the [planetary] ruler, Uranus, which is now balsamic-conjunct Pluto, and in the middle of that balsamic conjunction stands the black Moon Lilith. It shouldn't be too difficult to figure out the causative factors of the traumas which have preceded the current life, and which will be recreated in the current life, have all been caused by the nature of the biological families and the types of societies—i.e. patriarchal—in which she was born. Families and societies that have had the effect of making psychological and emotional imprints that have been utterly self-destructive to her.[112]

There is no uncertainty here. Statements are made about past lives and the causes of family conditions and psychological patterns in this life with complete assurance. There is no trace of doubt. Jung's statements that 'Symbols are tendencies whose goal is as yet unknown' and, 'Symbols are

---

[112] Steven Forrest and Jeffrey Wolf Green, *Measuring the Night: Evolutionary Astrology and the Keys to the Soul* (Boulder CO: Daemon Press, 2000), p. 63.

the best possible formulation of an idea whose referent is not clearly known', are not represented in such astrology.[113]

Such certainty is also evident in the work of the founder of Green and Forrest's spiritual-evolutionary astrology, the British theosophist, Alan Leo (1875–1917). We could consult, for example, Leo's descriptions of the sun in zodiac signs: Sun in Libra, 'is rather popular and generally liked', Sun in Scorpio is, 'unfavorable for the parents', Sun in Sagittarius is, 'naturally religious and sincere', and Sun in Capricorn is, 'ambitious and aspiring, desirous of power and fame'.[114] Alan Leo was the most influential of those fin-de-siècle astrologers who rejected the interpretative fabric of medieval horoscopic astrology in favour of an astrology of spiritual aspiration. Yet he no more suffered from doubt than did any medieval practitioner. His spiritual, 'esoteric' astrology was an astrology of fact, and his certainty was located in a certain absolutism inherent within theosophy. H. P. Blavatsky, founder of the Theosophical Society, had this to say about astrology:

> Astrology is a science as infallible as astronomy itself, with the condition, however, that its interpreters must be equally infallible....In astrology one has to step beyond the visible world of matter, and enter into the domain of transcendent spirit.[115]

Blavatsky's assertion of astrological infallibility was written just seven years after Papal infallibility was finally theologically defined at the first Vatican Council of 1869–70. Infallibility was thus a topic of discussion while she was composing her major literary works. If astrology, in the theosophical cosmos, is infallible, astrologers can say anything and be absolutely correct. This is what enables Green and Forrest to make certain statements about past lives. However, theosophy leads in different directions: to astrology as the identification of fact, as in Green and Forrest, or to astrology as the reading of symbol, as advocated by Rudhyar. The problem of the boundary between the two,

---

[113] Jung, 'Mysterium Coniuntionis', ¶668 and n. 54.

[114] Alan Leo, *How to Judge a Nativity* (1903; repr., London: Modern Astrology, 1922), pp. 76–77.

[115] H. P. Blavatsky, *Isis Unveiled* (1877; repr., Pasadena, CA: Theosophical University Press, 2 Vols., 1976), I.259.

though, is a partly a matter of the text. Theosophical astrology may indeed depend on the reading of symbols, but its aspiration is to describe a universe which is entirely knowable, down to its merest details.

For Cassirer everything is symbolic: he refers to 'the universal symbolic function'.[116] He adds, 'It is a common characteristic of all symbolic forms that they are applicable to any object whatsoever'.[117] Cassirer's conclusion is of significance for astrology for, if everything is symbolic, there can be nothing that is literal, and hence non-symbolic. As Donald Verene put it, 'In speaking about symbols, we are accustomed to make a distinction between what is symbolic and what is literal. Cassirer's epistemology rejects this distinction. What might be defined as literal is, in fact, in Cassirer's view, symbolic'.[118] But, if everything is symbolic then a new problem is raised for astrology: the claim to special status inherent in the claim that 'astrology is a symbolic language' disappears for everything is a symbolic language. Except, of course, that in Platonic terms the appearance of symbols is mathematically regulated and can therefore, theoretically, be projected indefinitely into the future.

CONCLUSION

The first conclusion is historical. The sceptics' argument that astrologers' claims of untestability is a response to scientific failure is rebutted on the grounds that such claims are located in a theoretical model which long precedes modern science. This does not mean that no astrologers advocate untestability as a result of negative testing of astrology, or as a defence against scientific scrutiny, but no research has been

---

[116] Ernst Cassirer, *The Philosophy of Symbolic Forms,* Vol. 3., The Phenomenology of Knowledge (1955; repr., New Haven, CT: Yale University Press, 1971), p. 1.

[117] Ernst Cassirer, *The Myth of the State* (New Haven, CT: Yale University Press, 1967).

[118] Donald Philip Verene, *The Origins of the Philosophy of Symbolic Forms: Kant, Hegel, and Cassirer* (Evanston, IL: Northwestern University Press, 2011), p. 14. See also Cyrus Hamlin and John Michael Krois, eds., *Symbolic Forms and Cultural Studies: Ernst Cassirer's Theory of Culture* (New Haven, CT: Yale University Press, 2004).

conducted in order to demonstrate this. The sceptical argument needs to be supported by evidence, rather than made as a statement of self-evident truth. However, the fact is that astrologers disagree on whether astrology's definition as a symbolic language suggests that astrology, as a single entity, is not testable. Simply, if it does not exist as a single entity, it cannot be tested as one. Only individual claims can be tested, but the results of such tests, either negative or positive, have no consequences for the field as a whole. Astrologers themselves disagree on the question of testability. Dane Rudhyar claimed that a symbolic astrology is removed from the domain of empirical science, but Jeff Mayo, John Addey and Charles Harvey disagreed.

The definition of judicial astrology as a symbolic language makes sense in the context of a Platonic cosmos, but it suffers from the same uncertainty as does Platonic cosmology when taken as a whole: is the cosmos to be understood as a set of ordered mathematical processes, or does it consist of shifting Ideas? Most astrological literature which uses the definition 'symbolic language' nevertheless relies not on the fluid uncertainty required by Rudhyar but the certainty typical of Hickey, without, however, making the inevitable leap advocated by Mayo: that if symbols are based on order, their existence must be testable. The second conclusion, then, is that the question, 'Is astrology a symbolic language?' can be answered in the affirmative, but the answer reveals nothing about astrology's nature, only about the claims made by astrologers.

The Rudhyarian paradox—that astrology can simultaneously lack any content or connection with astronomical order, yet make definite statements which are justified by that order—is largely ignored by practicing astrologers. Most astrologers simultaneously regard the rules of astrological interpretation as having no fixed value, yet insist on the truth of their statements to clients; between 70% and 97% agree with the proposition that the efficacy of astrological techniques in interpretation depends on the astrologer.[119] This doctrine was well-known in medieval astrology, yet is no less problematic for that. The implications are serious for those astrology schools and textbooks which teach systems which are the same for all students,

---

119 Campion, *Astrology and Popular Religion*, p. 181.

irrespective of the student. Yet, if astrology operates as a symbolic language, much can be misconstrued in translation. Rudhyar himself was aware of the difficulties, and sounded a cautionary note:

> ...of course the thing you have to be careful of when you are dealing with so-called occult, or esoteric ideas is that you never know too much how much you are dealing with symbols and how much you are dealing with what you might call a higher reality. In some ways you are dealing with both. But to what extent you do from one to the other, to what extent the idea that a different plane has exactly the same kind of reality as the physical plane, involves how much is true or not....When you are dealing with intra-atomic physics and with particles which are no longer only particles, but which are waves, not only waves of something but waves of probabilities.[120]

The third conclusion, then, is that if astrology is a symbolic language, it has not two forms, but three. Two of these are overtly Platonic and assume that the symbol is embedded in the essential fabric of the cosmos; the first is as a language of metaphor in which the astrologer's interpretation is primary, and the second is a language of order, dependent on an independently existing state of affairs. The third form is derived from theosophy but moves into the strictly representational, in which it is not necessary for the astrological symbol to be embedded in the Platonic essence. However, these are theoretical positions which are not necessarily represented in practical astrology. To the statement 'astrology is a symbolic language', then, the retort must be made 'what kind of symbolic language?'

---

[120] Sheila Rayner, ed., *Dane Rudhyar: Interviewed by Sheila Finch Rayner, Clare G. Rayner and Rob Newell* (Long Beach, CA: California State University Library, 1977), p. 99, cited in Ertan, *Dane Rudhyar*, p. 84.

ART, ASTRONOMY AND SYMBOLISM IN THE AGE OF SCIENCE

Gary Wells

ABSTRACT: Symbolic representations of celestial phenomena have always had an important role in art. Such representations have been a way of picturing the invisible, of expressing abstract concepts, and of embodying a personal and collective understanding of astronomical objects and events. But this long iconographical tradition contrasted with the empirical, scientific view of nature that artists came to regard as the proper domain of a modern art after the Enlightenment. The divide between the 'real' and the symbolic was symptomatic of this new view of art and of science. Yet, the persistence of symbolic associations, metaphorical references, and multilayered iconography in artistic representations of the sky suggests that such symbols were (and are) still meaningful and relevant.

Symbolic representations of celestial phenomena have always had an important role in art. Such representations have been a way of picturing the invisible, of expressing astronomical concepts, and of embodying a personal and collective understanding of astronomical objects and events. But this long iconographical tradition contrasted with the empirical, scientific view of nature that artists came to regard as the proper domain of a modern art after the Enlightenment. The perception of a divide between the 'real' and the symbolic was symptomatic of this new view of art and of science. Yet, the persistence of symbolic associations, metaphorical references, and multilayered iconography in artistic representations of the sky suggests that such symbols were still meaningful and relevant during the age of science.

I begin with the precursor to one of the most familiar modern images of the sky. Vincent van Gogh's *Starry Night Over the Rhône* of 1888 (fig. 2.1) is less well known than the artist's later *Starry Night* of 1889 (Museum of Modern Art, New York). In this painting, van Gogh brought together his two primary concerns as artist—the emotional and sensory life of humanity, and the longing for spiritual connection to nature. The couple who stroll along the river front are at the visual entry point to the painting. They are us; they represent the world of human relationships and emotions; their relationship as a couple is a reminder of the artist's own contrasting loneliness. We visually ascend by crossing the river, literally and metaphorically, to the starry sky above.

Fig. 2.1: Vincent van Gogh, Starry Night over the Rhône, *1888. Musée d'Orsay, Paris.*

Van Gogh thought of the sky in spiritual, if not religious, terms, a realm both remote and yet potentially accessible. His understanding of the stars was shaped symbolically: they remind us of our own limits and our mortality. The lights of the town and their reflections on the waters mediate between the earthly and celestial realms. The artificial lights are imitations of the stars in the sky. Their sparkling reflections on the rippling waters of the Rhône are the earth-bound recreation of the twinkling of those stars. The technology of artificial illumination contrasts with the natural stars. Light, both natural and artificial, is also an emblem of the sense of sight, the chief concern of artists and thus the essential metaphor of visual art.[1]

[1] For a full investigation of Van Gogh's approaches to painting the night sky, see Sjaar van Heugten, Joachim Pissarro, and Chris

Just a few months before *Starry Night over the Rhône* was painted, Van Gogh speculated about the relationship of art to the stars. In a letter to his sister, he observes:

> It often seems to me that the night is even more richly coloured than the day, coloured in the most intense violets, blues and greens. If you look carefully you'll see that some stars are lemony, others have a pink, green, forget-me-not blue glow. And without labouring the point, it's clear that to paint a starry sky it's not nearly enough to put white spots on blue-black.[2]

In a letter to the artist Eugène Boch, he describes this painting in particular:

> And lastly, a study of the *Rhône*, of the town under gaslight and reflected in the blue river. With the starry sky above—with the Great Bear—with a pink and green sparkle on the cobalt blue field of the night sky, while the light of the town and its harsh reflections are of a red gold and a green tinged with bronze. Painted at night.[3]

Van Gogh emphasized this last point—the scene is not one of the imagination or of memory, but painted directly from observation.

Van Gogh mused about the existence of other worlds, and what those worlds might offer visually for an artist. And while he does not directly suggest that looking to the stars is looking at heaven, he felt that death is the link between the stars and us. Death was transformative, a process akin to that of a caterpillar changing into a butterfly, as he writes to fellow artist Emile Bernard:

> That existence of painter as butterfly would have for its field

---

Stolwijk, *Van Gogh and the Colors of the Night* (New York: The Museum of Modern Art, New York, 2008).

[2] No. 678 (*Brieven 1990* 681, *Complete Letters* W7). From: Vincent van Gogh, To: Willemien van Gogh Date: Arles, Sunday, 9 and about Friday, 14 September 1888. Location Amsterdam, Van Gogh Museum, inv. nos. b707 a-b V/1962.
Accessed at http///www.vangoghletters.org/vg/letters/let678.

[3] No. 693 (*Brieven 1990* 696, *Complete Letters* 553b). From: Vincent van Gogh, To: Eugène Boch Date: Arles, Tuesday, 2 October 1888. Location Amsterdam, Van Gogh Museum, inv. nos. b598 a-c V/1962.
Accessed at http///www.vangoghletters.org/vg/letters/let693.

> of action one of the innumerable stars, which, after death, would perhaps be no more unapproachable, inaccessible to us than the black dots that symbolize towns and villages on the map in our earthly life. Science—scientific reasoning—seems to me to be an instrument that will go a very long way in the future.[4]

Van Gogh's working methods were based on direct observation. Like a scientist, he pursued the visual facts, rather than the conventions of representation. Indeed, he attempted to paint his night scenes at night, directly on the spot, rather than at the studio from memory. But he also wanted his audience to think beyond the observation about the bigger significance. Observation alone is not sufficient. The act of choosing and depicting a specific subject is to explore its significance. His speculations about the afterlife, transformation, and the connection between earth and the heavens are the conceptual framework for his observations. This makes the conclusion to his note to Bernard all the more unexpected: that science may be a means of guiding us, or guiding our imagination, toward a kind of spiritual mapping of the stars. This is not the simple cartographic mapping of one of the most readily identifiable constellations in the northern hemisphere. I read this as a rich metaphor for the symbolic representation of place and experience. It was Van Gogh's way of suggesting that the overlapping realities of earth and space have meaning for us even if one of those realities remains unknown. Science, perhaps, may remove some of that mystery.

Van Gogh's optimism about science reflected the nineteenth century's belief that the universe was ultimately knowable. The popular view of the night sky was being transformed by new discoveries in astronomy. Access to popular publications about the sky contributed to a general rationalizing and demystifying of the sky and its phenomena. Explanations of eclipses, comets, and phases of the moon, as well as new discoveries, were standard fare in education, public lectures, textbooks, and the popular press. While there is no doubt that the phenomenon of stellar

---

[4] No. 632 (*Brieven 1990* 635, *Complete Letters* B8). From: Vincent van Gogh To: Emile Bernard Date: Arles, Tuesday, 26 June 1888. Location New York, Thaw Collection, The Morgan Library & Museum.
Accessed at http///www.vangoghletters.org/vg/letters/let632.

colour noted by Van Gogh was based on his own observations, the same effect was noted more impersonally in astronomy texts, as seen in this diagram from Amédée Guillemin's book *Le Ciel* of 1866 (fig. 2.2).

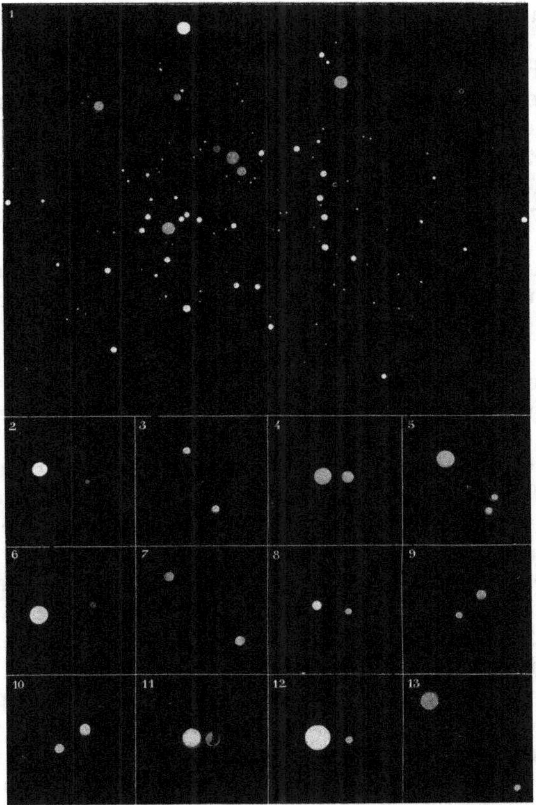

Fig. 2.2: Amédée Guillemin, 'Étoiles colorées', from Le Ciel, *Paris: 1866.*

The traditional understanding of and the personal encounter with the sky, with its meaningful signs, mythical narratives, and symbolic associations of stars and planets, was being rapidly lost in the modern era. The rise of the city simply meant that fewer people looked up at the stars. The meaning of the sky, its entire symbolic character, was changing. Fewer people needed to use the sky as a timepiece, or to determine planting times or discern the weather from observations of the sky. The irony is that as direct contact with the natural sky was being lost under the glare of streetlights, its abstract appearance became more familiar than the real appearance, through the many images, artistic and scientific, that were accessible to the public. Direct observation was replaced by the visual reproduction.

It is difficult to underestimate the impact of these images upon the popular imagination. Much as the telescope had changed fundamental perceptions of the sky beginning in the seventeenth century, the dissemination of visual information during the nineteenth century transformed those perceptions further. Photography, for example, revealed a more complicated sky than had been visually apparent even in the largest telescopes. Photography was revealing the sheer magnitude of the cosmos, the seemingly infinite number of stars. But it also revealed the hidden reality of phenomena barely glimpsed or even invisible to the eye. In this cultural context, van Gogh was no naïve stargazer, but a product of an increasingly science-literate public. His creative struggle was symptomatic of the times—the problem of interpreting the sky through art in an age of modern science.

Van Gogh's later study of the night sky, the famous *Starry Night* of 1889, comes about nine eventful months after *Starry Night over the Rhône*. This is not the place to retell the story of those nine months, but it is important to point out that the symbolic construction of the 1888 painting has been concentrated and intensified in the 1889 painting. As Albert Boime demonstrated, this painting still stays faithful to the direct observation of the night sky, such that we can identify its moment of creation based on the exact configuration of stars, moon and the planet Venus.[5] But two elements separate this work from its predecessor. First, the human presence is implied but not directly represented. The landscape dominates, and the distant town is the only marker of human habitation. Second, the element of abstraction, or at least of imagination or speculation, is also not a part of the earlier work. The great sweep of light that links the moon and the bright light of Venus in the centre of the composition, is not on the celestial map, nor is it something visible in the

[5] Albert Boime, 'Van Gogh's *Starry Night*: A History of Matter and a Matter of History', *Arts Magazine* 59, no. 4 (December 1984): pp. 86–103.

predawn hours as Van Gogh painted this canvas outdoors. It is, however, an artifact not of imagination so much as of memory.

His addition to the image of the sky may reflect contact with the science of the day and apparent knowledge of the running debate about the so-called 'spiral nebulae'. Van Gogh professed that he could only paint what he saw, not what he imagined, yet here memory may have guided him to embellish his painting in an uncharacteristic departure from observed reality. Numerous reproductions of Lord Rosse's drawings of spiral nebulae, invisible except through powerful telescopes, spread the idea that the sky contained unseen objects that only the instruments of science could reveal (fig. 2.3). The debate about the spiral structure of certain nebulae was resolved when photography imaged these elusive structures and 'proved' their reality.

Fig. 2.3: R. Dale, 'The Great Spiral Nebula', after 1850 (?), Institute of Astronomy Library, University of Cambridge.

What Van Gogh was confronted with and what he seems to have addressed in this painting was, in some ways, the same problem that had posed itself from the beginning of the age of science—the reconciliation of observed and the intuited, of the object in nature and the imaginative or creative symbolic reality. Science could present visualizations of an unseen reality, but artists had trafficked in the unseen realities of memory and imagination for centuries. While the word crisis is perhaps too strong a term, we might say that there was a reevaluation of the role of representation and symbolism in art under the presence of science in the nineteenth century. We might also suggest that this reevaluation of representation drove the modernist innovations in art of the subsequent twentieth century. How, then, did an artist so literal as van Gogh confront the problem of an unseen nature when he gazed up at the June sky from the olive groves and wheat fields of southern France? I want to consider the question in a more general fashion as we look at the intersections of art and astronomy in the nineteenth century, an age shaped by science.

Astronomy and art, in their modern forms, each propose a certain vision of the sky, the one driven by data, the other by personal vision and interpretive expression. Both aspire to a representation of truth. But at the points where these two intersected, troubling questions sometimes arise. The division between science and art is sometimes blurred by technological mediation, a fact that we encounter today in the vastly expanded visual regime of astronomical imagery. The impulse to aestheticise the presentation of data, to 'make art' from information, is a desire to make that data appear compelling to an audience that may be asked to support the costs of doing big science. Today, when we admire or perhaps critique the heavily processed, colourised, digitally enhanced images from instruments like the Hubble Space Telescope, we can address the formal aspects, the 'style', but we rarely ask what it means, in the sense of personal or cultural meaning. And we almost never regard these images as intrinsically or consciously symbolic. We have divided the regimes of the visual into symbolic and non-symbolic realms. But not so long ago, this division was at best ambiguous, and a symbolic meaning to nearly all images was precisely what viewers expected and asked.

One narrative of the history of art tells us that the changes sweeping through European art during the nineteenth century represented a rejection of past traditions of style and iconography in favour of a new vision of art, especially in the relationship of art and nature. The classic-romantic divide of the early nineteenth century gave way to a greater naturalism by mid-century. The concept of an art driven by the same empirical observations performed by science, and which placed direct experience at the heart of artistic practice, placed art along a parallel path with the sciences.

The faith in art as the exercise of rational observation helps explain the century's fascination with the limits and boundaries of such art. The painter John Constable would express the opinion in 1836 that the observation of nature took precedence over the invention of forms, and that art and science intersected:

> I am anxious that the world should be inclined to look to painters for information on painting. I hope to shew [*sic*] that ours is a regularly taught profession; that it is *scientific* as well as *poetic*; that imagination alone never did, and never can, produce works that are to stand by a comparison with *realities*.[6]

Constable continued in a subsequent lecture:

> Painting is a science and should be pursued as an inquiry into the laws of nature. Why, then may not a landscape be considered as a branch of natural philosophy, of which pictures are but the experiments?[7]

Constable's statement does not imply that the pursuit of nature through a kind of painterly science is an end in itself, nor must it exclude the possibility of meaning beyond the observed facts. The optimism that scientific inquiry would lead to the truth of nature had a strong appeal, even when it threatened to strip away the mystery and emotion of a more imaginative art. But the idea that a deep symbolic structure was essential to the very existence of art was firmly entrenched in the nineteenth century. But this belief would spill over to other forms of visual representation, opening up the possibility of a symbolic life in even non-artistic images, a fact that will become important to us as we look at the paths of science and art through this period. Bluntly stated, there was no work of art, and perhaps no image, produced in the nineteenth century that did not carry at least the potential for symbolic value for its contemporary audience.

Carl Gustav Carus, in his 1831 *Nine Letters on Landscape Painting*, said:

---

[6] C. R. Leslie, *Memoirs of the Life of John Constable, Esq, R.A.* (London: Longman, Brown, Green, and Longmans, 1845), p. 333.

[7] Leslie, *Memoirs*, p. 355.

> For it is only in this way that man can reveal himself as a totality, and although art and science can be separated from one another in analysis, they can never be wholly separated from one another in actuality. The presentation of science can therefore never succeed without art...and the production of the work of art will equally remain impossible without science.[8]

Carus' close friend Caspar David Friedrich would seek to carry out this depiction of the totality of the human experience. His images of men and women contemplating the moon in the landscape were metaphorical constructions, intuitive reactions that personalized the experience of the sky (*Two Men Contemplating the Moon*, c. 1830, fig. 2.4). But they were also grounded in representational fact, based on close observation of nature, and then framed with the emotional and symbolic elements that pushed science to one side in favour of the art. Friedrich's repeated use of the moon as an object of Romantic yearning and contemplation was part of the larger fascination in the early nineteenth century with the meaning, rather than the reality, of moonlight. Arthur Schopenhauer described this meaning as a personal connection with how the moon moves us, rather than what it actually is: '...It is sublime, in other words, it induces in us a sublime mood, because, without any reference to us, it moves along eternally foreign to earthly life and activity, and sees everything, but takes part in nothing'.[9]

---

[8] Charles Harrison, et al., ed., *Art in Theory 1815–1900* (Oxford: Wiley-Blackwell, 1998), p. 105.

[9] Arthur Schopenhauer, *The World as Will and Representation*, vol. 2, trans. E. F. J. Payne (New York: Dover, 1958), p. 374. See also Sabine Rewald, *Caspar David Friedrich: Moonwatchers* (New York: Metropolitan Museum of Art, New York, 2002), p. 12.

Fig. 2.4: Caspar David Friedrich, Two Men Contemplating the Moon, *c. 1830. The Metropolitan Museum of Art, New York, Wrightsman Fund, 2000 (2000.51).*

The heritage of the romantic view of sky and earth was a search for meaning through the contemplation of nature. There was, as in Van Gogh's works, a desire to connect, or reconnect, an increasingly demystified sky with human affairs and history. Alfred Stevens' image of a woman gazing out of a window at a moonlit seascape, *The Milky Way* (1885–86), reveals how nineteenth century viewers saw two superimposed layers of meaning (fig. 2.5). One layer was defined by the representation of the familiar world and the surface narrative structure of the painting's subject. The other layer was a latent, parallel, and partially hidden symbolism. The woman who gazes from the window, a common trope of European Romanticism throughout the century, represented the contrast of interior and exterior states, be it the soul and the material world, the private and the public, or the feminine and the masculine. The contemporary viewer would have wanted, in fact insisted, that the narrative of the picture synchronise these two layers. Here, a woman is awaiting the return of her lover/husband from some voyage, but the extended metaphor is one of the longing soul confined within an earthly prison. Unlike Van Gogh's paintings of the night sky, which divert narrative in order to develop a deeply idiosyncratic iconography, Stevens' more conventional painting presented the night sky within the traditional symbolic language of the day. The starry sky and the moon are emblems of travel and navigation, of distant places, of romantic longing, and of lunar femininity and beauty.

Fig. 2.5: Alfred Stevens, The Milky Way, *1885–86. Galerie Patrick Derom, Brussels.*

American painter Jasper Cropsey's view of an ancient world lapsing into twilight, *Evening in Paestum*, 1856, was also painted within the conventions of the symbolic language of the nineteenth century (fig. 2.6). Here, the narrative layer points us to historical, rather than individual, possibilities.

Fig. 2.6: Jasper Cropsey, Evening in Paestum, *1856. Frances Lehman Loeb Art Center, Vassar College, Poughkeepsie, NY.*

Such images of distant lands, of Greece, North Africa, the Middle East, were partly reveries of the Romantic poet-traveller. Much like Friedrich's moon watchers, it is a painting that invites a calm, contemplative view of sky and landscape. Twilight was the moment of transition, the fugitive interface between night and day. The twilight nocturne was by convention calm, thoughtful, and melancholy. But there is also a layer of significance that united sky and earth. In juxtaposing the moon and stars with the ruins of Paestum, Cropsey relies upon a common visual trope of the era—the new moon and emerging stars over ancient ruins was widely seen in the nineteenth century as an emblem of the new history, the modern world, superseding the decaying history of the past.

Even as art moved toward an ostensibly realist agenda, the images of sky and moon remained potent symbols. The numerous twilight moons of Jules Breton, often coupled with peasant women returning from the fields, suggested an ageless tradition preserved and perpetuated as a relic of lost knowledge. A certain nostalgia permeated such imagery, a different kind of yearning for return or for reconnection with the sky. It is not surprising that Van Gogh was strongly drawn to Breton's art, as he was to that of Jean-François Millet. Millet's *Starry Night* (c. 1850–65) might have been the specific painting that inspired Van Gogh to paint his own night skies. Van Gogh in his own way sought to emulate the directness of Millet's methods and the elder artist's humility before nature. Millet, like Van Gogh, reminded the viewer of the duality of appearance and meaning, those two layers by which the audiences of the nineteenth century responded to the work. A rural road leads upward to the horizon, also leading to the star-filled sky, in an obvious attempt to force upon us a semi-narrative of ascent and transcendence. But it is the residual symbolism of the image, absent a strong overarching narrative, that appealed to Van Gogh. The path to the stars, like Van Gogh's river of reflections, serves as a metaphorical bridge between earth and sky, life and death, matter and idea.

The expectation that all images were meaningful and potentially symbolic animated the public debate about nineteenth century art. Observation stripped of meaning was mere documentation; symbolism without nature verged on decadence. The idea of the sky was changing rapidly as the century progressed, assaulted by artificial lights and by new insights into the nature of celestial phenomena. But not all images of the sky were pictorial, and the venerable tradition of allegory represented an alternative, albeit increasingly old fashioned, means for communicating abstract concepts about nature. Thus, works like Louis-Ernest Barrias' *Nature Unveiling Herself Before Science,* 1899, are delightful anachronisms, reminding us of the limitations of a pictorial art in expressing certain subjects. How does one visually represent the concept of 'science', or a phenomenon like 'electricity'? The seriousness with which allegorical imagery was taken, not to mention the sheer quantity of works produced, even at the turn of the century (and the 1900 Universal Exposition in Paris was the dramatic swansong of this chapter of art), suggest that there was still a place for the language of personification in the absence of new approaches to an art of symbolic ideas.

While the Symbolist movement in art recast the allegorical mode to more personal and eccentric aims, the early decades of the twentieth century highlighted the failure of allegory to remain relevant in the emerging modernist aesthetic. Allegory as a symbolic language could only become ironic, or at best self-reflexive. For example, Mikalojus Ciurlionis' Zodiac series might have a 'modern' reductive and abstract style, but the basis of his iconography, the astrological rather than astronomical aspects of the constellations, pulls his paintings back into the past rather than forward to the future. It would be for later movements and artists, particularly among the Surrealists, to pick up this tradition and revitalize it. The expressive gap between Ciurlionis and Max Ernst or Joan Miró is far greater than the chronological separation.

Another important aspect in the symbolic visualization of the sky in the nineteenth century, although not immediately apparent nor initially perceived as such, was the science illustration. By the mid-nineteenth century, the basic astronomical facts about the visible phenomena of the sky were a part of nearly every child's education, and broadly disseminated in publications and pictures. The scientific illustration, and later the astronomical photograph, were primarily vehicles of explanation, not expression. The aesthetic aspects of scientific and astronomical illustration in the nineteenth century were important primarily for their popular appeal. Increasingly lavish illustrated publications demanded that artists work with, or sometimes as, scientists in order to give visual interest to data. The view of the Milky Way over Paris from Amédée Guillemin's *Le Ciel,* is strikingly similar to the artists' representation of the night sky (fig. 2.7).

Fig. 2.7: Amédée Guillemin, 'Le ciel de l'horizon de Paris vu à Minuit le 20 Juin', from Le Ciel, *Paris: 1866.*

This is the urban vision of the night sky, the horizon dotted with the spires and domes of readily recognizable Paris monuments. This specificity goes beyond the scientific needs of the star chart—it is an aesthetic choice that adds human interest to the scene and gives context to the view, much like the connecting elements of Van Gogh's or Millet's paintings link earth and sky. At the same time, it is an idealized view, free of smoky plumes from chimneys and the glare of gaslights. As an illustration that intersects art, it emphasizes the new reality of the stargazer, the educated urban dweller whose connection to the sky is more tenuous and restricted than that of his rural ancestors. These star charts were the closest that some people would come to scrutinizing the actual night sky.

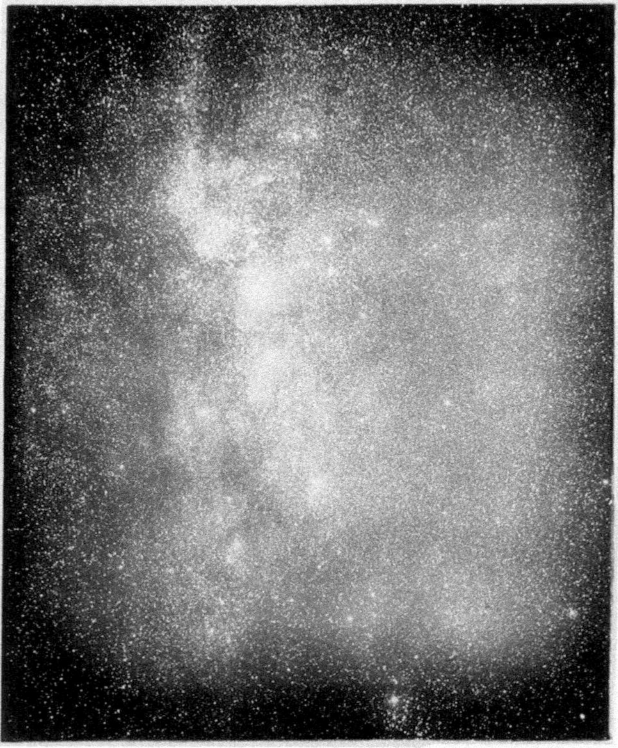

Fig. 2.8: Maximilian Wolf and Johann Palisa, The Milky Way, *c. 1900. San Francisco Museum of Modern Art. Foto Forum purchase, 2006.23.*

It is precisely this context that is stripped from an end-of-the-century photograph of the same subject by Maximilian Wolf and Johann Palisa (fig. 2.8). Removing what artists call the 'staffage' of people or foreground objects, the starry sky is here independent of human relationships, self-sufficient and presented as a thing in itself. This is the modern vision of the starry night, the end result of the empirical arc of the nineteenth century, without the supporting framework of Friedrich's thoughtful watchers, Romanticism's ancient monuments, or Van Gogh's twinkling town lights. Fully purged of humanity, it strikes us as the essence of a future modernist vision, an abstract construction rather than a document.

Now the aesthetic and, to some degree, the symbolic aspects of science illustration in the nineteenth century (and certainly in the twentieth as well) were dependant upon the viewer's familiarity of certain conventions of representation. The famous images of the moon created by John Russell (Birmingham Museums and Art Gallery) just before the beginning of the nineteenth century were products of the intersection of the artistic and the scientific. Russell was a portraitist and professional artist as well as a keen telescopic observer. He brought his skills as an artist to the task of capturing the nuances of the lunar surface in some of the most visually accurate images of the moon in the pre-photographic era. But these were clearly illustrative, not expressive, works. Russell segregated his art and his science by isolating the image, much as we would see it if we were looking through the restricted field of view of a telescope. This narrow view emphasizes the fact of the moon's existence and appearance, without romantic or symbolic connotations. In this manner, he anticipated the dispassionate photography of the moon that would, in the later nineteenth century, rationalize and thus demystify it. But other 'hybrid' artists, including photographers, sought to connect the expressive and even symbolic aspects of art to the needs and demands of science in illustrations and photographs that straddled conventional boundaries. Examples range from Eugène Viollet-le-Duc's alpine sketch of a solar eclipse[10] to Carleton Watkins' photo *Solar Eclipse from Mount Santa Lucia*, January 1, 1889, to James Nasmyth's sculpted models of the lunar surface.[11] The common

[10] Pierre A. Frey and Lise Grenier, *Viollet-le-Duc et la montagne* (Grenoble: Editions Glénat, 1993), p. 69.

[11] James Nasmyth, *James Nasmyth, Engineer: An Autobiography* (New York, 1883), p. 336.

denominator was the contextualization of the sky—comparing and contrasting the earthly and familiar landscape with all its latent meaning, to the phenomena of the heavens.

Even some scientists, at least in more popular forums, sought to return a sense of context and imagination to their illustrations of the moon. Camille Flammarion commonly included 'artistic' images in his popular astronomy books, like views imagined from the moon or alien planets. The success or drama of such an image depends entirely on a visually literate audience that recognized those conventions and appreciated the clever inversion of the earthly landscape. The collaboration of artist and scientist, or the overlapping talents of the two disciplines (both relied, at least in the pre-photographic era, on careful observational and drawing skills), meant there was a segment of this mode of science illustration that invited more direct comparisons to the art of the era. Flammarion used a number of talented graphic artists to supply his images. Some are generic (in a few cases poached from other sources), while others are clearly to spec, illustrating some novel theories and insights by the astronomer. Antoine Pralon's speculative image, 'What painter could imagine the strange light in a world illuminated by four suns and four moons?' from Camille Flammarion, *Astronomie populaire*, of an alien world pushes us to the boundaries between science and a visionary art (fig. 2.9). We see Van Gogh's coloured stars from a different perspective. But even here, the prevailing conventions of contemporary landscape painting give this alien vision an oddly familiar feeling. It could not be otherwise. Too abstract an image would defeat the purpose of the illustration, calling attention to itself rather than the concept for which it stands. The artist might have felt that the landscape of a distant planet was really no different than that of some unusual place on earth. It was more convenient to repurpose the existing imagery of dramatic terrestrial landscapes than to invent something totally new. But did this appropriation of the conventions of landscape art also bring with it the symbolic associations of those landscapes? There are few contemporary responses to guide us.

Fig. 2.9: 'What painter could imagine the strange light in a world illuminated by four suns and four moons?' from Camille Flammarion, Astronomie populaire, *Paris: 1880.*

The data of astronomy in the nineteenth century was cast into a multitude of graphic forms (fig. 2.10). These design solutions to the problem of visualizing data have a relevance to our own contemporary world, where the torrent of information that we must digest, scientific or otherwise, requires creative visual solutions to make that information accessible. The presentation of science data has been in recent years studied as evidence of the broader visual culture of the past. James Elkins, Barbara Stafford, and others have tried to understand the special context in which scientific illustrations operate as aesthetic and symbolic experience.[12]

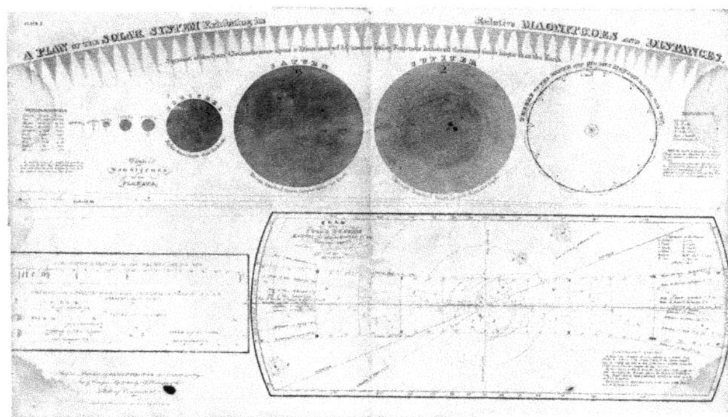

Fig. 2.10: 'A Plan of the Solar System Exhibiting it's Relative Magnitudes and Distances', Published by F. J. Huntington, Cleveland: 1835.

These scholars have noted the organizing design principles, the decisions made about size, shape, colour, or composition, that transcend the mere presentation of data and nudge these images into some indeterminate relationship to high art. Few scholars contend, however, that there was a symbolic dimension to these diagrams, charts, and schematic maps.

But more recent approaches to the scientific illustration perhaps changes that attitude. I believe that there is symbolic potential in some of these images that gives a richer experience to the contemplation of the sky in our own time. Beyond the data that such illustrations present, their aesthetic and symbolic life is a function of the associations and experiences we bring to them from the past as well as the present. I suspect that Jason Rowe's diagram of planetary candidates from the Kepler mission may provide the same

[12] James Elkins, *Six Stories from the End of Representation: Images in Painting, Photography, Astronomy, Microscopy, Particle Physics, and Quantum Mechanics, 1980-2000* (Stanford: Stanford University Press, 2008). Barbara Maria Stafford, *Artful Science* (Cambridge, MA: MIT Press, 1994). Simon Schaffer, 'On Astronomical Drawing', in Caroline A. Jones and Peter Galison, *Picturing Science, Producing Art* (New York: Routledge, 1998), pp. 441–74.

kind of sublime experience for some that Van Gogh had when looking at the night sky over a century earlier (fig. 2.11). The colors of the stars that Van Gogh noted have been systematized, the repetitive multitude of stars a metaphor for both methodical examination and expansive vistas, and the alien planets themselves, those that Van Gogh imagined as the destinations of souls, are black dots on a highly abstracted map.

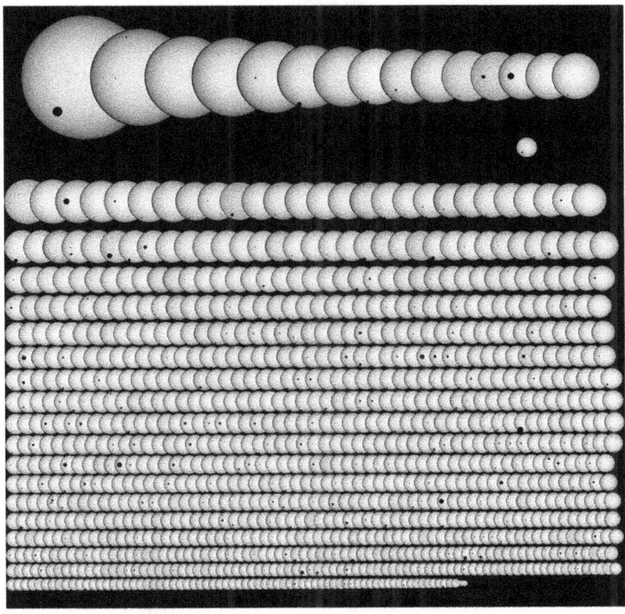

Fig. 2.11: Jason Rowe, 1,235 planet candidates from the Kepler Mission, *2011. Credit: Jason Rowe, NASA Ames Research Center and SETI Institute.*

I suggest that our interest in these images as aesthetic and symbolic objects is filtered through the revolutions in the visual arts of the twentieth century. The triumph of abstraction and non-objectivity have made these illustrations appealing on an aesthetic level that was not fully formed, or at least not fully articulated, when they were originally created. But I also suggest that there is, like in other nineteenth century images, always a latent symbolic potential to any image, no matter the source or purpose. If the common expectation of the nineteenth century was that all (art) images were significant, we would need to explain the demise of such expectations for more recent images.

If our musings about the twentieth and twenty-first century heritage of nineteenth century art and science lead us to rethink the relationship of celestial symbolism and the visual image, it is because we are open to possibilities that simply did not exist for earlier artists and scientists. The nineteenth century witnessed the end of one pictorial era and the beginning of another. The consequences of the new visual culture, shaped by photography and mass reproduction of images, are still with us today. We also accept that science is often about the invisible—the problematic aspect of the new technologies of the nineteenth century that forced artists to reconsider their pictorial strategies. We can now explore the full range of the electromagnetic spectrum outside the narrow window of human sight and code the invisible to make a pseudo-image. For some, that coded visualization carries the same symbolic weight today as the colours of the night did for Van Gogh one hundred and twenty years ago. In the early twentieth century, the possibilities of a radically transformed art led artists to the invisible as well. Abstraction and non-objectivity distilled the essence of nature down to the simplest, one might say elemental, forms. This kind of reductivism resonates strongly with the attitude expressed by scientists and their data. But art must carry the burden of history. As much as the modernist turn of the twentieth century was in part a stripping away of the conventions of iconography of the past, there remains that residue of significance that clings like a thin layer upon the surface of modern art.

A modern sublime, a kind of neo-Romantic image of natural forces and human experience, remain a powerful thread in contemporary art. The presence and meaning of the sky in the post-Apollo, postmodern era is yet more complex and vexed than ever before. The symbolic connection between viewer and sky is both richer and yet more mediated than at any time in history. Art and science have both given us a vast repertoire of images of every aspect of the sky, visible and invisible. New perspectives have displaced the old speculative visions of space. The views of the earth from the moon returned by the Apollo astronauts reignited a nostalgia for the traditions of iconography. It became a multipurpose symbol of technological triumph, environmental fragility, global unity, and existential loneliness. The images of the space age made concrete the centuries of speculation about the appearance of the earth, but it was the emotional reaction, the poetics of that image, that have come to define its place in the pictorial history of the sky. It is this symbolic reality that I believe Van Gogh would have understood.

PART TWO:

Ancient, Medieval, and Early Modern Expressions

THE BURNING SUN AND THE KILLING RESHEPH: PROTO-ASTROLOGICAL SYMBOLISM AND UGARITIC EPIC

Ola Wikander

ABSTRACT: In the mythological texts from Ugarit in modern Syria (probably fourteenth century BC), the motif of the deadly, drought-inducing sun plays a large part: the burning sun casts its destroying rays as a result of the power and influence of death, especially the god of death himself, Mot. However, the solar goddess Shapshu is in other cases portrayed as quite a benevolent entity. In this paper, I discuss this discrepancy in the light of an astronomical omen text that appears to mention the dangerous properties of the sun when it appears together with Resheph, the god of feverish, hot illness—probably representing the planet Mars. The presentation touches on the possibility of these motifs representing early stages of proto-astrological symbolism.

The roots of the most well known system of 'sky and symbol', i.e. horoscopic astrology, lie in the astral divination of ancient Mesopotamia. Understandably, this fact has led to a considerable scholarly interest in (particularly Babylonian) sky-based prediction, the study of what the Babylonians themselves referred to as the *šiṭir šamê*, 'the writing of heaven'. There is, however, another source of astral symbolism in the Ancient Near East that may, as I hope to show in this paper, have an important bearing on the history and development of the celestial bodies as carriers of symbolic meaning. This is the religious mythology and ideology of the ancient North West Semitic cultural sphere, especially typified by the writings from the city of Ugarit, on the Syrian coast, from the fourteenth and thirteenth centuries BC.[1]

[1] An earlier embryo of some of the ideas underlying this article was published in popular form in Swedish as Ola Wikander, 'Solen och pestguden i Ugarit', *Aorta* 27 (2011): pp. 17–21. On the general questions of the relationship between sun and the powers of the netherworld at Ugarit as well as many of the philological intricacies into which I do not have the space to enter here, I refer to my

These texts have been unearthed in the form of alphabetically inscribed clay tablets in a language today known as Ugaritic. A large number of such alphabetic cuneiform tablets have been found since the modern rediscovery of Ugarit in 1928, and this corpus constitutes the most comprehensive collection of pre-Christian North West Semitic textual material outside of the Hebrew Bible, with which the texts often display remarkable and in certain cases literal correspondences. Amongst other materials, the Ugaritic corpus includes mythological and ritual material relating to the cults of 'Canaanite' deities such as Baal, El, Anat, and it is to these texts that we must turn in order to gain an understanding of the ancient Ugaritians' view of the symbolic language of the sky, in this case, especially concerning the sun. It is today an established fact of scholarship that the Ugaritic texts form an invaluable collection of background material against which to understand the development of the Hebrew Bible, which gives their contents a direct relevance for the history of Abrahamic religion as a whole. The Ugaritic texts are our best source for the religious culture preceding and surrounding the nascent Israelite religion, and although our knowledge of Ugaritic religion and cosmology is far from perfect, the textual material from Ugarit still constitutes a treasure trove of information on ancient North West Semitic religion. In this paper, I intend to discuss a Ugaritic omen text concerning the possibly destructive role of the sun in the light of this narrative religious material. I will begin by giving a summary of relevant parts of these texts.

In the main mythological narrative recovered from Ugarit, the so-called Baal Cycle, the sun goddess Shapshu (whose name literally means 'sun' in Ugaritic) plays a multifaceted, somewhat ambivalent role. The Baal Cycle tells the story of how the young storm god Baal gained kingship of the gods and the universe by defeating his two cosmic enemies Yamm ('Sea') and Mot ('Death'). It is his conflict with the latter that concerns us here.[2]

---

doctoral dissertation *Drought, Death and the Sun in Ugarit and Ancient Israel.*

[2] The interested reader can find a good, modern English rendering of the Baal Cycle in Mark Smith, 'Baal', in Simon B. Parker, ed., *Ugaritic Narrative Poetry*, SBL Writings from the Ancient World 9 (Atlanta, GA: Scholars Press, 1997), pp. 81–176. This translation has the added benefit of including the original text on each facing page.

The Ugaritic tablet *CAT* 1.4 has recounted in great detail how Baal builds his palace, the symbol of his kingship, without which he cannot rule gods and men and send fertility to the earth in the form of his life-giving rains. The young god throws a house-warming party of sorts, inviting various gods to feast. One god does, however, not attend—Mot, the god of Death, who lives deep in the earth. At the beginning of tablet 1.5, he instead sends an invitation of his own, instructing Baal to descend to the land of the dead. Mot actually manages to intimidate Baal and coerce him into descending into the gullet of the dread ruler of the netherworld. Baal, the bringer of fertility, rain and life, has perished.[3] It is in this situation that the role of the sun becomes pivotal. Three times during the course of the Baal myth, the poet reiterates the same formulaic expression of the result of Mot's rule:

nrt ilm špš ṣḥrrt	The sun, the lamp of the gods, is burning,
la šmm b yd bn ilm mt	the heavens are powerless in the hands of Death, the divine one.

(*CAT* 1.6 II 24–25)[4]

This is the result of Death's rule: a terrible, all-consuming drought, sent out through the medium of the sun. The same phenomenon is described in a number of places in the Cycle:

---

[3] There has been a great deal of discussion concerning Baal's descent into the netherworld—what it means, and if the motif is even really present in the text. For a thorough survey of the various points of view and a compelling and convincing argument for Baal's death as a quite authentic part of the Cycle, see Tryggve Mettinger, *The Riddle of Resurrection: 'Dying and Rising Gods' in the Ancient Near East*, Coniectanea Biblica Old Testament Series 50 (Stockholm: Almqvist & Wiksell International, 2001), pp. 55–81.

[4] The Ugaritic texts are quoted according to the standard edition *CAT*, i.e. M. Dietrich, O. Loretz and J. Sanmartín, eds., *The Cuneiform Alphabetic Texts from Ugarit, Ras Ibn Hani and Other Places (KTU: Second, Enlarged Edition)*, Abhandlungen zur Literatur Alt-Syrien-Palästinas und Mesopotamiens 8 (Münster: Ugarit-Verlag, 1995). Translations of the Ugaritic texts are my own; in the interests of space and readability, the translations are mostly given 'as is', without the thorough philological discussions and references of the sort that would be necessary in a publication aimed at a purely Ugaritological audience.

'the furrows of the fields are parched', the text says.[5] This terrible calamity affects all of nature and is the main manifestation of Death's power in the world. In the basically agricultural society of Bronze Age Syria, drought was a threat like no other: indeed, one of the proposed contributing reasons for the sudden demise and collapse of Ugarit shortly after 1200 BC is intense and unremitting drought—maybe Mot won after all?[6]

The role of the burning drought and the terrible glare of the sun is apparent already in the threats directed at Baal before the hero's descent. Mot says:

tṯkḥ ttrp šmm	[...] The heavens will grow hot, they will shine,
krs ipdk ank	for I shall tear you asunder and swallow you.[7]
(*CAT* 1.5 I 4–7)	

Here, the calamity befalling the skies is expressly associated with the actions of the god of Death and his rule. The influence of the netherworldly powers and the terrible heat of the heavens are directly linked to one another.

At the end of the story, however, Baal somehow returns to life and confronts Mot; he cannot destroy him (Death never truly dies), but the sun goddess declares Baal the winner of the contest. The text actually ends with a hymnal praising of Shapshu, the sun, which describes her as the ruler of the shades of the dead.

The sun goddess Shapshu is not in any way an 'evil' character in the divine drama of the Baal Cycle: she is on the

---

[5] *CAT* 1.6 IV 1–3: *pl ʿnt šdm*

[6] The view that drought contributed significantly to the demise of Ugarit is held by Claude Schaeffer, 'The Last Days of Ugarit: Drought, Famine, Earthquakes and, Ultimately, Fire Ended Civilisation at Ugarit', *Biblical Archaeology Review* 9 (1983), pp. 74–75. However, the view that a military invasion by 'Sea Peoples' was involved is also common. For evidence of this, including contemporary letters from Ugarit possibly referring to such a military threat, see Michael C. Astour, 'New Evidence on the Last Days of Ugarit', *American Journal of Archaeology* 69, no. 3 (1965): pp. 253–58.

[7] My translation of this philologically complicated double-line is informed by the insights of J.A. Emerton, 'A Difficult Part of Mot's Message to Baal in the Ugaritic Texts', *Australian Journal of Biblical Archaeology* 2 (1972): pp. 50–71.

side of Baal and his friends and help them in many cases. At the end of the Cycle, she is the one who declares Baal the king of the gods. But she has an ambivalent part to play: because of the daily motion of the sun beneath the horizon, she was thought to descend into the land of the dead, and it is my view that this connection with death and dying is in many ways of great importance to understanding her symbolic associations. She helps in bringing Baal's body back from the netherworld, no doubt due to her connection with death and the underworld.[8] But the relationship goes further: it is through her that the god of Death manifests his rulership by sending out drought. She is a liminal character, whose association with the land of the dead gives vastly different results depending on whether the ruler of life (Baal) or the ruler of death (Mot) currently has the upper hand in the story. There is even a text that appears to describe a royal funeral, in which Shapshu is called upon to 'be hot' in order to open the way between the worlds of the living and the dead, an element that I interpret as a beneficial version of her 'burning' connection with the drought-inducing lord of the netherworld seen earlier.[9]

The idea of a solar deity connected to the netherworld is not unheard of in ancient Near Eastern Religion. One parallel example can be found in the so called *Epic of Liberation*, preserved as a bilingual in Hittite and Hurrian found at the Hittite capital of Hattusa, in the Hittite version of which the Storm God meets the sun deity by going 'down into the dark earth' (*kattanta tankuwai takni* in Hittite).[10] The sun living beneath the earth was not as strange as it may seem to modern people.

The dual and almost contradictory symbolic role of the sun in the Baal Cycle merits some attention: Shapshu is both the kingmaker, the symbol of the royal family—and the scorcher, the one who transcends the boundary between life and death, both for good and for ill. Already at this juncture,

[8] *CAT* 1.6 I 8–18.

[9] The text in question is *CAT* 1.161, with the relevant verb appearing (twice) in line 18.

[10] The *Epic of Liberation* has been published in Erich Neu, *Das Hurritische Epos der Freilassung I: Untersuchungen zu einem hurritisch-hethitischen Textensemble aus Ḫattuša*, Studien zu den Boğazköy-Texten 32 (Wiesbaden: Harrasowitz, 1996). The expression occurs at KBo XXXII 13, l. 10 (p. 221 in Neu's edition). The title of this Hittite solar deity is even 'the sun goddess of the earth' (*taknāš* dUTU-*uš*).

we may note the interesting parallelism between this dual view of the sun and the role of the heavenly body in the fully developed system of horoscopic astrology, with the sun both being regarded as the natural significator of kingship and as the bringer of combustion, 'burning' and negatively influencing planets close to it.

If we search outside the purely mythological corpus of Ugaritic texts, we may find even more illuminating illustrations of these dual symbolic natures of Shapshu/the sun. As a first step, let us look at some examples from the numerous ritual texts recovered from Ugarit. In a number of these texts, which to a large extent consist of lists of sacrifices made to various gods, Shapshu appears next to another, highly conspicuous, divine character, namely Resheph, the god of battles, plague, fever and heat, sometimes identified at Ugarit with a Mesopotamian counterpart, namely the chthonic god Nergal, a figure with similar gloomy connotations.[11] Resheph is a god well known from the Ancient Near East—his name even appears a few times in the Hebrew Bible, sometimes traditionally interpreted as a term for fire or arrows, for example Deuteronomy 32:24 (*mĕzê rāʿāb ûlĕḥûmê rešep*, 'The wasting of hunger, and the devouring of the fiery bolt' [Jewish Publication Society Bible]), and Habakkuk 3:5 (*lĕpānāyw yēlēk dāber wĕyēṣēʾ rešep lĕraglāyw*, 'Before Him goeth the pestilence, and fiery bolts go forth at His feet' [JPS]). In these cases, the word *rešep* was originally the name of the North West Semitic deity.[12]

The appearance of these deities (Shapshu and Resheph) next or close to each other in the sacrificial lists is a sign they were in some respect regarded as being associated with one

[11] Shapshu and Resheph (or Rashpu, as he was probably known in Ugaritic) occur together in the ritual text *CAT* 1.41/1.87, and also in the sacrificial list *CAT* 1.148, which has a syllabically written equivalent in RS 92.2004. All these texts can found in translation and original in Dennis Pardee (ed. Theodore J. Lewis), *Ritual and Cult at Ugarit*, SBL Writings of the Ancient World 10 (Atlanta, GA: Scholars Press, 2002). Resheph is identified with the Mesopotamian Nergal in the Ugaritic god-list RS 20.024.

[12] Important studies on Resheph are William J. Fulco, S.J., *The Canaanite God Rešep*, American Oriental Series 8 (New Haven, CT: American Oriental Society, 1976) and Edward Lipiński, *Resheph: A Syro-Canaanite Deity*, Orientalia Lovaniensia Analecta 181, Studia Phoenicia XIX (Leuven: Uitgeverij Peeters en Departement Oosterse Studies, 2009). The latter work differs from general opinion by holding (p. 262) that Resheph was not an essentially chthonic deity.

another. This may seem surprising: the goddess of the sun and the ruler of pestilence and fever would seem to be strange bedfellows indeed. But it is probably no coincidence: Resheph may, amongst other things, have been a god of the netherworld, the world below the horizon into which the sun was thought to descend every night, and we know that his Mesopotamian analogue, Nergal, was many times specifically associated with scorching heat and drought. In fact, an alternate name of Nergal is Erra, which was shown by J. J. M. Roberts to be derived from the Proto-Semitic verbal root *ḥrr* ('to scorch').[13] This means that this name could etymologically be interpreted as meaning 'Scorcher'. Nergal was also identified with the planet Mars in Babylonian astronomy. The possibility that the netherworldly role of the sun may be in focus in these sacrificial lists is underscored by the fact that one of the lists which mention Shapshu and Resheph together expressly describes the solar deity as *špš pgr*, 'Shapshu of the corpse'.[14]

This state of affairs gives us an indication that the symbolic role of the solar deity at Ugarit was thought of in terms of her connectedness not only with the god of Death himself (i.e., Mot) but also with lesser but still dangerous divinities such as Resheph, whose character goes together excellently with Shapshu's dangerous effects as sender of drought and terrible heat.

All of which brings us, at long last, to the text I specifically mean to discuss in this paper:

bṯṯ ym ḥdṯ ḫyr
ʿrbt špš
ṯġrh ršp
kbdm tbqrn
skn

(*CAT* 1.78)

[13] J.J.M. Roberts, 'Erra: Scorched Earth', *Journal of Cuneiform Studies* 24, no. 1/2 (1971): pp. 11–16. Roberts most closely associated the 'scorching' with the devastation connected to fire and war—this need not, however, be the only type of burning relevant.

[14] *CAT* 1.39:12, 17 and 1.102:12. In 1.102, Resheph occurs only two lines before this designation.

> The day of the new moon of Ḫiyar was shamed (or: on the sixth day of the new moon of Ḫiyar),
> the sun 'went in'.
> Her gatekeeper was Resheph.
> Two livers were studied:
> danger!

This text is one of the few astral omen texts we possess from Ugarit. It is quite a short piece of proto-astrological literature, and it involves a description of an astronomical event, together with a sort of micro-delineation of the meaning of that event. Exactly what astronomical reality the text is referring to is not entirely clear: how are we to view the significance of the word *ʿrbt* (literally 'she went in', the subject being Shapshu, the sun)? This word some researchers have interpreted simply as reference to the sun setting (that is, going into the netherworld), while others have seen in it a description of a solar eclipse—that she 'went in' during the day and not in the night time, as usual.[15]

Connected to this is a similar philological problem in the first line of the text: are the initial three letters a dating ('on the sixth day' or 'in the sixth hour') or a verb ('the day was shamed')? The latter interpretation is especially favoured by those seeing the text as a solar eclipse oracle. If, indeed, the text does describe an eclipse, one would be able to imagine possible dates for it, and there have, of course, been suggestions (such as 3 May 1375 BC and 5 March 1223 BC, dates put forward by Sawyer/Stephenson and de Jong/van Soldt, respectively).[16]

What concerns us here, however, is not the linguistic

---

[15] A convenient digest of various translations of this short but enigmatic text can be found in Nicholas Wyatt, *Religious Texts from Ugarit* (2nd ed.), The Biblical Seminar 53 (New York: Sheffield Press, 2002), pp. 366–67.

[16] J. F. A. Sawyer and F. R. Stephenson, 'Literary and Astronomical Evidence for a Total Eclipse of the Sun in Ancient Ugarit on 3 May 1375 BC', *Bulletin of the School of Oriental and African Studies, University of London* 33, no. 3 (1970): pp. 467–89; T. De Jong and W. H. van Soldt, 'Redating an Early Solar Eclipse Record (KTU 1.78): Implications for the Ugaritic Calendar and for the Secular Accelerations of the Earth and Moon', *Jaarbericht Ex Oriente Lux* 30 (1987–1988): pp. 65–77. The Sawyer/Stephenson interpretation presupposes that the 'Resheph' of the text is not the actual planet Mars but rather the star Aldebaran (which was very close to the sun) mistaken for that planet (p. 471).

conundrum of whether or not the text talks of an eclipse, but what is said about Shapshu/the sun in the text. The main statements are that she 'went in' and that Resheph was her 'gatekeeper' (*ṯāġiru*, cognate with the Hebrew *šôʿēr*). Given what we know of Resheph and his being identified with the Mesopotamian Nergal, it is possible to regard this as an expression of what we would refer to with modern terminology as a conjunction between Mars and the sun. If we, for the sake of argument, assume that this is the central message of the text and that *ʿrbt* simply means what it appears to do (that the sun set), the expression takes on an even greater import. The chthonic deity Resheph is Shapshu's gatekeeper, the one guarding her path into the netherworld, the land of the dead. The conjunction (?) with Resheph is then a central message of the text, regardless of whether the sun was eclipsed or not. However the case may be, the divinatory specialist apparently regarded Resheph as important enough to write about in the remarkably short report.

As a result of this astronomical observation, the Ugaritians appear to have resorted to another form of divination much favoured in the Ancient Near East: haruspical analysis of liver omens. And the result of these various techniques of divination applied in this case are laconically and succinctly stated at the end of the text: 'danger' (*skn* in Ugaritic).

The question is now: what does this mean? Why does the sun 'going in' together with Resheph/Mars constitute such a threat? In previous studies of this omen text, it has been common to regard the calamity as purely astronomically mechanistic in action, especially when the text is viewed as an account of an eclipse. An eclipse leads to danger, full stop. But why then mention Resheph specifically? Is the text only a report on an unusual astronomical observation and nothing more?

I would like to widen the perspective somewhat, and look at the omen in the light of what we know of Shapshu from the mythological texts, and of the symbolism attached to the sun in Ugaritic religion. After all, if a complex and multi-faceted collection of ideas are attached to the sun in the narrative literature from Ugarit, there is no methodological reason to expect less of a more technical text, such as an astronomical omen. There is no self-evident need to separate what we know of the Ugaritians' view of the sun in the context of religious epic and theology from their ideas concerning that heavenly body in the realm of astral divination. Indeed, it is a common feature of exponents of such divination to associate their ideas with the mythological and theological ideas of the surrounding religious culture.[17] The role of the diviner was one of the pre-eminent career paths of the ancient Near Eastern scribes, and, as a matter of fact, the scribe Ilimilku, responsible for writing down the Baal Cycle, is described in the colophon at the end of *CAT* 1.6 as being a disciple of a personage by the name of Attēnu, with the epithet *prln*, which may in fact mean 'diviner', showing the close connection between literary scribal activity and divinatory practice at Ugarit.[18]

So, we know the following: Shapshu is connected with the land of the dead. She functions as a messenger between this world and the chthonic realms. She can make kings and dispense justice, but she can also fall under the sway of the powers of the netherworld, which leads to her scorching the world with her drought-inducing rays. Her 'negative' characteristics come into play when she is ruled by the forces of Death, when her connection with the afterlife is so to speak 'perverted'. And she is associated with the destroying Resheph, the inducer of feverish heat, not least when she is referred to as 'Shapshu of the corpse'. Note the close conceptual correspondence between the ideas of burning drought and killing fever.

And what does the omen text tell us? It says that Shapshu had been seen together with 'Resheph' (probably Mars), that he was her gatekeeper on the way to the netherworld. He, the chthonic deity of disease and fever, had symbolically taken hold of her, which caused 'danger'. What danger could this have been, if not the very danger ascribed to Shapshu in the Baal Cycle itself—the perversion of her netherworldly connotations, the power of Death in the world, and the destroying heat itself, the glare of the sun identified with the

[17] For one salient example of this process (in the context of Indian astrology, and Western appropriation of the same), see Martin Gansten, 'Reshaping Karma: An Indic Metaphysical Paradigm in Traditional and Modern Astrology', in Nicholas Campion, ed., *Cosmologies: Proceedings of the Seventh Annual Sophia Centre Conference 2009* (Ceredigion, Wales: Sophia Centre Press, 2011), pp. 52–68.

[18] On the meaning of this word, see W. H. van Soldt, ''*atn prln*, "'attā/ēnu the Diviner"', *Ugarit-Forschungen* 21 (1989): pp. 365–68.

fever of Resheph? The contact between the sun and Mars was, I propose, read in a wider symbolic context, vividly expressing the danger inherent in the sun. The killing Resheph was interpreted as taking out the worst in the sun, as turning her from kingmaker into scorcher as he opened her way into the realm of the dead. The very fact that Shapshu is mentioned as 'going in' may be an indicator that her chthonic characteristics are here in focus. Just as the above-quoted little refrain from the Baal Cycle talks of Shapshu burning and destroying when she is (literally) 'in the hands of Death', having Resheph as her gate-keeper may well be imagined to have a similar, devastating result.

Read in this way, the little text *CAT* 1.78 becomes more than a report on an astronomical condition, whether an eclipse or not. It becomes part of an elaborate symbolic structure concerning the sun, the powers of Death and their relationship with one another, that touches upon the same dual associations connected with the sun that come to light in the classical system of astrology. The sun is the symbol of rulership, but also of potential destruction. The sun is used as a metaphorical vehicle with which to express apparently disparate concepts concerning life and death, kingship and calamity, lordship of the universe and the drought that kills all nature. When the sun was on the side of the storm god Baal, she was kingship personified, but when she was in the hands of Death or the burning and smiting Resheph/Mars, her character was regarded as much more unfortunate, connected to drought and destruction—combustion, if you will. Whether or not a conception such as this one was a direct influence on the classical astrological view of being under the rays of the sun is, of course, not self-evident, but the possibility is certainly there, and, in any case, the Ugaritic texts prove that such a conception of the contradictory roles of the sun was prevalent in the Ancient Near Eastern milieu in which astrological tradition has its earliest roots. Whether the influence was direct or not, the Baal Cycle and *CAT* 1.78 show us (when read together) the calamities that could be associated with the solar deity in the minds of the people of ancient Ugarit, a burning, scorching power from the forces of the netherworld, possibly 'channelled' through the medium of Resheph/Mars. Such, I argue, was the nature of that which the Ugaritic divinatory experts scribbled down at the end of their tablet:

'Danger!'

FROM BABYLON TO JERUSALEM: THE ROOTS OF JEWISH ASTROLOGICAL SYMBOLISM

Andrea D. Lobel

ABSTRACT: The history of astrology in early Judaism is complex, and is inextricably connected with the astrological traditions of Greece and Mesopotamia. Indeed, this history is one that is rife with ambivalence, and even outright suspicion, toward both astrology and Hellenism on the part of Jewish rabbinical authorities, particularly during the Late Antique period. However, at the same time, we note the usage—and later, the Jewish transformation—of the Greek zodiacal signs in literary, interpretive, and artistic works alike.

In this paper, I will briefly describe the origins of the Greek signs of the zodiac, set against the ancient background of Babylonian astrology. I will then discuss the ways in which Babylonian and Greek symbolism found their way into early Jewish zodiacal imagery—from Hebrew Bible interpretation, mosaic synagogue floor designs and Jewish mystical texts in the ancient world to contemporary writings on the topic of Jewish astrology. Throughout, we will examine the trajectory taken by astrological symbolism within Judaism, with the dual aim of negotiating its many historical transitions, and gaining a better understanding of the Jewish integration of elements previously considered foreign.

INTRODUCTION: ON TEMPORALITY, TERMINOLOGY AND THE TEXT

As Douglas Adams wrote in *The Hitchhiker's Guide to the Galaxy*, 'Time is an illusion. Lunchtime, doubly so'.[1] Where daily life in the ancient world was concerned, however, time and time-keeping devices—whether based on astronomical observation, sundials,[2] or fixed in clocks and calendrics—regulated societal, agricultural, and state functions, as well as

[1] Douglas Adams, 'The Hitchhiker's Guide to the Galaxy', in *The Hitchhiker's Trilogy* (New York: Harmony Books, 1979), p. 20.

[2] A sundial was, in fact, according to VanderKam, unearthed by R. de Vaux in 1954 at Khirbet Qumran (89).

religious observances. Moreover, timekeeping and calendar also served to demarcate and distinguish social and religious groups one from the other. As Sacha Stern has pointed out, a calendar provides such groups with 'an essential point of reference for interpersonal relations and time-bound communal activity. It determines how time is lived and utilized in the community, and sometimes shapes the community's distinctive identity'.[3]

This was certainly the case within early Judaism. While we know very little about the calendar used in Iron Age Judaism during the composition of the earliest books of the Hebrew Bible, more data exists from the Second Temple period, circa 538 BCE–70 CE. There is, in fact, strong evidence for the presence of both a solar and lunar calendar within Judaism. These included the solar calendar so common in sectarian Jewish literature during the Hellenistic period (e.g., within *Jubilees* and *1 Enoch*), and a lunar calendar, which was very likely influenced by the Babylonian lunar calendar.[4]

This leads us to the topic at hand—namely, the history of astrology in early Judaism, and more specifically, the question of historical and cultural provenance. For in the case of Judaism, as we will see, calendrics and astrological symbolism are strongly linked, particularly in the synagogue art of the Late Antique period (ca. the first through eighth centuries CE).

James Charlesworth has challenged the notion that astrology did not gain popularity in Judaism until the medieval era, and that prior to that time, it was simply considered pagan.[5] Whereas earlier scholars had been hesitant to address the question of magical or divinatory practices within Judaism, Charlesworth put this question to rest with his exposition of the numerous references to the zodiac within Judaism during the Second Temple and Late Antique periods.[6]

---

[3] Sacha Stern, *Calendar and Community: A History of the Jewish Calendar Second Century BCE–Tenth Century CE* (Oxford: Oxford University Press, 2001), p. v.

[4] Ibid., pp. 28–30.

[5] James H. Charlesworth, 'Jewish Interest in Astrology during the Hellenistic and Roman Periods', in *Aufstieg und Niedergang des römischen Welts* Vol 2., no. 20.2 (Berlin: de Gruyter, 1987), pp. 926–27.

[6] James H. Charlesworth, 'Jewish Astrology in the Pseudepigrapha, the Dead Sea Scrolls, and Early Palestinian Synagogues', *The Harvard Theological Review* 70, no. 3–4 (1977), pp. 183–200.

Given the broad historical scope of this paper, we must limit our field of inquiry to two key questions. First, what evidence do we have demonstrating the emergence of Babylonian and Hellenistic Greek astrology in Second Temple and Late Antique Judaism? And secondly, given the biblical prohibitions against idolatry and foreign practices, how did astrology—with its Babylonian and Greek origins—become entrenched within the common lexicon of Judaism?

We will begin with a brief overview of the admonitions against astrology found in the Hebrew Bible, its mixed reception in the Second Temple and Late Antique periods, and then summarize the transitions from earlier Mesopotamian celestial divination to horoscopic astrology in first century BCE Babylonia and Hellenistic Greece. Evidence will then be presented for the incursion of several of these astrological motifs into selected examples of Greco-Roman and rabbinic literature, as well as zodiacal symbolism found in Late Antique synagogues—the very seat of Jewish holiness and prayer. Finally, we will note the ongoing presence of astrology within Judaism despite the historical ambivalence and earlier biblical admonitions.

ASTROLOGICAL AMBIVALENCE: HEBREW BIBLE TO TALMUD

> *'Thus said the Lord: Do not learn to go the way of the nations,*
> *And do not be dismayed by portents in the sky;*
> *Let the nations be dismayed by them!' (Jer 10:2)*

Within the Hebrew Bible and in extra-biblical literature, we note numerous references to celestial objects, from the sun and the moon to stars, planets, and phenomena that suggest the observation of solar and lunar eclipses, comets, and meteors. Indeed this astronomical imagery has lent itself to a multiplicity of interpretations and uses, both within the biblical texts themselves and in later rabbinic literature. For example, we note numerous astrological debates within the Babylonian Talmud. However, in this, there is a tension, for the examination of astrology brings with it a complex of allied topics including magic (*kishuf*), idolatry (*avodah zarah*), and following foreign customs (*chukat goyim*). Practices expressly forbidden by the Hebrew Bible included divining, soothsaying, and necromancy,[7] and this encompassed divination by means of astral bodies, which came to be conflated with astrology in later interpretive sources.

Nevertheless, later musings on astrology eventually found themselves expressed in both Second Temple and Late Antique Jewish writings. It must first be mentioned that during the Second Temple period, there was not one monolithic form of Judaism, but many Judaisms, including sectarian forms such as the Jewish group that formed the Qumran community. During this period, it is sometimes not easy to identify precisely which group espoused particular astrological beliefs due to the difficulty of pairing writings with their rightful author/s. In Late Antique Judaism, however, sectarian groups are fewer, and we are able to identify many rabbinic texts with their authors, or certainly, with their communities.

We note the beginnings of early Jewish ambivalence toward astrology as a prognostic art during the Second Temple period in the sense that sources are both positively and negatively valenced, and this is the case whether or not the sources are sectarian in nature. During this period, Jewish thought was influenced by Hellenistic popular culture, which favoured divination, dream interpretation, and astrology. Although the authorities sometimes viewed divination and magic and related practices with suspicion, particularly during the Roman period, astrology was popular in the Greco-Roman world, and was—alongside its occult associations—held to be a scientific pursuit.

Indeed, despite Second Temple sectarian concerns regarding idolatry that had their origins in the admonitions of the Hebrew Bible, the earliest Jewish horoscopes are attested at Qumran. These comprise two types: The first is exemplified by 4Q318—also called a *brontologion*—which serves to determine the most auspicious time to perform a ritual or action based on the sound of thunder. The second is represented by 4Q186,[8] a genethlialogical, or fate-based, horoscope, which aims to determine the astrological destiny,

---

[7] This is particularly true within Deuteronomy 18 and the Deuteronomistic History. We also note specific prohibitions against the worship of astral bodies in Deut. 4:19 and 17:2–3, among other biblical verses.

[8] In 4Q186, we read 'A man who is born [. . .] under the influence of Taurus [. . .] will have long and thin thighs and toes'. Charlesworth, 'Jewish Interest', p. 938.

or *mazal*, of an individual by describing an astrological sign's influence upon his physical appearance.[9]

Also important is *1 Enoch* 72–82, also known as the *Astronomical Book (AB)*—a third century BCE apocalypse that describes the biblical character Enoch's tour of the heavens, complete with references to the zodiac. Four copies of this work were found among the Qumran scrolls, and were designated as 4Q208–11. The *AB* contains both astronomical and astrological imagery employed to support the book's apocalyptic agenda.[10] Within the text, the heavens are portrayed as part of a grand plan in which those who are on the side of goodness are rewarded, and those who are not are punished—for example, with the explosion of the sun and the veritable implosion of the cosmos prior to a 'new creation' instituted by God. Most telling for our purposes, however, is the astral ambivalence demonstrated in chapter 80, in which, 'it is said that the natural order will change in "the days of the sinners" who are characterized as worshippers of the stars'.[11] We note similar condemnations of star worshippers in *1 Enoch* 8:3, and in the second century BCE *Sibylline Oracles* which make reference to being 'deceived by the predictions of Chaldean astrology.'[12] Perhaps the central polemic against astrology in the Pseudepigrapha is found in *Jubilees* 12, which 'was directed against the astrological claim that the zodiac determines the yearly rainfall.[13]

By contrast, in *Judaica* and *Peri Ioudaion* (second/third centuries BCE) Artapanus extolled the virtues of astrology, describing Moses as possessing magical powers and astrological prowess that allowed him to triumph in his liberation of the Jews from Egypt.[14] Similarly, for Pseudo-Eupolemus (ca. first century BCE), Abraham is described as

---

[9] Francis Schmidt, 'Ancient Jewish Astrology: An Attempt to Interpret 4QCRYPTIC (4Q186)', in Michael E. Stone and Esther G. Chazon, *Biblical Perspectives: Early Use and Interpretation of the Bible in Light of the Dead Sea Scrolls*. (Leiden: Brill, 1998), p. 189.

[10] James C. VanderKam, *From Revelation to Canon: Studies in the Hebrew Bible and Second Temple Literature* (Leiden: Brill, 2000), p. 363.

[11] James C. VanderKam, *Enoch and the Growth of An Apocalyptic Tradition* (Washington, DC: The Catholic Biblical Association of America, 1984), p. 78.

[12] Charlesworth, 'Jewish Astrology', p. 188.

[13] Ibid., p. 189.

[14] James H. Charlesworth, ed., *The Old Testament Pseudepigrapha*. Vol. 2 (New York: Doubleday, 1985), p. 889.

having been born in Babylonia as per the biblical account, and as being an expert in astrology, spreading 'astrological lore' on his travels to Phoenicia.[15]

Indeed, both supportive and critical views of astrology come to us from Second Temple Jewish sources, and there is no consensus on the matter. Given the pervasive Greco-Roman contexts and the popularity of astrology in these cultures, by the fifth-sixth centuries CE, during the time of the Babylonian Talmud, astrological beliefs and imagery had become assimilated into the fabric of Judaism.[16] However, this too was not without ambivalence.

In Late Antiquity, for example, we note the presence of Talmudic debates regarding the concept of *mazal* (literally, 'constellation'), in which the destiny of an individual depends on the dominant positions of the astronomical bodies at the time of his or her birth. This astrally-mediated determinism was heavily debated in the Talmud and in later Jewish sources.[17]

The crux of the debate, of course, was the monotheistic imperative and the will of God set against a form of predestination in which the astral bodies served as intermediaries between humanity and Divine agency. Indeed, during the rabbinic era, we note competing views regarding astrology, from Talmudic statements stating that an individual's *mazal* is writ large in the stars, to rabbis expressing concern that *mazal* is tantamount to *avodah zarah*,[18] as well as *chukat goyim*. Some of these views are steeped in the biblical admonitions that we have noted, while other rabbinical authorities are supportive of astrological practices.

This disagreement comes into sharp relief in the *Babylonian Talmud* (*BT*) 156a–b, which presents a debate regarding the validity of astrology and the concept of *mazal* for Israel. That is, they inquire, is Israel immune from astrological influences, or is it not? Here, we see a debate presented between R. Hanina and R. Johanan Bar Nappaha (ca. 180–279), in which R. Hanina argued that Israel is influenced by astrological *mazal*, whereas R. Bar Nappaha

[15] Ibid., p. 873.

[16] Charlesworth, 'Jewish Interest', p. 930.

[17] Joshua Trachtenberg, *Jewish Magic and Superstition: A Study in Folk Religion* (Philadelphia: University of Pennsylvania Press, 2004), p. 250.

[18] This is also sometimes referred to as *avodat ha-kokhavim umazalot*—the worship of the stars and constellations.

did not believe in the existence of *mazal* for Israel.[19] As further evidence for the anti-astrological camp, some Babylonian rabbis during this period went so far as to forbid consultations with 'Chaldeans' (*BT Pesachim* 113b), a term with clear linkages to the Babylonian astrologers of the Hebrew Bible. Indeed, despite the existence of rabbis supportive of astrology, in practice, astrologers typically had 'bad reputations in Rabbinic sources, and they were not allowed to predict concerning Jews'.[20] Indeed, these foreign elements were held at arm's length in rabbinic Judaism, and were imbued with the taint of Babylonian idolatry. And yet, while there are tractates such as *Sanhedrin* 68a that demonstrate some 'discomfort with magic, including rabbinic knowledge of it',[21] the rabbis of the Talmud either practiced magic, including astrology, or were familiar with its use.[22] Indeed, this ambivalence toward astrological and other magical practices can be shown to exist throughout rabbinic sources in Late Antique Judaism.[23]

These mixed views scattered throughout Second Temple and Late Antique Jewish literature and the discourses that led to them ultimately, then, bring us to the question of Babylonian provenance, which colours much of the negatively-valenced astrological discourse we have noted. It is here that we turn to the evidence for a linkage between Babylonian astrology, its later Hellenistic counterpart, and Late Antique Jewish zodiacal symbolism.

On Providence and Foreign Provenance: From Babylon to Beit Alpha

The practice of celestial divination was performed in Mesopotamia from the Old Babylonian period onward by priestly scribes called *tupsar Enuma Anu Enlil* to divine the

---

19 Charlesworth, 'Jewish Astrology', p. 186.

20 Meir Bar-Ilan, 'Astrology In Ancient Judaism', in *The Encyclopaedia of Judaism*, Vol. 1, ed. J. Neusner, A. Avery-Peck and W.S. Green (Leiden: Brill, 2005), p. 132.

21 Kimberly Stratton, 'Imagining Power: Magic, Miracle, and the Social Context of Rabbinic Self-Representation', *Journal of the American Academy of Religion* 73, no. 2 (2005): p. 369.

22 Ibid., p. 366.

23 J. Magness, 'Heaven on Earth: Helios and the Zodiac Cycle in Ancient Palestinian Synagogues', *DOP* 59 (2005): p. 34.

future and secure the well-being of king and state.[24] We know of omen lists, or omen texts, dating to the second millennium BCE extending through the Seleucid period in the 300s BCE. These eventually came to influence the later development of astrology in the Hellenistic era.[25] The majority of these extant lists, however, date to the first millennium BCE, and most of these were from the seventh century BCE, from Assurbanipal's library. The omen records were presented in pairs including a protasis and an apodosis, forming an 'if-then' causal sequence.[26] For example, in the section on planetary omens in *Enuma Anu Enlil*, the forecast is that 'If Venus rises in Month VIII: [hard times will seize] the land'.[27]

The very earliest signs of transition from celestial divination and omen lists to individual horoscopic astrology took place at some point during the eighth-seventh centuries BCE, with the rise of the conquering Assyrians, who collected the earlier astrological materials, including all of the omen listings. These celestial omens were rarely related to individuals prior to the fifth century BCE, during the rise of horoscopic astrology, which was, in turn, adapted by the Persians upon conquering Babylon in 539 BCE, then further modified by the Greeks during the Hellenistic period and finally, widely disseminated.[28]

We note the presence of birth charts from 410 BCE onward. According to David H. Kelley and Eugene Milone as well as Nicholas Campion, the zodiacal signs were first noted in an astronomical diary in the fifth century BCE, and this was followed by the earliest Greek examples shortly thereafter.[29]

---

[24] Francesca Rochberg, *The Heavenly Writing: Divination, Horoscopy, and Astronomy in Mesopotamian Culture* (Cambridge: Cambridge University Press, 2004), p. 182.

[25] J. C. Greenfield, J. C. & M. Sokoloff, 'Astrological and Related Omen Texts in Jewish Palestinian Aramaic', *JNES* 48 (1989): pp. 201–2; Rochberg, *The Heavenly Writing*, p. 169.

[26] Rochberg, *The Heavenly Writing*, p. 170.

[27] Ibid.

[28] E. Reiner, *Astral Magic in Babylonia*, Transactions of the American Philosophical Society 85.4 (Philadelphia: American Philosophical Society, 1995), p. 13.

[29] David H. Kelley and Eugene F. Milone, *Exploring Ancient Skies: An Encyclopedic Survey of Archaeoastronomy* (New York: Springer Science + Business Media, Inc., 2005), p. 498; Nicholas Campion, *Dawn of Astrology: A Cultural History of Western Astrology: The Ancient and Classical Worlds* (London: Continuum International Publishers,

The names of the majority of these zodiacal constellations were given in Mesopotamia, and soon found themselves transformed into twelve 30-degree wide signs by the Greeks—becoming *zodiacos kuklos*, a circle of animals.[30] As Campion explains 'by approximately the first century CE, the twelve signs of the zodiac were, at long last, fixed in the forms we recognize today'.[31]

And it was precisely during this period that another transition was taking place—namely, that from the era of the Second Temple priests during the Greco-Roman period to rabbinic Judaism, upon the destruction of the Second Temple by the Romans. What we know from this era comes to us from a broad cross-section of texts, many of which were extra-biblical in nature.

What is clear from the available evidence is that there are numerous references to the Babylonian provenance of astral wisdom in early Jewish literature. Indeed, Adolfo Roitman notes this in his examination of Judith 5:6, which contains the statement 'This people are descendants of Chaldeans'. Roitman points to the association of Jews with Chaldeans in this verse, and this, in turn, emphatically highlights the Mesopotamian origins of the patriarch Abraham in Ur Kasdim, or Ur of the Chaldees.[32] As Roitman explains it:

> In sum, the biblical as well as the Hellenistic sources show that the term 'Chaldeans' was transferred during the Hellenistic and Roman age from its narrow denotation as the name of a nation [. . .] to a more general class: all the persons who had studied or practiced the Chaldean sciences (namely, astrology, astronomy, divination, and so on.[33]

Annette Reed notes that other Second Temple and Late Antique sources—such as *Jubilees* 12:17–18, Philo, *On Abraham* 69–70, the Midrash in *Bereshit Rabbah* 44:12, and *BT Shabbat* 156a—explicitly describe Abraham's newfound monotheism as 'a revolutionary departure from Meso-potamian beliefs'.[34] This discourse became a shorthand for

---

2008), p. 80.

[30] Kelley and Milone, *Exploring*, p. 26.

[31] Campion, *Dawn of Astrology*, p. 183.

[32] Adolfo D. Roitman, 'This People are Descendants of Chaldeans' (Judith 5:6): Its Literary Form and Historical Setting', *Journal of Biblical Literature* 113, No. 2 (1994): pp. 255–56ff.

[33] Ibid., p. 256.

[34] Annette Yoshiko Reed, 'Abraham as Chaldean Scientist and

associations with paganism, polytheism, and non-Jewish religion in later periods.[35] Indeed, if Abraham was seen as Chaldean, it was far easier to encompass all of Israel as well.[36] It must also be noted that the ambivalent tendencies mentioned earlier also found their way into the portrayal of Abraham in the *Antiquities of the Jews* 1.154–168, by Josephus, in the first century CE, for here, he is shown both having rejected 'Chaldean science' in favour of monotheistic belief while at the same time later having 'transmitted astronomy/astrology to Egypt, appealing to the positive connotations of this art for apologetic aims'.[37]

We also note evidence of celestial divination and astrology with a clear Babylonian stamp in a document brought to light by J. C. Greenfield and M. Sokoloff, who describe an astrological omen text composed in Jewish Palestinian Aramaic. As they state, 'the only other Aramaic text of this type known from Jewish literature is a fragmentary *brontologion* from Qumran Cave IV giving the signs of the zodiac for the days of the month, followed by predictions that can be drawn from thunder'.[38] To explain the existence of such texts, they point to the great repository of Babylonian omen literature, and in particular, to the wide dissemination of the *Enuma Anu Enlil* omen series.[39]

Greenfield and Sokoloff also note the influence of Babylonian celestial divination upon the Ugaritic omen texts, many of which were nearly identical in content and structure to their earlier counterparts.[40] Paolo Xella further enhances our understanding of this relationship by making it clear that Mesopotamian celestial divination was well established in Ugarit, where 'Mesopotamian tradition was followed quite slavishly' with only minor local adaptations.[41] Supporting this contention of transmission and influence are copies of Mesopotamian celestial omen texts have been found at

Father of the Jews: Josephus *ANT*. 1.154–168, and the Greco-Roman Discourse About Astronomy/Astrology', *Journal for the Study of Judaism* 35, no. 2 (2004): pp. 125–26.

[35] Ibid., p. 126.

[36] Roitman, 'Chaldeans', p. 258.

[37] Reed, 'Abraham as Chaldean Scientist', pp. 120ff.

[38] Greenfield, J.C. and M. Sokoloff, 'Astrological and Related Omen Texts', p. 202.

[39] Ibid., p. 201–2

[40] Ibid., p. 202.

[41] Paolo Xella, 'The Omen Texts', in *Handbook of Ugaritic Studies*, ed. Wilfred G. E. Watson and Nicolas Wyatt (Leiden: Brill, 1999), p. 353.

Ugarit, particularly at Ras Shamra. These include an alphabetic cuneiform tablet containing observations of the sun and the moon, and displaying reliance upon *Enuma Anu Enlil.*[42]

Given the many pieces of sometimes conflicting data from cultures spanning the region, scholars do not always agree about the precise nature of the influences upon the culture/s of ancient Israel by its neighbors and earlier cultures in the ancient Near East (ANE), including Ugarit. However, as J. Edward Wright expresses it, of one thing we are certain—the history of ancient Israel 'was closely intertwined with the histories of the great powers surrounding it and intermittently dominating it'.[43] As such, we do note a number of celestial motifs common to ancient Israel and other ANE societies.[44] These include cosmologies and celestial cosmographies, including the structure of heaven itself, which was, in ancient Israel, conceived[45] in tripartite form, with the heavens at the top, earth in the middle, and the netherworld of Sheol below.[46] This is similar to the conceptions that existed in other regions of the ANE, such as the Mesopotamian view of the universe being divided into 'heaven, earth, and netherworld.'[47]

Clear evidence of Mesopotamian provenance is present in the *Astronomical Book of Enoch* as well. Like *Jubilees* and many other scrolls found at Qumran, *1 Enoch* espouses a solar calendar, and this is mirrored by the age of Enoch in the Hebrew Bible in Genesis 5:21–24—that is, 365 years old. *1 Enoch* begins with a list of Sumerian kings who ruled before the Great Flood.[48] The seventh king is known as Enmeduranna or Enmeduranki, and he was linked with Shamash, the sun god, who, according to some texts, taught the seventh king the divinatory arts. It is notable that Enoch too is listed as the seventh biblical patriarch after Adam, the first human being, a point often highlighted by scholars as

---

[42] Ibid.

[43] J. Edward Wright, *The Early History of Heaven* (Oxford: Oxford University Press, 2000), p. 3.

[44] Ibid., p. 52.

[45] One of the main perceptions of the heavens in ancient Israel. Indeed, as Wright states, there has never been 'a single biblical version of heaven' (Wright, 56).

[46] Ibid., p. 53.

[47] Ibid., pp. 54–55.

[48] This was preserved by Berossus, among others.

being strongly suggestive of Mesopotamian influence.[49]

Finally, in a similar vein, we note the presence of these Babylonian and Greek elements in Late Antique synagogue art—specifically in the zodiacal motifs on synagogue mosaic floors found at Sepphoris, Hammath-Tiberias, Huseifa, and Na'aran, and other locations. Perhaps the most striking example, however, may be found in the sixth century CE mosaic floor at *Beit Alpha*, which combines Jewish and Pagan motifs. Here, the sun god Helios appears within the central panel in his chariot, surrounded by the stars and the moon, as well as the twelve signs of the zodiac. These are combined with images of the four Jewish *tequfot*, or seasons, one appearing in each corner.[50]

While some scholars have asserted that it is impossible to ignore the deeper syncretism of the mosaic floor at *Beit Alpha*, particularly given the central presence of the image of the Greek sun god Helios driving his quadriga through a starry firmament, the astrological imagery has been dismissed by others as merely decorative. Erwin Goodenough minimized the Hellenistic motifs, arguing that the image of Helios was not pagan, but represented the God of Judaism.[51] Reinforcing the linkage between timekeeping and communal regulation, however, Rachel Hachlili and several other scholars interpret of the zodiac mosaic floors as visual calendars[52] that would have been 'employed as a significant framework for the annual synagogue rituals'.[53] Indeed, for Hachlili, the roles of the astrological mosaics are far from being decorative or pagan—instead, they are Hellenistic, and ultimately, Babylonian—motifs *appropriated* within a Jewish context for timekeeping, ritual and liturgical purposes.

Indeed, it should be noted that in addition to its immediate Greco-Roman provenance, many of these zodiacal images bear striking resemblances to those found on Mesopotamian cylinder seals and in other astrological iconography found in the fertile crescent.[54] While it is clear to

---

[49] Bar-Ilan, 'Astronomy In Ancient Judaism', p. 144.
[50] Rachel Hachlili, 'The Zodiac in Ancient Synagogal Art: A Review' *Jewish Studies Quarterly* 9 (2002): pp. 248–49.
[51] Erwin R. Goodenough, *Jewish Symbols in the Greco-Roman Period*, Volumes 1–12, Bollingen Series XXXVII (New York: Pantheon Books, 1953-65), Vol. 8, p. 177.
[52] Hachlili, 'The Zodiac', p. 234.
[53] Ibid., p. 237.
[54] Gavin White, *Star-Lore: An Illustrated Guide to the Star-lore and*

most scholars today that much earlier and ultimately foreign provenance exists for many key astrological texts and symbols within early Judaism, this was not always the case within early Jewish communities. Whereas the astronomy and astrology found in many of the key texts and symbols of early Judaism, including *1 Enoch 72–82, Jubilees,* and the zodiac synagogue floors, were Mesopotamian, and later, Greek in origin, the religious authorities, asserts Meir Bar-Ilan, often simply 'did not admit it'.[55]

CONCLUSION: GIFTS OF THE MAGI

We have noted some of the ways in which astrology and its symbolism found their way from the Babylonian and Hellenistic Greek contexts into early Judaism. Indeed, the linkages between the Mesopotamian material and the early Jewish documents are highly suggestive—from the connection between the patriarch Abraham to Ur Kasdim in the Hebrew Bible itself to the references to the Mesopotamian lists of kings found in *1 Enoch* 72–82. Additionally, we cannot underestimate the significant historical link of Judaism with Babylon during the Babylonian Exile, which began circa 586–587 BCE, leaving Judaism with a Babylonian legacy that is now inextricable from its cultural and theological fabric. When one fully considers Sacha Stern's view that calendar and community are strongly connected, it becomes quite evident that the Babylonian month names[56] that became many of the Jewish months were but one expression of Babylonian influence upon the development of early Judaism. Moreover, the Babylonian lunar calendrical influence upon the Jewish lunisolar calendar have not only had an impact upon Judaism, but transformed it into something quite new—a religion that differed substantially from the Israelite culture and religion that preceded the exile.

Throughout Jewish history, we note a tendency toward what I might term making foreign material kosher. That is to say, there has been a tendency to appropriate and absorb

---

Constellations of Ancient Babylonia (London: Solaria Publications, 2008).

[55] Bar-Ilan, 'Astrology', p. 13.

[56] These included: Nissanu (Nisan), Ayaru (Iyar), Simanu (Sivan), Du'uzu (Tammuz), Abu (Av), Ululu (Elul), Tashritu (Tishrei), Arahsamnu (Marcheshvan), Kislimu (Kislev), Tebetu (Tevet), Shabatu (Shevat) and Addaru (Adar). (Campion, 2000, p. 524)

external imagery to reinforce the identity of a very different community. As James Charlesworth tells us, 'zodiacal symbols were sometimes demythologized from non-Jewish religions and remythologized in Jewish categories'.[57] In Late Antique Jewish synagogues, for example, the zodiac imagery of the foreign host culture coexisted with Jewish calendrical symbolism to reinforce communal ritual and serve as a visual reminder of communal identity in time and space.

Even so, Late Antique Judaism, in its evolving forms, continued to wrestle with the dichotomy between the monotheistic imperative and its admonitions and whatever it might have been that continued to entice those who wrote history to engage in the belief in and practice of astrology. And yet, what might these attractive elements have been? I would suggest that astrology and its symbolism provided Late Antique rabbis with several incentives to reinterpret or perhaps even reframe astrological pursuits as being compatible with rabbinic Judaism.

I would suggest that one of the most likely reasons for the trend toward acceptance of Babylonian and Greek astrological elements, including zodiacal imagery, was their association with *power*. Indeed, according to Jacob Neusner, the rabbi was considered to be 'a holy man' with magical abilities in Sasanian Babylonia during the third century BCE.[58]

As Neusner writes, 'The two indigenous sciences of Babylonia, astrology and medicine, were studied by the rabbis'.[59] As a wonder-worker, the rabbi became 'a master of ancient wisdom both of Israel and of his native Babylonia and privy to the occult'. He could also cast horoscopes even as he wielded expertise in Torah. Despite the early rabbi's facility in astrology and magic, however, God was still considered to be the supreme ruler who could override the horoscope and astral influences at any time.[60] Whereas the foreign Magus might have been considered a sorcerer, 'the Iranians may well have seen the rabbi as a Judaized Magus'. Their domains were very similar, and both included astronomy/astrology, theology, and healing among their areas of expertise.[61] Although we see connections between

---

[57] Charlesworth, 'Jewish Astrology', p. 184.

[58] Jacob Neusner, 'Rabbi and Magus in Third-Century Sasanian Babylonia', *History of Religions* 6, no. 2 (1966): p. 170.

[59] Ibid., p. 171.

[60] Ibid., p. 172.

[61] Ibid., p. 173.

the magic and 'the dangerous other' in the Babylonian Talmud, Kimberly Stratton too underscores the point that the Babylonian context and its valuation of the Magus and his power would have created a more positive valence associated with the magical arts.[62]

A second possible incentive for the absorption of Babylonian and Hellenistic astrological symbolism and practice into rabbinic Judaism was that the discourse of astronomy/astrology was strongly linked with *antiquity*.[63] Given the fact that the rabbis had only recently emerged as a new force in Judaism after the destruction of the Second Temple, a connection to an ancient practice would likely have been seized upon as a marker of the rabbinic claim to the wisdom and authority of the past. It is then, I would suggest, the claim to both magical power and to the authority of the past that may have provided rabbinic Judaism with sufficient incentive to allow astrology to be integrated piecemeal into Judaism despite the debate and ambivalence surrounding its incorporation.

Bringing the Babylonian and Hellenistic astrological symbolism into the contemporary context, it is of no small import that astrology has become extremely popular within Ultra-Orthodox and Hasidic Jewish communities. Like the rabbis of antiquity, they too have found ways of absorbing astrological content and rendering it kosher. Unsurprisingly, Orthodox and Hasidic usage of astrology in their literature shows no evidence of being aware of its ultimate origins in the Babylonian and Greek worlds. Indeed, the Babylonians—i.e., the Chaldeans—and the Hellenizers are considered to be ancient cultures antithetical to Judaism and its theology, and this is reiterated time and again in contemporary confessional Jewish sources.[64]

---

[62] Stratton (p. 366) believes that the negative valence associated with magical arts may have been regional in nature, associated with the Jerusalem Talmud in Palestine and the negative view of magic by the Romans. By contrast, the Babylonian Talmud was keen to present rabbis as magical experts due to the Sasanian valuation of the Magi. Hence, magical prowess would have served to 'enhance rabbinic stature and authority'. For Pliny, for example, magic is linked to 'the threat of foreign invasion'. Romans are, for him, liberators from what he labels 'monstrous rites'. (Reed, 'Abraham as Chaldean Scientist', p. 155)

[63] Reed, 'Abraham as Chaldean Scientist', p. 142.

[64] Matityahu Glazerson, *Above the Zodiac: Astrology in Jewish Thought* (Northvale, NJ: Aronson, 1997); Gad Erlanger, *Signs of the Times: the*

Nevertheless, interest in astrology persists in these communities. While the symbols used in the Jewish appropriation of the zodiac are identical to the ones with which we are familiar, their names have changed. Instead of using Aries, Taurus, Gemini, and the like, Orthodox and Hasidic Jews have given them Hebrew names.[65] In so doing, these communities have—mirroring the example of the Late Antique synagogues—remythologized the zodiac signs within a Jewish context, associating each sign with persons and events found in the Hebrew Bible, or with either Jewish mystical concepts or exhortations to spiritual growth. For example, pointing to the transition from Aries (*Taleh*) and the Hebrew month of Nissan (formerly Nisanu in the Babylonian context) to Taurus (*Shor*) and the month of Iyar (Ayaru), Eliyahu Glazerson states that 'From a lamb one becomes like a bull carrying the burden of Torah'.[66] Of particular interest—and lacking no irony given the presence of this commentary within a Jewish astrological work—is one of the contemporary descriptions of Keshet, the bow, also known as Sagittarius. Given its placement in the Jewish month of Kislev, in which Chanukah takes place, the apologetics are particularly compelling. Here, as Gad Erlanger writes, the shooting of the bow is emblematic of 'the bow of Judaism that fights Greek thinking'.[67] Although the Hellenistic Greeks tried to 'wound' Jews in the matter of their identity,[68] the Jews ultimately prevailed over Greek culture, leading to the celebration of Chanukah.[69]

Reiterating the thesis of Sacha Stern on calendar and community boundary maintenance, these astrological elements are, in contemporary Jewish writings, often presented in conjunction with references to the Jewish calendar and its holiday cycle. It is noteworthy that the foreign elements that were once held so strongly at bay within Judaism have emerged and re-emerged throughout

---

Zodiac in Jewish Tradition (New York: Feldheim, 2001).

[65] Erlanger, *Signs of the Times*, p. 12. The Judaized/'kosher' signs include Taleh (the ram); Shor (the bull); Teomim (the twins); Sartan (the crab); Aryeh (the lion); Betulah (the virgin); Moznaim (the balance scales); Akrav (the scorpion); Keshet (the bow); Gedi (the kid); D'li (the pail); and Dagim (the fish).

[66] Glazerson, *Above the Zodiac*, p. 23.

[67] Erlanger, *Signs of the Times*, p. 162.

[68] Ibid., p. 187.

[69] Ibid., p. 176.

Jewish history within the holiest venues and the most stringently religious communities.

Somehow, astrology has not only found itself remythologized within Jewish categories, but it has also overcome biblical prohibitions and rabbinic ambivalence to be considered part and parcel of Judaism and its very theology. This appropriation, or koshering, of the astrological symbolism of the ancient other, with or without apologetics, speaks volumes regarding the flexibility of religious groups to accommodate new iconography and ideas over time. In the case of Babylonian and later Hellenistic Greek astrological symbolism, its incorporation into Jewish thought and art did not take place without controversy; however, its endurance and acceptance within Judaism reveals the ongoing power of communal and religious needs to transform the forbidden into normative practice.

THE PERUGIA FOUNTAIN: AN ENCYCLOPAEDIA OF SKY, CULTURE AND SOCIETY

Darrelyn Gunzburg

ABSTRACT: This paper focuses on the Fontana Maggiore, a fountain located in the centre of Perugia in Umbria in central Italy, built in 1278 and sculpted by Nicola and Giovanni Pisano. The sculptural narrative of this urban monument contains not only religious imagery but also civic and allegorical themes and symbols, including the Labours of the Months, and the zodiac. It is the purpose of this paper to demonstrate that physically through the carvings on the lower basin, the fountain depicts in stone a narrative that reflects the heavens, as well as containing celestial symbolism within its orientation. It considers the location of the fountain within the town of Perugia and the historical circumstances that led to its construction and the commissioning of the stone sculpture work as a way of understanding its functional importance. It then examines the orientation of the images of the lower basin and their reflection of the sky. Finally the paper concludes that this monument and its sculptural narrative can be considered to be an omphalos linking the individual and their community to sky, earth and seasonal rhythms, and is offered as an example of the human desire to build structures that occupy liminal spaces, which link the townspeople to the sky.

HISTORICAL BACKGROUND

The story of the Fontana Maggiore begins with the economic revival emerging in the tenth century that encouraged the proliferation of towns. The increasing dominance of merchants and bankers begat a flourishing industrial and mercantile middle class with growing political importance.[1] As a result, the Italian communes emerged in the late eleventh and thirteenth centuries, in part the result of the struggle between the Empire and the Papacy and the

---

[1] J. K. Hyde, *Padua in the Age of Dante* (Manchester: Manchester University Press, 1966), p. 13.

ensuing political deterioration of both powers.[2] The communes established themselves as municipal bodies with limited sovereignty, engaged in the business of local government yet still dependent upon external seigneurial or monarchical authority for economic, military and trade relations, and crimes where the death penalty was incurred.[3] Those communes that were able to turn limited into full sovereignty became known as city-states. In 1186 Henry VI, *rex romanorum* and future emperor, gave diplomatic recognition to the Perugian consular government. Pope Innocent III (Pope from 8 January 1198 until his death in 1216) followed this decision, acquiescing that Perugian civic practices carried the force of law.[4]

Perugia is a city built on a hilltop far from the mountains and with meagre freshwater springs of its own. Thus in 1275-76, when the Perugian commune determined its plans for the following decades, they included not only new and enlarged residences for the government and the *Consiglio del Popolo*, improved road access, and a new cathedral but also the construction of an aqueduct and fountain to bring water to the city centre.[5] To provide this water, a new aqueduct needed to be built from Monte Pacciano three miles outside the town. This project also required the construction of a fountain which was to be the terminus of the fresh water for the Perugians and located between the Duomo (cathedral) and the Palazzo Comunale (town council hall), also known as the Palazzo dei Priori.[6] This was the site where five newly-built roads leading from each of the five ancient city gates converged, thus forming a centre point to the town, an omphalos (fig. 5.1). This fountain took a little over a year to complete and was known as the Fontana Maggiore.

---

[2] Helene Wieruszowski, 'Art and the Commune in the Time of Dante', *Speculum* 19, no. 1 (1944): p. 14.

[3] S. R. Epstein, 'The Rise and Decline of Italian City States', *Working Paper* No.51/99 (1999): p. 1.

[4] Raffaele Rossi, Attilio Bartoli Angeli, and Roberta Sottani, *Perugia*, vol. 1 (Milan: Elio Sellino Editore, 1993), pp. 120–40.

[5] Trevor Dean, *The Towns of Italy in the Later Middle Ages*, trans. Trevor Dean, Manchester Medieval Sources Series (Manchester and New York: Manchester University Press, 2000), pp. 6–7.

[6] William D. Wixom, 'A Glimpse at the Fountains of the Middle Ages', *Cleveland Studies in the History of Art* 8 (2003): pp. 12–13.

Fig. 5.1: Aerial view of the Fontana Maggiore from Google Earth showing the location of the fountain between the Duomo and the Palazzo dei Priori.

So important was the fountain as the town's water supply that, in its *Statuti di Perugia* of 1342, the Perugian commune set down in writing prohibitions that protected the water of the Fontana Maggiore from being polluted or soiled. These included embargos against washing, laundry, drawing water to make lime, work leather, prepare parchment, or allowing animals to drink from it. The 1342 statutes also made precise reference to the sculptures when it listed 100 lire as the penalty 'for anyone who with stone, iron or wood broke... any of the images carved there'.[7] The alternative for the offender unable to pay the fine was the loss of his right hand. With the Fontana Maggiore being central in both its location within the city and in the city's development plans, the commune undertook the task of commissioning the decoration of the fountain with its rich and complex stone narrative.

Prior to the emergence of communal self-governorship the commissioning of buildings and the decoration of altar-pieces, carved statuary, reliquaries, illuminated manuscripts, and stained glass was chiefly the dominion of the church, kings and princes. This changed as the communes gained their independence and stood alongside church, kings and princes as patron, utilising this position of strength to

[7] Francesca Bocchi, 'Regulation of the Urban Environment by the Italian Communes from the Twelfth to the Fourteenth Cenutry', *Bulletin of the John Rylands Library* 72 (1990): pp. 75–76.

reinforce statements of communal pride.[8] Thus it was that the Perugian commune commissioned Nicola Pisano (c.1220 / 1225–c.1284) and his son Giovanni (c.1250–c.1315) to carry out the sculptural work on the Fontana Maggiore. John White declared that these sculptors comprised two of the three men working as sculptors in late-thirteenth-century Italy whose work gave rise to a pictorial revolution (the third being Arnolfo di Cambio). By creating an innovative sculptural language that employed antique prototypes, as well as seeking new representational means through naturalistic forms, Nicola Pisano and his son, Giovanni, brought a previously unseen vividness and immediacy to sculptural expression.[9] Anita Fiderer Moskowitz calls attention to the fact that, for such a fundamentally functional civic monument, there was no close precedent for the scale, the intricacy of design, and the lavish iconographic program. She describes it 'as original in its form and conception in the history of fountains as are Nicola Pisano's pulpits and the tomb of St. Dominic in relation to the preceding pulpit and tomb traditions'.[10]

THE NARRATIVE IN THE IMAGES

The resulting Fontana Maggiore of Perugia was and still is a three-tiered fountain, polygonal in shape and built using an alternation of pink and white marble (fig. 5.2). The lowest level is set on a ring of steps from which rises a twenty-five-sided basin. Each face of the twenty-five sides forms a diptych, two images joined by a central link, embellished with sculptured reliefs and articulated by colonettes, small thin columns used for decoration. Each diptych is separated from the next by a tripartite compound pier of twisted or patterned marble. Inside the circle of this lower basin are set twenty-four external and thirty-four internal columns that support a smaller twelve-sided basin. This second basin, formed by twelve plain-sided concave diptychs, is embellished with twelve figures placed within each diptych and twelve figures placed at the angles where each diptych

[8] Wieruszowski, 'Art and the Commune in the Time of Dante', p. 17.
[9] John White, *Art and Architecture in Italy 1250–1400*, Pelican History of Art (New Haven: Yale University Press, 1993), p. 73.
[10] Anita Fiderer Moskowitz, *Italian Gothic Sculpture c.1250–c.1400* (Cambridge: Cambridge University Press, 2001), p. 43.

meets the next one, thus reversing the scheme of the lower basin which contains images on the faces only but without figures at the angle corners. Rising from the centre of this middle basin is a single column on top of which sits a bronze basin supporting three female caryatids, three draped female figures replacing what would normally be a column.

Fig. 5.2: The Fontana Maggiore showing the three basins with the Palazzo dei Priori behind it. Photo: *D. Gunzburg.*

The fountain has undergone two major reconstructions since its completion—one after an earthquake in 1349, and the second in 1948–49 after the Second World War. The complication for modern restorers was to recover the correct order of the reliefs of the middle basin. However, while White argued that the inscriptions and figures of the middle basin had been incorrectly transposed, in contrast he noted that the order of the reliefs in the lower basin could not be altered due to 'small variations in dimension which meant that the surrounding cornices had been individually tailored to fit each particular relief'.[11]

---

[11] John White, 'The Reconstruction of Nicola Pisano's Perugia Fountain', *Journal of the Warburg and Courtauld Institutes* 33 (1970): p. 71.

The contents of the sculpture work are varied. In the lower basin sit the Liberal Arts, the story of Romulus and Remus as the progenitors of Perugia, the gryphon, the symbol of Perugia, and the Guelph emblem of the lion, scenes from The Old Testament such as David and Goliath, and Samson and Delilah, along with the Labours of the Months and the zodiac. The twenty-four figures of the middle basin contain abstract personifications, such as Ecclesia Romana and Victoria Magna, figures representing cities and localities, such as Rome and Chiusi, figures from The Old Testament, such as Moses, David and Solomon, Saints such as John the Baptist and Benedict, Eulistes the founder of Perugia, and men who were alive at the time of the fountain's construction and assisted with the completion of the fountain, namely the *podestà* Matteo da Correggio, and *Capitano del Popolo* Ermanno da Sassoferrato.[12] Francis Ames-Lewis describes the fountain as Perugia's 'manifesto of civic identity'.[13] This combination of religious with civic components is also reflected in the location of the fountain in the Piazza Maggiore between the Duomo (cathedral) and the Palazzo Comunale (town council hall), as cited earlier.[14]

The Fontana Maggiore was constructed and sculpted at a time when Scholasticism, a method and a system that aimed to reconcile the Christian theology of the Church Fathers with the Greek philosophy of Aristotle and his commentators, was the dominant form of theology and philosophy in the Latin West. It was also a period when, as Diana Norman clarifies, the new urban-centred religious orders, such as the Dominicans and Franciscans, encouraged a more direct relationship between ordinary mortals and God and his saints.[15] Moskowitz observes that Nicola Pisano's two polygonal pulpits in Pisa and Siena that pre-

---

[12] John Pope-Hennessy. *Italian Gothic Sculpture*. Vol. 1, An Introduction to Italian Sculpture (London: Phaidon Press Limited, 2000), p. 232; White, 'The Reconstruction of Nicola Pisano's Perugia Fountain', p. 72.

[13] Francis Ames-Lewis, *Tuscan Marble Carving 1250–1350: Sculpture and Civic Pride* (Aldershot: Ashgate, 1997), p. 214.

[14] William D. Wixom, 'A Glimpse at the Fountains of the Middle Ages', p. 8.

[15] Diana Norman, 'Change and Continuity: Art and Religion after the Black Death', in *Siena, Florence and Padua: Art, Society and Religion 1280–1400, Volume I: Interpretive Essays*, ed. Diana Norman (New Haven: Yale University Press, 1995), p. 181.

date the fountain may be amongst the earliest visual expression of the spread of Scholasticism, a view also held by Ames-Lewis, White, and John Pope-Hennessy.[16] These two pulpits depict narratives of biblical history interwoven with representations of the Virtues and Vices, Liberal Arts and pagan prophetesses of Antiquity and inform the viewer about the natural world and the nature of religious experience.[17] This naturalistic expression of Scholasticism is also reflected in the fountain through the images of the Labours of the Months and signs of the zodiac, clearly visible around the lower fountain, and situated along with the religious and allegorical themes, civic statuary, the liberal arts, and moral stories of the fountain as a whole.

LABOURS OF THE MONTHS

James Carson Webster's study of the Labours of the Months, identified in sculptured doorways of churches and cathedrals from antiquity to the medieval period, revealed a recurring pattern involving two elements: the first stemmed from astronomy and the 'desire to make more memorable the movement of time, to introduce some sense of order, of commemorate recurrence'; the second was 'humanistic, reflecting man's life on earth'.[18] According to Webster, these Labours of the Months emerged in sculptured friezes as 'illustrations', possibly dating from the second or first century BCE, initially as personifications of the month, 'human figures in constant quiet pose',[19] distinguishing season and climate by their clothing and employing attributes to suggest occupation. With the appearance of the Chronograph of 354, a late Roman codex, there was the beginnings of the movement away from passive personification towards active representation of scenes of contemporary life. Webster's research revealed a transition of image representation between the fourth and ninth centuries,

---

[16] Ames-Lewis, *Tuscan Marble Carving 1250–1350*, pp. 89–95; White, *Art and Architecture in Italy 1250–1400*, pp. 85–88; Pope-Hennessy, *Italian Gothic Sculpture*, pp. 12–13.
[17] Anita Fiderer Moskowitz, *Pisano Pulpits* (London: Harvey Miller Publishers, 2005), p. 8.
[18] James Carson Webster, *The Labors of the Months*, Northwestern University Studies in the Humanities, no. 4 (New York: AMS Press, 1938), pp. 93–94.
[19] Ibid., pp. 5, 94.

a metamorphosis from the assertion of nature—'the simple idea of ripening fruits, of growing grain, of blooming flowers, of birds which appeared in the spring'[20]—to the human capacity to capitalize on this bounty of nature, emphasizing 'daily activity which fed him and played a part in his salvation'.[21]

David M. Robb found that the earliest sculptured zodiac in Romanesque art in Italy occurred in the decoration of the Porta dello Zodiaco of the Sacra di S. Michele in the Frazione di S. Ambrogio dated c.1114, and the earliest example of the Labours of the Months as calendar illustration in Italian sculpture occurred in the decoration of S. Zeno Cathedral, Verona, dated c.1120.[22] By the twelfth century these representations of active occupation, the Labours of the Month, were well established and found in carvings on the archivolts, door jambs, facade reliefs and floors of many Romanesque and Gothic churches, as well as in medieval stained glass windows, misericords, and in many illuminated manuscripts. These images were regionally based, with the Italian labours of June illustrated by reaping and of July by threshing, whilst in France, at a higher latitude, labours of July were illustrated by reaping and by threshing in August.

The sculptures of the lower basin of the Fontana Maggiore continue this tradition of the Labours or Occupations of the Months connected with their appropriate astrological zodiac sign, representing the cycle of the year and synthesising seasons and calendar. Each image of the Labours of the Months along the lower basin of the Fontana Maggiore is depicted as a single human figure engaged in an activity, a seasonal labour. The left image of each pair is inscribed with the month carved above it and contains a sculptured sign of the zodiac within it. The right image of each pair is inscribed above with either 'Uxor' (wife) or 'Socius' (companion) (see fig. 5.3 for an example of one month).

---

[20] Ibid., p. 102.
[21] Ibid., p. 103.
[22] D. M. Robb, 'Niccolò: A North Italian Sculptor of the Twelfth Century', *Art Bulletin* 12, no. 4 (1930): pp. 381, 410.

Fig. 5.3: Februarius and Pisces, Fontana Maggiore, Perugia, Italy. On the left a fisherman sits on a rock with a rod in his right hand, a basket on his left next to him. The fishes of Pisces are carved in the top right hand corner. In the right hand scene ('socius') the fisherman carries the fish home. Photo: *D.Gunzburg.*

According to Jonathan Alexander, it was at this time that society formed itself into three orders—those who prayed, those who fought, and those who laboured. This was reflected iconographically as the Labours of the Months represented by the non-religious orders of nobility (those who fought) and peasant (those who laboured).[23] Alexander notes that, by appearing to support the sacred narrative of the Bible in their placement on church facades, chosen and controlled by the third order—those who prayed—the Occupations 'complete the Church's spiritual and physical existence'.[24] Whilst not commissioned by the church, nevertheless the images of the lower basin of the Fontana Maggiore also appear to be supporting a sacred narrative, flanked as they are by scenes from the Old Testament, and overlooked by the figures of St Paul, St Lawrence, St

---

[23] Johnathan Alexander, 'Labeur and Paresse: Ideological Representations of Medieval Peasant Labor', *Art Bulletin* 72, no. 3 (1990): p. 437.

[24] Ibid., p. 438.

Herculanus, St Benedict and St John the Baptist above them on the middle basin. Thus the Labours of the Month are a seasonal expression of the roles of the nobility and the peasantry in the order of the world.

ASTROLOGICAL IMAGES ALONG THE LOWER BASIN

On the lower basin of the Fontana Maggiore, the Labours of the Months and their zodiacal imagery follow each other consecutively in an anticlockwise direction beginning with the month of Januarius and the zodiac sign of Aquarius and completing with December and the zodiac sign of Capricorn. In the northern hemisphere the ecliptic, the apparent path of the sun through the heavens, is viewed to the south. In the scheme of the Fontana Maggiore the zodiac images have all been placed to the south of the scheme, thus facing the arc of the sun every day of the year. In this circular fountain where the images for the months of the year and their occupations could have been placed in any orientation, they have been placed to face the rising and the setting of the sun.

Furthermore the month of Januarius and the zodiac sign of Aquarius is placed at a point that is south of due west (azimuth 237^{0} / declination -23^{0} 32′) and the month of December and the zodiac sign of Capricorn is placed at a point that is north of due east (azimuth 57^{0} / declination 23^{0} 32′) (fig. 5.4).

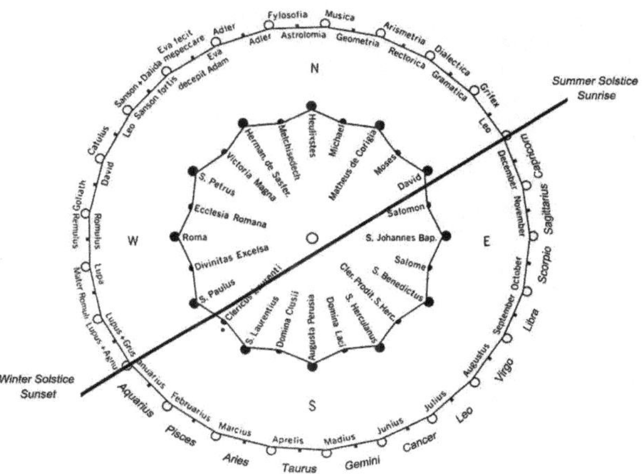

Fig. 5.4: Plan view showing the lower (outer ring) and middle (inner ring) basins of the Fontana Maggiore, after John White's reconstruction of the figures of the middle basin, and indicating the Labours of the Months scheme with Januarius placed at the Winter Solstice sunset point and completing at the Summer Solstice sunrise point.

These two points correspond to the position on the horizon of the sun at the winter solstice sunset in the south west (Januarius/Aquarius) and to the position on the horizon of the sun at the summer solstice sunrise point in the north east. The winter solstice sunset point corresponds with traditional images connected with the Labours of the Month at that time of the year. A nobly dressed man is seated in front of a roaring fire. A whole roast chicken on a plate rests on his knee. He holds a wine cup in his left hand. The image of the water bearer of Aquarius is carved in the top right hand corner such that it appears to be pouring water into the man's wine cup. His wife sits on the other side of the roaring fire, supporting a mug on her knee in her left hand and a plate in her right hand (fig. 5.5). This beginning point of the monthly/astrological cycle coincides with the beginning of the liturgical year—the birth of Christ.

Fig. 5.5: Januarius and Aquarius, Fontana Maggiore, Perugia, Italy. On the left a nobly dressed man is seated in front of a roaring fire. A whole roast chicken on a plate rests on his knee. He holds a wine cup in his left hand. The image of the water bearer of Aquarius is carved in the top right hand corner such that it appears to be pouring water into the man's wine cup. In the right hand scene ('uxor') his wife sits on the other side of the roaring fire, supporting a mug on her knee in her left hand and a plate in her right hand. Photo: *D.Gunzburg.*

Whilst one may never know if this was the intention of those who constructed and sculpted this scheme, it appears as if these images have been placed to reflect visually the orientation of the ecliptic to the south of the observer, and placed so that they reflect two important points of the calendar: the beginning of the year oriented to the winter solstice sunset and carrying in its Christian symbolism the birth of Christ; and the final image of the zodiac scheme oriented to the extreme north-east horizon position of the sun at the summer solstice sunrise.

GEARS AND SEASONS

There is an additional area in which the Fontana Maggiore of Perugia reflects celestial symbolism and this concerns its architectural design. The chronicles of the time, such as those written by Florentine Giovanni Villani (c. 1276 or 1280–1348), Albertino Mussato (1261–1329), and Rolandino of Padua (1200–1276), emphasized the causal relationship between communal triumphs and accomplishments, particularly military victories, and astrology.[25] Villani reflected the characteristic attitude of his day towards astrology in comprehending that both natural phenomena and human actions were influenced by the movements of the heavenly bodies, that God bestowed life and movement to humanity through a primary action which passed down through the concentric spheres of stars and planets to the motionless realm of the earth beneath the moon. It was then the province of the astrologers or natural philosophers to interpret the movements of the planets in order to understand the manifestations of God's will on earth.[26] Jonathan B. Riess notes that in his writing, Mussato, one of Padua's leading literary and political thinkers, perceived that

> history, as made up of cycles that repeated themselves in regular and predictable patterns, implied that history could be brought to the service of the state. History could thus be 'scientifically' employed for social and political ends. The past had lessons to teach. Among these lessons was that of vigilance in the preservation of republicanism because the alternative, tyranny, would be abhorrent. Thus, far from locking man into a deterministic or mechanical universe, the particular form of astrology that flourished at the moment made it possible for man to come to grips with the past and to influence, as did the planets, the present and the future.[27]

The physical layout of lower and middle basins, with 25 diptychs in the lower basin and 12 in the middle basin, do

---

[25] Hyde, *Padua in the Age of Dante*, pp. 304–6.

[26] Sharon Dale, Alison Williams Lewin, and Duane J. Osheim, eds., *Chronicling History: Chroniclers and Historians in Medieval Renaissance Italy* (University Park, PA: The Pennsylvania State University Press, 2007), pp. 122–23, 133–34.

[27] J. B. Riess, *Political Ideals in Medieval Italian Art: The Frescoes in the Palazzo dei Priori, Perugia (1297)* (Ann Arbor: UMI Research Press, 1981), p. 73.

not create an exact 1:2 correspondence. Seen from above, offset to each other, this deliberate displacement of basins suggest the cogs of a gear, wheels with teeth that transmit torque and aids motion. White likens this to the way many Gothic windows contain varying counterbalanced radiating parts which also suggest movement and he applies this idea of discordance to the fountain which when observed from eye level, White suggests, stimulates the viewer to move around it.[28] One could add that such a disjunction of precise agreements offers the stone equivalent of musical or literary syncopation, directing the viewer to keep in constant motion. The three-dimensional sculptural work in the middle basin further aids this motion by drawing the viewer to engage with it from a multiplicity of angles. Of sculpture work in general Charles Harrison noted that

> … it is of the essence of the experience of sculpture that we tend to find ourselves in motion in its presence, seeking optimal positions and angles of view, and that different lighting conditions will tend to bring out different aspects and effects.[29]

On the Fontana Maggiore one could argue that this gear-like design produces a regular flow of rhythm that never offers a point of rest and reflects the cycles of the seasons, forever in motion, thus reproducing the phenomenological experience of the seasons and cycles.

Additionally one can argue that this replicates the hierarchical model of an Aristotelian cosmology, all moved by gears. The top of the fountain with its the three caryatids (the Trinity) suggests the unmoved mover or primary cause of all motion; below this is the middle basin with its mixture of Biblical images and saints, the sphere of the fixed stars; and finally the lower basin which contains the heavenly bodies and the Labours of the Months of the sublunar world.[30]

---

[28] White, *Art and Architecture in Italy 1250–1400*, pp. 88–89.

[29] Charles Harrison, *An Introduction to Art* (New Haven: Yale University Press, 2009), p. 187.

[30] Aristotle, *Metaphysics*, XII:8, in Johnathan Barnes, ed., *The Complete Works of Aristotle*, Bollingen Series LXXXI.2 (Princeton, NJ: Princeton University Press, 1984), Vol. 2, pp. 1695–98.

FONTANA MAGGIORE AS LIMINAL SPACE

A third area where celestial symbolism is expressed in the sculptural narrative of the Fontana Maggiore is in its symbolic and metaphoric representation of a liminal space that connects the viewer to the sky as heaven. Pope-Hennessy advocated that in shape the fountain reflected or quoted the Pistoia holy-water basin of Giovanni Pisano, carved in the early 1270s, with its three caryatid figures forming the column on which the basin rested.[31] Moskowitz proposes that the form of the whole fountain would have been one that was inherently recognized by medieval Perugians, physically by mirroring liturgical fittings and fixtures within churches such as two– and three–basined baptismal and cloister fonts, pulpits and altar ciboria; and spiritually by linking the fountain with theological concepts such as the *fons vitae* (the fountain of life), and baptism as rebirth.[32] Nicola Coldstream terms this 'micro-architecture', the phenomenon which 'emerged from a process of exchange between objects and media, in which the meaning and significance of one was adopted by another, only to be reflected back to the source'.[33] So just as the baptismal font and the water it contained reminded worshipers every time they entered the church of their initiation into the Christian religion via the rite of baptism and their relationship to God and the saints in heaven, so by its similar shape and the water it contained, the Fontana Maggiore contained the same attributes as a holy water or baptismal font.

One could argue that the sculptural narrative of the Fontana Maggiore, set in the open, in the public square, performs a similar function as the religious narratives on the pulpits in Pisa and Pistoia in educating the urban population about the specific role of the commune. Yet this fountain was also a place where the community came at least once if not several times a day to collect water. One can therefore see the fountain as a place where two worlds cross, that which constitutes what today Mircea Eliade would define as 'the paradoxical place where those worlds communicate, where passage from the profane to the sacred world becomes

---

[31] Pope-Hennessy, *Italian Gothic Sculpture*, p. 232.

[32] Moskowitz, *Italian Gothic Sculpture c.1250–c.1400*, p. 43.

[33] Nicola Coldstream, *Medieval Architecture*, Oxford History of Art (Oxford: Oxford University Press, 2002), pp. 162–63.

possible'[34] thus defining a liminal space. In this way the routine event of collecting water would have merged with these religious recognitions and formed, for the people of Perugia, a sacred space between heaven and earth, satisfying both physical and spiritual thirst.

CONCLUSION

The Fontana Maggiore can be understood as a practical civic monument that represents, via its sculptural narrative, the place of the commune and its governing bodies within a cosmological scheme. In its three tiers it articulates a comprehensive understanding of the world, judicially, theologically, and astrologically. It was also constructed in a way that fused the basic necessity of water with the familiar shape of religious ritual, thus connecting the earthly need for water with spiritual thirst. Furthermore by its placement at the centre of the city, between the main civic and religious buildings and situated at the confluence of five roads, the fountain became an omphalos, one which linked the individual and their community through sculpted imagery to sky, earth, and through the Labours of the Months which are oriented towards the arc of the sun, seasonal rhythms. It is an illustration of the human desire to build structures that occupy liminal spaces, one that physically grounded them and at the same time metaphorically linked them to the sky.

---

[34] Mircea Eliade, *The Sacred and the Profane: The Nature of Religion*, trans. Willard R. Trask (Orlando: Harcourt Inc., 1957), p. 25.

THEOSIS, VISION, AND THE ASTRAL BODY IN MEDIEVAL GERMAN PIETISM AND THE SPANISH KABBALAH

Elliot Wolfson

ABSTRACT: This paper will explore the notion of the astral body in the works of the Rhineland Jewish Pietists and the Spanish Kabbalists of the twelfth to thirteenth centuries. Specifically, it will focus on contemplative practices in the two streams of medieval Jewish esotericism that involved ascetic renunciation of the body as a means to cultivate a vision of the celestial image, culminating in the angelification and divinisation of the human.

The history of astrological speculation and practice in Judaism is a long and variegated one that stretches back to Hellenistic Late Antiquity and likely has even earlier roots in the Near Eastern context of ancient Israel.[1] Obviously, it lies beyond the scope of this essay to supply an adequate survey of this complex subject. Suffice it to say, however, that in post-biblical Jewish sources astrology was considered predominantly a component of esoteric wisdom, that is, a wisdom considered secretive and thus limited to a small coterie of initiates. Even the rabbinic sages who flatly denied the relevance of astrology to Judaism—the most famous articulation thereof is the dictum *ein mazzal le-yisra'el*, that is, Israel is not subject to astrological influence[2]—did not repudiate its efficacy and importance. Particularly significant in this regard are techniques of astral magic employed to draw down the force of spirituality from above through

[1] James H. Charlesworth, 'Jewish Astrology in the Talmud, Pseudepigrapha, the Dead Sea Scrolls, and Early Palestinian Synagogues', *Harvard Theological Review* 70 (1977): pp. 183–200.

[2] Babylonian Talmud, Shabbat 156a, Nedarim 32a. See Gregg Gardner, 'Astrology in the Talmud: An Analysis of Bavli Shabbat 156', in *Heresy and Identity in Late Antiquity*, ed. Eduard Iricinschi and Holger M. Zellentin (Tübingen: Mohr Siebeck, 2008), pp. 314–38.

specially designed physical localities or talismanic effigies. The effectiveness of such magical procedures is predicated on the scientific assumption that stellar emanations exert an influence upon processes in the material world. I concur with the view of Kocku von Stuckrad that the analogical correspondence between the celestial and terrestrial realms can be considered the tenet of astrology that stands at the core of the alchemical-magical worldview. Rather than applying to astrology the theologically laden and implicitly pejorative labels of superstition or apostasy, it is more accurate to consider it the central esoteric discipline that emerges from hermetic discourse. To cite his own words:

> Instead of assuming a causal and mechanistic influence of the stars astrologers try to establish analogies and symmetric correspondences between the planetary zone and the earth—hermeticism's famous 'as above, so below'. Hence, a denotation of astrology would read as follows: Astrology is a concept of interpretation describing the quality of a given time, i.e. the essence of simultaneously and synchronically occurring events which are connected to inherent symbols and meaning.[3]

Viewing astrology principally as a hermeneutical practice, as a semantic system of interpretation based on symbolically decoding mundane events in light of their heavenly signs, provides the key to comprehending the role of astrology in medieval Jewish esotericism. The crucial, but by no means only, source that influenced later astrological works authored by Jews was *Sefer Yeṣirah*, a text that is dated anywhere from the third to the ninth centuries. Properly speaking, the work should not be described as a single composition but rather as an aggregate of distinct literary strands that have been woven together through a complicated redactional process whose stages are not clearly discernible.[4] What is clear, however, is that this relatively small treatise began to have an inordinate influence from the tenth century when the different versions of the text were established more firmly and commentaries on it were written in various parts of the

---

[3] Kocku von Stuckrad, 'Jewish and Christian Astrology in Late Antiquity: A New Approach', *Numen* 47 (2000): p. 5–6.

[4] For a review of the problem and reference to the various scholarly approaches, see Elliot R. Wolfson, 'Text, Context, Pretext: A Review Essay of Yehuda Liebes's "Ars Poetica in Sefer Yetsirah"', *Studia Philonica Annual* 16 (2004): pp. 218–28.

Jewish diaspora. Especially important was the commentary composed in Byzantine Southern Italy by Shabbetai Donnolo (913–c. 982), the *Sefer Ḥakhmoni*, a work that deals mainly with astrology. For our purposes I will mention only that Donnolo greatly expands the microcosmic correlation of the universe and the human body enunciated in the second part of *Sefer Yeṣirah*, a principle that he affirms in conjunction with the presumption that astrophysical forces are mirrored in and therefore can manipulate the anatomical life of human beings.[5] Another composition from this region and time that embraced astrology explicitly as a legitimate Jewish enterprise was the *Megillat Aḥimaaz* composed by Aḥimaaz ben Paltiel in Capua in 1054.[6] Both texts maintain a distinction between astronomy as the science that measures the movements of celestial bodies and observes their impact on the natural world, on the one hand, and astrology as the technique of interpreting the influence of these bodies on future events, on the other. The former was accorded an elevated status in the economy of religious law, since it was the principle vehicle for temporal calculation and establishing the calendar correctly, whereas the latter (in the technical classification of medieval Christian sources, judicial astrology) presented a far more ambivalent—even if not outright heretical or illegal—picture, inasmuch as the art of prognostication and the determinism linked to the stars and planets seemingly challenged the claims to God's omnipotence and human free will, two pillars of the

---

[5] See Andrew Sharf, *The Universe of Shabbetai Donnolo* (New York: Ktav, 1976); Elliot R. Wolfson, 'The Theosophy of Shabbetai Donnolo, with Special Emphasis on the Doctrine of *Sefirot* in his *Sefer Ḥakhmoni*', *Jewish History* 6 (1992): pp. 281–316; and the recently published comprehensive study of Piergabriele Mancuso, *Shabbatai Donnolo's Sefer Ḥakhmoni: Introduction, Critical Text, and Annotated Translation* (Leiden: Brill, 2010), pp. 2–76.

[6] Concerning this work, see Robert Bonfil, 'Between Eretz Israel and Babylonia', *Shalem* 5 (1987): pp. 1–30, esp. 10–12, 19 (Hebrew); Robert Bonfil, 'Myth, Rhetoric, History? A Study in the *Chronicle of Ahima'aẓ*', in *Culture and Society in Medieval Jewry: Studies Dedicated to the Memory of Haim Hillel Ben-Sasson*, ed. Menahem Ben-Sasson, Robert Bonfil, and Joseph R. Hacker (Jerusalem, 1989), pp. 99–136, esp. 103–7 (Hebrew); Joseph Marcus, 'Studies in the *Chronicle of Ahimaaz*', *Proceedings of the American Academy of Jewish Research* 5 (1933/34): pp. 85–91; Steven D. Benin, 'The Chronicle of Ahimaaz and its Place in Byzantine Literature', *Jerusalem Studies in Jewish Thought* 4 (1984/85): pp. 237–50.

monotheistic worldview as it crystallised in medieval theological treatises.

Noteworthy in this regard is the fact that Judah Halevi and Abraham Ibn Ezra, two prominent Andalusian Jewish thinkers of the twelfth century, utilised astral magic to explain the true intent of certain ritual observances.[7] This has far-reaching implications in narrowing the gap between what is often considered to be magic and the accepted understanding of religious custom. I note, in passing, that both of these figures were informed by Islamic sources that had preserved the older mix of magic, astrology, divination, and alchemy evident in the Hermetic literature of the Hellenistic East. Based on these currents of Greco-Roman philosophy, the Muslim thinkers developed a corpus of what they considered to be occult science. Texts contained in this body of literature are often referred to as *sir* or *sir al-asrār*, respectfully rendered into Latin as *secretum* and *secretum secretorum* and into Hebrew as *sod* and *sod ha-sodot*. As William Eamon pointed out, the 'philosophical foundation' of the Arabic secret sciences 'was the doctrine that the world was a network of hidden correspondences and a reservoir of powerful occult forces. This perspective infiltrated Islamic natural philosophy'.[8] It is no exaggeration to say that the principle of hidden correspondences is one of the rudimentary ideas—we could even call it a ground concept—that informed the various trends of Jewish esotericism that evolved in the twelfth and thirteenth centuries. We should also bear in mind that the cosmological conception of dual reference—each event relates externally to nature and internally to the supranatural—has a parallel in the

[7] Shlomo Pines, 'On the Term *Ruḥaniyyot* and its Origin and on Judah Halevi's Doctrine', *Tarbiz* 57 (1988): pp. 511–40, esp. 524–30; Dov Schwartz, *Studies on Astral Magic in Medieval Jewish Thought*, trans. David Louvish and Batya Stein (Leiden: Brill, 2005), pp. 1–26, 103–13. See also Y. Tzvi Langerman, 'Some Astrological Themes in the Thought of Abraham ibn Ezra', in *Rabbi Abraham ibn Ezra: Studies in the Writings of a Twelfth-Century Jewish Polymath* ed. Isadore Twersky and Jay M. Harris (Cambridge, MA: Harvard University Press, 1993), pp. 28–85; Shlomo Sela, *Astrology and Biblical Exegesis in Abraham Ibn Ezra's Thought* (Rama-Gan: Bar-Ilan University, 1999) (Hebrew).

[8] William Eamon, *Science and the Secrets of Nature: Books of Secrets in Medieval and Early Modern Culture* (Princeton, NJ: Princeton University Press, 1994), p. 40.

hermeneutical belief that every word of Scripture has an apparent and a latent connotation. The parallel lines meet in the portrayal of the universe as a cosmic text: every facet of the world can be seen as a mirror through which the hidden God is disclosed. On this account, natural phenomena were treated like a transparent symbol whose inner meaning points to a spiritual reality that is invisible. The esoteric interpretation of the exoteric text served as the basis for the figurative study of the world as the book of nature that requires decoding. The preponderance of these themes is another indication of the complex interweaving of the threads of philosophy, science, and mysticism in the religious outlook developed by Muslims and Jews on the basis of Hermetic and Neoplatonic sources.

Here it is worthwhile recalling the argument proffered by Moshe Halbertal in his *Concealment and Revelation: Esotericism in Jewish Thought and its Philosophical Implications* that the three major trends in Jewish thought that evolved in the twelfth and thirteenth centuries—the period that Halbertal calls 'the age of esotericism and its disclosure'[9]—kabbalah, philosophy, and astrology, all laid claim to having the key to decode the secrets of the Torah. Halbertal is to be given credit for discerning a unity in the 'esoteric structure' of these disparate forms of intellectual expression,[10] even if I might quibble with his understanding of esotericism.[11] What protagonists of the three orientations share is the presumption that sacred scripture has an internal meaning that complements the external; the latter addresses the needs of the community at large and the former is thought to be exclusive to a select body of individuals. Appeal to the medium of the concealed allowed the various Jewish thinkers to absorb and to incorporate the diverse cultural influences of Aristotelianism, Neoplatonism, and Hermeticism, and thereby transform the tradition by promoting foreign elements in the guise of the hidden and deeper import of the canonical text.[12]

---

[9] Moshe Halbertal, *Concealment and Revelation: Esotericism in Jewish Thought and its Philosophical Implications,* trans. Jackie Feldman (Princeton, NJ: Princeton University Press, 2007), p. 5.
[10] Ibid., p. 6–7.
[11] See my book review in the *Journal of Religion in Europe* 2 (2009): pp. 314–18.
[12] Halbertal, *Concealment and Revelation,* pp. 39–41, 138–39.

From the standpoint espoused by Halevi and Ibn Ezra, the biblical-rabbinic law is the vehicle to secure the emanation of spirit from above, and hence it is not merely the case that astrology was viewed as a legitimate scientific tool to analyze history and especially to calculate the advent of the messianic era based on planetary conjunctions, as we find, for example, in the fifth chapter of the twelfth-century treatise *Megillat ha-Megalleh* composed by one of Ibn Ezra's teachers, the Catalan mathematician, astronomer, and philosopher, Abraham bar Ḥiyya (1065–1145), but it was treated as an essential component of Judaism, indeed, its hidden essence. Contrary to the opinion that the astrologer's assumption that the motions of the celestial bodies incite corresponding changes in the sublunar world constitutes an assault on human freedom and divine omnipotence, Halevi and Ibn Ezra, leading intellectuals of their time, maintained that through astrology one can unlock the deepest mysteries, indeed the secrets that constitute, in Halevi's telling formulation, 'the root of faith and the root of heresy'.[13]

Dov Schwartz has argued that within the framework of Jewish rationalism the occult traditions surrounding astral magic developed in an 'esoteric climate' because of their similarity to the characterization of idolatry—drawing down stellar spirituality upon effigies was viewed as analogous to the worship of idols. The esotericism, on this score, would be essentially political in nature, that is, matters that might be perceived as dissenting had to be concealed. He also notes, however, that the esotericism bespeaks the fact that these intellectuals sincerely believed that astrology was an arcane science cultivated by Jews from antiquity.[14] Given the proximity of astral magical practice to the kabbalists' theurgical explanation of the commandments, that is, the presentation of ritual law as the means to draw down the divine efflux, which both maintains the unity within the Godhead and sustains the different worlds that constitute the links in the ontic chain, it should come as no surprise that when Jewish mysticism flourished historically on the European continent in the twelfth and thirteenth centuries, many of the theosophic doctrines and occult activities would be expressed precisely in astrological terms. The harsh

[13] Judah Halevi, *Sefer ha-Kuzari*, trans. Yehuda Even Shmuel (Tel-Aviv: Dvir, 1972), I.77, p. 23.
[14] Schwartz, *Studies on Astral Magic*, p. 9.

critique leveled against astrology, and especially astral magic, as a form of idolatry or pseudo-science (in contrast to the legitimate science of astronomy) on the part of Moses Maimonides in his letter to the rabbis of Lunel as well as other places in his written work—following the lead of his Muslim predecessor al-Fārābī—proved to be an aberration (in its own time and for several centuries until the modern period).[15] Various post-Maimonidean currents of Jewish mysticism indicate that this practice and the conceptual worldview whence it arose were considered a legitimate part of Jewish theology.

I will limit my reflections in this article to one particular motif that has figured prominently in astrological speculation, the astral body.[16] Even more limitedly, I will be focusing on the role of this symbolic notion in the contemplative practices of two streams of medieval Jewish esotericism in the thirteenth century, the Rhineland Jewish Pietism represented by Eleazar ben Judah of Worms and the prophetic kabbalah promulgated by Abraham ben Samuel Abulafia of Spain. I am especially interested in the promotion of the ascetic renunciation of the physical body as a means to facilitate a vision of the celestial image, culminating in the angelification of the human. The aetheral or subtle body can also be designated the 'imaginal' body, a term that denotes that the somatic form, which is configured in the imagination of the visionary, corresponds to the angelic image in the *'ālam al-mithāl* or the *mundus imaginalis*, expressions utilised by Henry Corbin to designate the intermediate plane—what Ṣūfis call *barzakh*—between spirit and matter. This world, the interval of the in-between, is not to be construed as unreal; it is, as Corbin put it, a 'third world halfway between the world of sensible perception and the world of intelligibility', realm of 'autonomous forms and images...preserving all the richness and diversity of the sensible world but in a spiritual state'.[17] But even these words are inadequate inasmuch as

---

[15] Ibid., pp. 27–54.

[16] For a detailed discussion of this theme in Jewish mystical sources, see Gershom Scholem, *On the Mystical Shape of the Godhead: Basic Concepts in the Kabbalah*, trans. Joachim Neugroschel, ed. and revised by Jonathan Chipman, foreword by Joseph Dan (New York: Schocken Books, 1991), pp. 251–73.

[17] Henry Corbin, 'The Visionary Dream in Islamic Spirituality', in *The Dreams and Human Societies*, ed. G. E. von Grunebaum and Roger Caillois (Berkeley: University of California Press, 1966), pp.

they still suggest a polarity between the spiritual and material, the very dyad that is undermined by the fact that this realm is characterised by the coincidence of opposites, including the presumed opposition between real and imagined. What is real is real as imagined precisely because what is imagined is imagined as real.

I will begin with the German Pietists or the Ḥasidei Ashkenaz. Although there were several different circles that are considered branches of this historical phenomenon, my comments relate to the main group of Pietists, the Qalonymide circle, which was active in the twelfth and thirteenth centuries in the cities of Mainz, Regensburg, Speyer, and Worms.[18] The three major figures of this school were Samuel the Pietist, Judah ben Samuel, and Eleazar ben Judah. According to the legend preserved by the Pietists, they trace their lineage back to the clan of the poet Qalonymous ben Qalonymous in Southern Italy, a region, as we have seen, that was rife with astrological speculation. The descendants of the Qalonymide family were supposed to have been transplanted from Lucca to Mainz by Charles the Bald, who lived in the second half of the ninth century, although some sources name Charlemagne, but his death in 914 renders this identification suspect.[19] Even though the efforts to discover the historical grain of truth behind this narrative have proved to be futile, most historians believe that there is no reason to doubt that there was an Italy-Rhineland route of Jewish immigration into northern Europe.[20]

406–7. For a more extended discussion of Corbin's notion of the imaginal, see Elliot R. Wolfson, 'Imago Templi and the Meeting of the Two Seas', *RES* 51 (2007): pp. 121–35.

[18] For discussion of the different circles and their literary compositions, see Joseph Dan, *The Esoteric Theology of Ashkenazi Hasidism* (Jerusalem: Bialik Institute, 1968), pp. 46–67 (Hebrew).

[19] Dan, *The Esoteric Theology*, pp. 14–20; Abraham Grossman, 'The Migration of the Kalonymos Family from Italy to Germany', *Zion* 40 (1975): pp. 154–85 (Hebrew).

[20] Joseph Dan, 'Ashkenazi Hasidim, 1941–1991: Was There Really a Hasidic Movement in Medieval Germany?', in *Gershom Scholem's Major Trends in Jewish Mysticism 50 Years After: Proceedings of the Sixth International Conference on the History of Jewish Mysticism*, ed. Peter Schäfer and Joseph Dan (Tübingen: J. C. B. Mohr, 1993), pp. 87–101; Ivan G. Marcus, 'History, Story, and Collective Memory: Narrativity in Early Ashkenazic Culture', *Prooftexts* 10 (1990): pp. 365–88; Ivan G. Marcus, 'The Historical Meaning of Hasidei

In the Rhineland, the Qalonymide family established itself again and rose to a position of rabbinic authority in competition with the other dominant model of authority in the Ashkenazi cultural milieu, the Tosafist school, headed by the grandchildren of the eleventh-century Northern French commentator and exegete, Solomon ben Isaac (Rashi). This school represented elitist intellectualism, placing primacy on the dialectical study of Talmudic literature as the primary mode of religious commitment; the Pietists presented an alternate paradigm of devotion, one based on prayer,[21] extreme submission of the finite human will to the infinite will of God,[22] and adoption of ascetic and meditative practices to attain a state of prophecy, which consists of seeing the glory and receiving the Tetragrammaton,[23] which is identified further as the mystical essence of Torah, a

Ashkenaz: Fact, Fiction or Cultural Self-Image?' in *Gershom Scholem's Major Trends in Jewish Mysticism 50 Years After*, pp. 103–14. For an attempt to explain the origins of Ḥasidei Ashkenaz in a 'proto-pietism' from the eleventh and twelfth centuries, which has parallels to developments in Christian society, see Robert Chazan, 'The Early Development of Hasidut Ashkenaz', *Jewish Quarterly Review* 75 (1985): pp. 199–211.

[21] Joseph Dan, 'The Emergence of Mystical Prayer', in *Studies in Jewish Mysticism: Proceedings of Regional Conferences Held at the University of California, Los Angeles and McGill University in April, 1978*, ed. Joseph Dan and Frank Talmage (Cambridge, MA: Association for Jewish Studies, 1982), pp. 85–120; Joseph Dan, 'Prayer as Text and Prayer as Mystical Experience', in *Torah and Wisdom-Studies in Jewish Philosophy, Kabbalah, and Halacha: Essays in Honor of Arthur Hyman*, ed. Ruth Link-Salinger (New York: Shengold Publishers, 1992), pp. 33–47; Joseph Dan, 'Pesaq ha-Yirah veha-Emunah and the Intention of Prayer in Ashkenazi Hasidic Esotericism', *Frankfurter Judaistische Beiträge* 19 (1991–1992): pp. 185–215 (the three essays are reprinted in Joseph Dan, *Jewish Mysticism in the Middle Ages* [Northvale: Jason Aronson, 1998], pp. 221–311); Ivan G. Marcus, 'Prayer Gestures in German Hasidism', in *Mysticism, Magic and Kabbalah in Ashkenazi Judaism: International Symposium Held in Frankfurt a. M. 1991*, ed. Karl Erich Grözinger and Joseph Dan (Berlin: Walter de Gruyter, 1995), pp. 44–59.

[22] Haym Soloveitchik, 'Three Themes in the *Sefer Ḥasidim*', *AJS Review* 1 (1976): pp. 314–15; Ivan G. Marcus, *Piety and Society: The Jewish Pietists of Medieval Germany* (Leiden: E. J. Brill, 1981), pp. 25–26.

[23] Dan, *The Esoteric Theology*, pp. 74–76; Elliot R. Wolfson, *Through a Speculum That Shines: Vision and Imagination in Medieval Jewish Mysticism* (Princeton, NJ: Princeton University Press, 1994), pp. 234–47.

fundamental axiom shared by the major currents of medieval Jewish esotericism, and one that we can reasonably assume has archaic roots.[24] There is thus a threefold identification at the core of the Pietist mystical orientation, the name = the glory = the torah (each of which, parenthetically, is associated with the number 32, which is the numerology of the word *lev*, or heart, the locus of the spiritual vision).[25] For Samuel and Judah, this ideal of piety was to be realised by a small fraternity of practitioners, who were recipient of the oral traditions that comprised both practice and lore; those who belonged to the group were distinguished as the elect of Israel, God's remnant, in contrast to the rest of the Jews who are designated as the 'wicked'. In the case of Eleazar, there is a relative democratization of the ideal reflected in his attempt to communicate the main Pietistic teachings to the wider Jewish community.[26]

To comprehend the concept of the astral body, we must begin with a brief description of the Pietistic view on the nature of prophecy and envisioning that which is ostensibly beyond visualization. As I delineated in great detail in the chapter on the German Pietists in my monograph *Through a Speculum That Shines: Vision and Imagination in Medieval Jewish Mysticism* (1994),[27] three different views on the nature of prophetic revelation are offered: according to the first opinion, God is invisible and therefore what the prophet sees are images (*dimyonot*) in his mind, a process compared to an illusion (*aḥizat einayim*) that may be conjured by magical means, a sleight of hand. The second opinion likewise maintains that God is invisible and thus what the prophet beholds is the glory that emanates from God. Insofar as the glory has a fixed dimension it is visible. However, the place of the attachment of the glory to God—referred to scripturally as God's face (Exodus 33:20)—is not visible. The third opinion agrees with the second but maintains that the glory is a created light outside of God.

---

[24] Elliot R. Wolfson, 'The Mystical Significance of Torah Study in German Pietism', *Jewish Quarterly Review* 84 (1993): pp. 43–77; Wolfson, *Through a Speculum That Shines*, pp. 247–54.

[25] Wolfson, 'The Mystical Significance', pp. 67–69; Wolfson, *Through a Speculum*, pp. 252–54.

[26] Marcus, *Piety and Society*, pp. 15–16, 55–74.

[27] Wolfson, *Through a Speculum That Shines*, pp. 195–214. My abbreviated discussion here is based on the far more textually documented analysis found in that monograph.

Critical to assessing these different views is the role of imagination and the status accorded the imaginary. The faculty by which one imagines, generally designated as *dimyon*, is not characterised by either of the two main functions attributed to the imagination in medieval Aristotelianism. That is, the activity of the *dimyon* is not construed as either representation through an image of an absent sensible object or as the presentation of that which is unreal based on the sensory impressions retained in memory. Perhaps reflecting a Neoplatonic understanding, the role of the imagination is to conjure visible forms of that which is invisible, the divine glory whose luminosity is intrinsically without shape. The images, accordingly, are not merely epistemological constructs but are ontological entities, in fact, the spiritual archetypes of earthly realities that are experienced concretely within the imagination.

In some of the pietist sources, these images are identified more explicitly as angelic forms (*mal'akhim*) that exist in the spiritual domain that is parallel to an individual's thoughts, the mirror reflection, as it were, the image of the image. The angels, in other words, are the theophanic forms that take shape within the mind. Consider the following passage from the *Sefer ha-Shem* of Eleazar: 'It is customary for God to clothe the thoughts of his decrees, to show [them] to the prophets so that they will know that God has set his decrees. The prophet knows his thoughts according to the vision that he sees. At times this vision is called an angel'.[28] In addition to explaining the prophetic envisioning of the glory, this idea is deployed to elucidate the proper intentionality in prayer. The angels are identified as the imaginal forms that the divine places in the mind/heart of the worshipper and by which the nonphenomenalizable is encountered phenomenally. Insofar as the images seen by the visionary are the visible representation of the invisible, it may be said that for the German Pietists the imagination is the faculty that mediates between the corporeal and spiritual. These images do not 'objectively' represent truth but neither are they entirely false or subjective. The imaginary is both real and illusory. The paradox is pithily expressed in the assertion preserved in

[28] MS British Museum 737, fol. 223a, translated in Wolfson, *Through a Speculum*, p. 206, and see other relevant sources cited there in n. 64.

several sources that *God is both outside and within the image*.[29] The glory can be imagined in concrete forms, including most importantly that of an anthropos, for God, the transcendent One, is not the image; yet, one can worship that very image because God is present therein. The positive value accorded the imagination is also underscored by the theological claim that it is the will of God that is responsible for placing these images within the prophet's mind.

The Pietists utilised this notion of a mental image to explicate the biblical idea that man is created in the divine image, which is connected with the conception of the celestial image (*ṣelem*) that corresponds, as Scholem has written, to a kind of astral 'archetype' occupying a 'sphere of non-corporeal, semi-divine existence'.[30] In the most detailed formulation of the latter idea in *Sefer Ḥasidim*, there is an implicit connection between the *ṣelem* and prophetic vision as the relevant context has to do with the revelation of the divine glory to Moses:

> It is written 'I will make all My goodness pass before you' (Exodus 33:19). The word 'all' (*kol*) [in the expression 'all My goodness', *kol ṭuvi*] numerically equals fifty [which corresponds to the fifty] gates of understanding.[31] Upon each and every gate there is appointed an angel. The expression 'all My goodness' (*kol ṭuvi*) numerically equals [the word] *mazzal*, [zodiacal sign], for [God] showed Moses the sign of their souls....This is the meaning of 'all My goodness before you'. It should have been written 'all My goodness before your eyes'. Rather [the use of the expression 'before you', *al panekha*] alludes to the fact that the face of Moses is above when He passes above over that very form, the angel comes down upon him and informs him....Thus it says, 'And God created man in His image, in the image of God He created him' (Genesis 1:27): one [image] above and one [image] below.[32]

---

[29] For references, see Wolfson, *Through a Speculum*, pp. 200–1, 207.

[30] Scholem, *Major Trends*, pp. 117–18. See also Scholem, *Origins of the Kabbalah*, ed. R. J. Zwi Werblowsky, trans. Allan Arkush (Princeton, NJ: Princeton University Press, 1987), pp. 112–14.

[31] Babylonian Talmud, Rosh ha-Shanah 21b, Nedarim 38a. The rabbinic concept had a considerable impact on various forms of medieval Jewish esotericism.

[32] Judah the Pious, *Sefer Ḥasidim*, ed. Jehuda Wistinetzki and Jacob Freimann (Frankfurt am Main: Wahrmann Verlag, 1924), §1514, pp. 369–70.

The scriptural narrative that occasions this comment is the request of Moses to see the divine glory. The voice of God responds that a vision of the face will be denied, since no human can survive such an experience, but he will be able to see the back as God causes all of his goodness to pass before him. The expression 'all My goodness', which contextually signifies the glory, is interpreted by the medieval German Pietist astrologically based on the numerological equivalence of *kol ṭuvi* and *mazzal*—both have the value of 77. What God showed Moses was the astral signs of all the souls including his own. This is an extraordinary recasting of the biblical text: the divine back that the prophet beheld is the angelic forms (*mazzalot*) engraved on the fifty gates of understanding, the archons that bear the likeness of the faces of each person. Most astonishingly, this astrological principle is employed to explain the seminal idea that Adam was created in the image of God, that is, the image below reflects the image above. Eleazar reiterates the same notion:

> Every angel who is an archon of the zodiacal sign (*sar mazzal*) of a person when it is sent below has the image of the person who is under it....And this is the meaning of 'And God created man in his image, in the image of God he created him' (Genesis 1:27). Why is [it written] twice, 'in his image' and 'in the image'? One image refers to the image of man and the other to the image of the angel of the zodiacal sign which is in the image of the man.[33]

In an earlier Pietistic text, which very likely was one of Eleazar's sources, the prophetic vision is similarly described in terms of the intermediary of the celestial image: 'Then the prophet sees the image above, and the sign (*mazzal*) which is the archon of that image in the image of (that) man....Thus he sees the image and the speech of that image...he sees the image of the archon which is his sign'.[34] What is apparently distinctive about Eleazar's interpretation is the claim that this celestial image is configured only within the mind of one who perceives it. I note, in passing, that the Pietistic notion is not fatalistic or deterministic, i.e., one's fate is not irrevocably

[33] Eleazar of Worms, *Ḥokhmat ha-Nefesh* (Benei Beraq, 1987), Ch. 48, 80. For discussion of the astral image in the thought of Eleazar of Worms, see Scholem, *On the Mystical Shape*, pp. 260–61.
[34] MS Oxford-Bodleian 1567, fols. 60a–b, quoted by Dan in *The Esoteric Theology*, p. 225 n. 8.

determined by the celestial form in whose image one is created. On the contrary, the morphological resemblance between earthly anthropos and the celestial form endows the former with theurgical significance over the latter.

By way of summary, the Pietists endorsed the idea that each person has a corresponding celestial angel (*sar mazzal*), a *Doppelgänger*, which assumes the image of that person upon descending to the world. This is how they explained the biblical idea that Adam was created in the divine likeness, that is, the human being is fashioned in the likeness of the angelic form that is one's own image. Even though that form is not immaterial, it is a subtle body that assumes concrete shape only within the imagination. In a manner consonant with the dream,[35] the prophetic visualization of the external image is a projection of the psychological state of the visionary, i.e., what appears outside the mind is, to borrow the locution of the Jungian analyst, Erich Neumann, a 'psychic image-symbol' that derives from the interior state of the individual and is projected upon the external world.[36] The visual image is, in effect, a symbolic depiction of an internal psychic condition. The astral body can be apprehended only through the image configured in the

[35] On the role of dreams in the thought of the German Pietists, see Monford Harris, 'Dreams in Sefer Hasidim', *Proceedings of the American Academy of Jewish Research* 31 (1963): pp. 51–80, and Monford Harris, *Studies in Jewish Dream Interpretation* (Northvale: Jason Aronson, 1994), pp. 15–38; Joseph Dan, 'On the Teaching Concerning the Dream in Ḥasidei Ashkenaz', *Sinai* 68 (1971): pp. 288–93 (Hebrew); Tamar Alexander-Frizer, *The Pious Sinner: Ethics and Aesthetics in the Medieval Hasidic Narrative* (Tübingen: J. C. B. Mohr, 1991), pp. 87–89 and 91–94; Tamar Alexander-Frizer, 'Dream Narratives in "Sefer Hasidim"', *Trumah* 12 (2002): pp. 65–78 (Hebrew); Moshe Idel, *Les Kabbalistes de la nuit* (Paris: Editions Allia, 2003), pp. 12–28 ; Moshe Idel, 'On Še'elat Ḥalom in Ḥasidei Aškenaz: Sources and Influences', *Materia Giudaica* 10 (2005): pp. 99–109; Annelies Kuyt, 'Hasidut Ashkenaz on the Angel of Dreams: A Heavenly Messenger Reflecting or Exchanging Man's Thoughts', in *Creation and Re-Creation in Jewish Thought: Festschrift in Honor of Joseph Dan on the Occasion of his Seventieth Birthday*, ed. Rachel Elior and Peter Schäfer (Tübingen: Mohr Siebeck, 2005), pp. 147–63. For my own investigation of this matter, see Chapter 3 in Elliot R. Wolfson, *A Dream Interpreted Within a Dream: Oneiropoiesis and the Prism of Imagination* (New York: Zone Books, 2011).

[36] Erich Neumann, *The Origins and History of Consciousness* (Princeton, NJ: Princeton University Press, 1973), p. 294.

imagination of the visionary, typically located in the heart according to both Islamic and Jewish gnosis.

A similar process is attested in the kabbalistic teachings of Abraham Abulafia, the Spanish visionary responsible for promulgating what he called 'prophetic kabbalah' or the 'kabbalah of the names', the implementation of meditative practices that lead to the unitive state of *devequt,* the disembodied conjunction of the human and divine intellects. Somewhat improbably, Abulafia was able to combine this basic tenet of Maimonidean religious philosophy, whose epistemological and ontological contours are elicited from Muslim sources wherein the Aristotelian and Neoplatonic currents are intertwined, with the esoteric doctrines and mystical practices (mediated chiefly through the works of the Rhineland Jewish Pietists, but also through select treatises of the Catalonian and Castilian kabbalists that either preceded or were contemporary with him) to produce his distinctive understanding of kabbalah as a path (*derekh*) to attain gnosis of the name (*yedi'at ha-shem*).[37] In a manner closer to Judah Halevi than to Maimonides,[38] he maintained that the knowledge of this name, which is the essence of the tradition, is not grasped by speculation shared universally by all nations, but by a prophetic vision unique to the people of Israel. Moreover, insofar as this name is equated with the Torah, and the Torah is composed of the twenty-two letters of the Hebrew alphabet, we may surmise that this knowledge is inherently linguistic in nature. With regard to this matter as well Abulafia deviates from Maimonides, who regarded all language, including Hebrew, to be determined by societal contrivance. Following a much older current in mystical and rabbinic sources, Abulafia maintained that the vital life force of all existence is linked inherently to the 'holy tongue' (*leshon ha-qodesh*), the 'mother of all languages' (*em kol ha-leshonot*),[39] which he thus considered to be 'natural' in contrast to the 'conventional' status of the all the other seventy languages.[40] To cite one of dozens of passages that

---

[37] Scholem, *Major Trends,* pp. 136–38; Moshe Idel, *The Mystical Experience in Abraham Abulafia* (Albany: State University of New York Press, 1988), pp. 14–24; *idem, Language, Torah, and Hermeneutics in Abraham Abulafia* (Albany: State University of New York Press, 1989), pp. 101–9.

[38] Halevi, *Sefer ha-Kuzari,* IV.15, p. 172.

[39] Abraham Abulafia, *Mafteaḥ ha-Ḥokhmot* (Jerusalem, 2001), p. 60.

[40] Idel, *Language, Torah, and Hermeneutics,* pp. 12–14, 16–27, 143–45 n.

illustrate the point from Abulafia's *Imrei Shefer*:

> Know that if all the languages are conventional [*heskemiyyot*], the holy language is natural [*ṭiv'it*]...for it is not possible that there not be a natural language [*lashon ṭiv'it*] whence all the languages derive, and it is like the matter for all of them, nor is it possible that there not be a natural script [*mikhtav ṭiv'i*] whence all scripts emerge in the image of the primal Adam from whom all human beings were created.[41]

Just as Hebrew is deemed to be the language of creation, so the Jewish people represent the ethnicity that embodies the human ideal most fully. This standing is connected more specifically to their possession of the divine name, which is expressed somatically as the inscription of the sign/letter of the covenant on the male organ and psychically as the envisioning of the name in the imaginal form of the divine anthropos. This possession, which Abulafia and other kabbalists considered unique to the Jewish people, facilitates the actualization of their angelic potentiality.[42] Regarding this matter, Abulafia is simply inconsistent: on the one hand, all languages are comprised within Hebrew, and hence secrets pertaining to the name can be found in every language and it is even possible to practice the art of letter combination in any language, but, on the other, Hebrew is privileged as the sole language that is essential and not contingent.

By means of the praxis of letter combination, one receives the name—the 'first of all the created beings', the entity 'upon which the holy spirit is engraved', the 'soul that comprises all the souls'[43]—and is thereby conjoined to the effluence of intellectual light, an act that Abulafia viewed as equivalent to attaining a state of prophecy. To cite one of many passages that illustrate the Maimonidean influence on Abulafia's understanding of prophecy. In *Sefer ha-Ḥesheq*, he remarks:

55, 146 n. 71; Elliot R. Wolfson, *Abraham Abulafia—Kabbalist and Prophet: Hermeneutics, Theosophy, and Theurgy* (Los Angeles: Cherub Press, 2000), pp. 58–59, 62–64.

[41] Abraham Abulafia, *Imrei Shefer* (Jerusalem, 1999), pp. 67–68.

[42] See Elliot R. Wolfson, *Venturing Beyond: Law and Morality in Kabbalistic Mysticism* (Oxford: Oxford University Press, 2006), pp. 58–73.

[43] Abulafia, *Imrei Shefer*, p. 46.

> All these matters emanate from the Active Intellect, which informs the person about the truth of the substance of his essence by means of the permutation of the letters [*ṣeruf ha-otiyyot*] and the mentioning of the names [*hazkarat ha-shemot*] without doubt, until the person is restored to the level of intellect so that he may be conjoined to him in the life of this world in accord with his capacity and in the life of the world to come in accord with his comprehension.[44]

In *Ḥayyei ha-Olam ha-Ba*, Abulafia writes that all the prophets, sages, and philosophers concur that

> the imaginary form [*ha-ṣiyyur ha-dimyoni*] is very physical [*gufani me'od*], and thus it is not possible to imagine [*leṣayyer*] anything of the intelligible matters [*ha-devarim ha-muskalim*] except by means of participation with comprehension of a corporeal form [*hassagat ṣiyyurit gashmit*]. Thus, even with respect to the prophet, in the moment that he comprehends the prophetic word that emanates from God, blessed be he and blessed be his name, through the intermediary of the Active Intellect in man by means of the permutation of letters contemplated in the heart, it is not possible that he will not imagine [*yeṣayyer*] what he has comprehended in the form of a body speaking a certain matter…on account of the reality of the imaginative faculty that is in him. But the prophet knows in truth that the body imagined by him [*ha-guf ha-mitdammeh lo*] in the moment of prophecy has no physical existence [*meṣi'ut gufanit*] at all, but it is an intelligible entity, entirely spiritual [*davar sikhli kullo ruḥani*], and it materialises in relation to him [*hitgashem eṣlo*] in the moment of his comprehension, since the prophet is a body that contemplates the truth in his spiritual intellect [*heyot ha-navi guf ha-maskil ha-emet be-sikhlo ha-ruḥani*] that is actualised in that moment.[45]

Abulafia obviously follows the philosophical approach of Maimonides, with the crucial difference that the means to attain prophecy is identified by him as the meditative technique of letter permutation. The content of the experience, however, is identical; prophecy consists of configuring the intelligible in a corporeal form as it is mediated through the imagination. The prophet knows that this form has no actual physical existence outside the mind; it

---

[44] Abraham Abulafia, *Sefer ha-Ḥesheq* (Jerusalem, 2002), p. 8.

[45] Abraham Abulafia, *Ḥayyei ha-Olam ha-Ba* (Jerusalem, 1999), pp. 48–49.

is a spiritual entity—in other passages identified more specifically with the angelic Metatron—that materialises as a human form in the imaginative faculty of the prophet,[46] whose body has been transfigured in light of comprehending the truth in the state of intellectual conjunction.[47] Referring to this process in the continuation of the passage from *Ḥayyei ha-Olam ha-Ba*, Abulafia writes:

> It is known that we, the community of Israel, the congregation of the Lord, know in truth that God, blessed be He and blessed be his name, is not a body or a faculty in a body, and he never materialises. But his overflow creates a corporeal intermediary, and it is an angel in the moment of the prophecy of the prophet.[48]

The corporeal image of the incorporeal substance configured in the imagination of the enlightened visionary (*maskil*) is the figure of Metatron, also identified as the 'angel of the Lord' (*mal'akh yhwh*), that is, the angel whose name is YHWH, the imaginal form of the Active Intellect, the incarnational presence that assumes the shape of an anthropos (the idealised Israel) in the imaginative faculty.[49] Alternatively, Abulafia depicts this ecstatic state in another passage in *Ḥayyei ha-Olam ha-Ba* as seeing the letters in the 'mirror of prophecy' (*mar'eh ha-nevu'ah*) as if they were the 'dense bodies' (*gufim avim*) of living angels that speak to the visionary mouth to mouth 'in accord with the abundance of

---

[46] On the vision of the human form and prophecy in Abulafian kabbalah, see Idel, *The Mystical Experience*, pp. 95–100.

[47] See Wolfson, *Language, Eros, Being*, pp. 234–42. An almost identical perspective to Abulafia is found in Ibn al-'Arabī's depiction of the manifestation of Gabriel in human form in his *Fuṣūṣ al-ḥikam*. See Muḥyīddīn Ibn al-'Arabī, *The Bezels of Wisdom*, translation and introduction by Ralph W. J. Austin, preface by Titus Burckhardt (New York: Paulist Press, 1980), p. 121: 'The appearance of the Angel to him as a man was also from the plane of the Imagination, since he [Gabriel] is not a man but an angel who took on himself human form. This [form] was transposed by the beholder with gnosis to its own true form'.

[48] Abulafia, *Ḥayyei ha-Olam ha-Ba*, p. 49.

[49] For discussion of this aspect of Abulafia's understanding of prophecy, see Idel, *Mystical Experience*, pp. 89–90, 100–4. With respect to the matter of the angelic body beheld in the prophetic vision, there is an interesting affinity between Abulafia's kabbalah and Islamic mysticism, especially Shī'ite esotericism. See Wolfson, *Language, Eros, Being*, p. 239.

the intelligible form [*ha-ṣiyyur ha-sikhli*] that is contemplated in the heart that converses with them'.[50] From here we can adduce a transposed materiality that is rooted in the belief that the body, at its most elemental, is constituted by semiotic inscription, the flesh that is the word garbed in the word that is flesh. The letters themselves, therefore, are the prophetic matter (*ḥomer ha-nevu'ah*) that assumes the shape of angelic interlocutors engaged dialogically with the soul.

Perhaps the most explicit formulation of this idea is found in *Sha'arei Ṣedeq*, a treatise composed by Abulafia's disciple, Nathan ben Saadiah Harar.[51] The final stages of the meditational practice of the 'way of the names' (*derekh ha-shemot*) that leads to prophetic experience, which the author equates, following the appropriation of the Maimonidean ideal on the part of his teacher, with the disembodied state of intellectual conjunction, identified as well as the worship of the heart predicated on knowledge of the name, is described as follows:

> By drawing forth words from thought [*maḥshavah*] he will force himself to come out from under the rule of his natural intellect [*sikhlo ha-ṭiv'i*], for if he wants not to think he cannot, and he should be lead initially in the phases of writing [*ketav*] and speaking [*lashon*], and also through the mouth, which is the form [*ṣiyyur*]. When he is to come out from under its rule, another effort is necessary, which consists of drawing thought gradually forth from its source until reaching the level that will compel him not to speak and concerning which he has no ability to overcome. And if he has it in his power to prevail and to continue drawing forth, then he will go out from his inwardness, and it will take shape in his purified imaginative faculty [*koaḥ ha-dimyoni ha-zakh*] in the form of the translucent mirror [*mar'ah zakhah*], and this is 'the flame of the encircling sword' (Genesis 3:24), the back rotating and becoming the front. He discerns his inmost being [*mahut penimiyyuto*] as something outside of himself... For when a form is not perfect it is detached from its essence until it is nullified and garbed in a purified imaginal form [*ṣurah dimyonit zakhah*] through which the letters are combined in a perfect, orderly, and adequate combination. It seems to me that this form is called by kabbalists 'garment'

---

[50] Abulafia, *Ḥayyei ha-Olam ha-Ba*, p. 159.

[51] See Idel, *The Mystical Experience*, pp. 91–92; Idel, *Studies in Ecstatic Kabbalah* (Albany: State University of New York Press, 1988), pp 73–89, 98 n. 18, 149 n. 42, 151 nn. 61–62.

[*malbush*].[52]

After having documented in graphic detail the arduous phases of the meditational practice of letter combination and the peculiar experiences ensuing therefrom, the author discloses the culminating state of contemplation whereby one passes beyond the threshold of thought and speech. The state of mindfulness devoid of concepts, images, and words, leads experientially to the breakdown of the perceptual distinction between inside and outside, as the external form beheld by the adept is the radiance of the internal light, which, in turn, is a reflection of the external form. As we read in a tradition reported in another text preserving the teachings of the same disciple of Abulafia, 'The enlightened sage (*he-ḥakham ha-maskil*), his honor, R. Nathan, blessed be his memory, said: "Know that the perfection of the secret of prophecy for the prophet is when suddenly he sees the image of himself standing before him, and he forgets himself and it disappears from him, and he sees his image before him speaking to him and informing him of future events". Concerning this secret the sages, blessed be their memory said,[53] "Great is the power of the prophets for they compare the form to its creator"'.[54] The peak of prophecy consists of the spontaneous dissolution of the distinction between inside and outside: the form that the prophet sees communicating information about the future is his own image.

To express the matter in a more contemporary phenomenological idiom, the noematic presence in consciousness inhering in some 'objective' substance or substratum, the representation of an outer sense, is given to consciousness as an image of the noetic manifold of an inner sense projected

---

[52] My translation is based on the Hebrew text published in Moshe Idel, *Le Porte della Giustizia, Sa'are Sedeq* (Milano: Adelphi Edizioni, 2001), p. 482. For alternative English renderings, see Scholem, *Major Trends*, p. 155; Idel, *The Mystical Experience*, p. 108.

[53] *Genesis Rabbah* 27:1, edited by Julius Theodor and Chanoch Albeck (Jerusalem: Wahrmann Books, 1965), pp. 255–56.

[54] I translate from the text transcribed by Gershom Scholem, 'Eine kabbalistische Erklärung der Prophetie als Selbstbegegnung', *Monatsschrift für Geschichte und Wissenschaft des Judentums* 74 (1930): p. 287. For an alternative English translation, see Scholem, *Major Trends*, p. 142. A version of the text was published in *Shoshan Sodot* (Korets, 1784), 69b, noted and cited by Idel, *The Mystical Experience*, pp. 91–92.

outward as an object that stands over and against the mind, but that projection is itself predicated on the presumption that the noematic presence is a 'subjective' projection of the noetic manifold. At the moment of vision, through the force of the imagination that has been purified by ascetic practice and the purging of all discriminate forms, the heart of the mystic becomes a translucent mirror, the screen/veil through which the internal is externalised and the external internalised, the seeing of one's inward form projected outward as the outward image propelled inward, the vision in which the difference of identity between seer and seen is overcome in the identity of their difference—the mind sees itself as the mirror reflecting the mind that mirrors, embracing thereby the paradox of visible invisibility of which we spoke above—and hence the imagination is referred to by the scriptural image of the 'flame of the encircling sword', that is, the sword that revolves in such a way that the back becomes the front in confronting the front becoming the back. In the polished mirror of imagination, the form detached from its essence, that is, the form that has no form, not even the form of formlessness, clothes itself in an 'imaginal form'. We are told, moreover, that through this form 'the letters are combined in a perfect, orderly, and adequate combination', which is identified with the kabbalistic notion of the garment, *malbush*.[55] I would suggest that the author is alluding here to the permutation of the Hebrew letters that constitute the Torah. If this conjecture is correct, then the point of the passage is that the Torah is the imaginal form through which the formless is envisioned. It is likely, moreover, that implied here is the identification of Torah and the Active Intellect or the angelic Metatron.[56]

Astrologically, Abulafia ascribes a dual comportment to Metatron, which is described in various ways, including his identification of Metatron as the angel of the moon with its dark and light side or as the Teli, the serpentine constellation,[57] which is linked to (or said to hang from) the Tree of

[55] Idel, *Le Porte della Giustizia*, pp. 245–50.
[56] Idel, *Language, Torah, and Hermeneutics*, pp. 34–41, 79–80, and 163 n. 33; Idel, *Absorbing Perfections: Kabbalah and Interpretation* (New Haven: Yale University Press, 2002), pp. 348–50.
[57] On the astrological symbol of the Teli in Abulafia, see Idel, *Studies in Ecstatic Kabbalah*, pp. 77–78; Wolfson, *Abraham Abulafia*, p. 145, n. 135. The connection of this image and Jesus has been explored most extensively by Robert Sagerman, 'Ambivalence toward Christianity

Knowledge of Good and Evil,[58] insofar as it comprises the attribute of compassion on the right and the attribute of judgment on the left, a theosophic truth also conveyed by the image of the head and tail of the astrological dragon, depicted as well by the two archangels Michael and Gabriel or as the male and female cherubim, upon which the *Shekhinah* is enthroned. In the theophanic incarnation, the coalescence of opposites is effectuated and the distinction between matter and form is blurred. In other contexts, the satanic and angelic are identified by Abulafia as two faces of Metatron, sometimes portrayed as the attributes of mercy and judgment, and thus we may assume that in grouping together human, angel, and Satan,[59] what he intends is that an individual can emulate either dimension of Metatron. This symbolism may relate as well to Abulafia's portrayal of Christianity as demonic, which he associates with its idolatrous nature, that is, the worship of the image (the term that Abulafia often uses is *demut*) of the divine body, a characterization that is based, in turn, on the assumption that the tenets of this Abrahamic faith originate in the imagination rather than reason.[60] In an ironic twist, the religion that dogmatically professes the incarnation of God in human form is placed on the level of Satan as opposed to Adam, whereas the religion upon whom the prophetic tradition has been bestowed expresses its adamic nature by actualising the capacity to conjure the angelic-astral body, the anthropomorphic configuration of the incorporeal, in the imagination. It is feasible, then, to surmise that the three terms, *adam*, *mal'akh*, and *saṭan*, signify the struggle on the

---

in the Kabbalah of Abraham Abulafia' (Ph.D dissertation, New York University, 2008), pp. 183, 259–68, 276–79, 281, 285–87, 294–96, 299, 302, 305, 307–13, 316, 318, 323, 327–31, 335–37, 340, 348–70, 421–22, and 461–64.

[58] Idel, *Studies in Ecstatic Kabbalah*, p. 52; Harvey J. Hames, *Like Angels on Jacob's Ladder: Abraham Abulafia, the Franciscans, and Joachimism* (Albany: State University of New York Press, 2007), pp. 80–81; Sagerman, 'Ambivalence', p. 107. See Abraham Abulafia, *Sitrei Torah*, p. 144; Abraham Abulafia, *Mafteaḥ ha-Sefirot* (Jerusalem, 2001), p. 85, where the Teli is connected to the 'copper serpent' (Numbers 21:9), whose power is magic; Abraham Abulafia, *Mafteaḥ ha-Tokhaḥot* (Jerusalem, 2001), p. 8.

[59] See Idel, *Studies in Ecstatic Kabbalah*, pp. 38–39.

[60] Idel, *Studies in Ecstatic Kabbalah*, pp. 45–61; Idel, *Messianic Mystics* (New Haven, CT: Yale University Press, 1998), pp. 62 and 97–99; Wolfson, *Venturing Beyond*, pp. 62–63.

psychological plane between the evil and good inclinations, which corresponds to the battle on the theological plane between Christianity and Judaism, Jesus of Nazareth and the Messiah of Israel, the seals of the sixth and seventh days of the week, the material Tree of Knowledge and the spiritual Tree of Life.[61] The threefold distinction can also be cast in temporal terms that were a commonplace in the cosmological order that Abulafia derived from Maimonides: the satanic corresponds to corruptible matter and it is thus subject to time, the angelic corresponds to the incorruptible intellect and it is thus not subject to time, and the human being is a composite of matter and intellect and it is thus both subject to and not subject to time, or, to put it in a different terminological register, the human being has the capacity to eternalise the temporal by temporalising the eternal.

To sum up: Mystical gnosis of the name, which is achieved as a result of the technique of letter-combination, entails a state of intellectual conjunction that Abulafia also designates by the rabbinic notion of eschatological felicity, the 'life of the world to come' (*ḥayyei ha-olam ha-ba*).[62] Although the latter retains something of its original connotation in Abulafia's scheme, he was far more interested in utilising the phrase to denote an interior state of spiritual transformation occasioned by the triumph of intellect over imagination, spirit over body, an orientation that is attested as well in other medieval Jewish philosophical exegetes, poets, and kabbalists. Abulafia does not go so far as to negate entirely the nationalistic aspects of the messianic ideal, but it is clear from his writings that his messianism is primarily psychic in nature.[63] Tactilely, the ecstatic experiences the illumination as being anointed with oil, and thus the one who is illumined is not only capable of being redeemed proleptically prior to the historical advent of the messiah, but such an individual noetically attains the rank of the messianic figure.[64] The anointment also denotes the priestly status of the illuminate;[65] indeed, in the unitive state, the

---

[61] Idel, *Studies in Ecstatic Kabbalah*, pp. 51–52.

[62] Scholem, *Major Trends*, pp. 131–35; Wolfson, *Abraham Abulafia*, pp. 54–55.

[63] For discussion of the spiritual and noetic dimension of the Messiah in Abulafia, see Idel, *Messianic Mystics*, pp. 65–79.

[64] Scholem, *Major Trends*, p. 142.

[65] Idel, *Messianic Mystics*, pp. 94–97.

ecstatic assumes the role of high priest,[66] the position accorded Metatron in the celestial Temple, the angelic viceregent summoned by Abulafia as the object of conjunction.[67] We may conclude, therefore, that the phenomenon of anointment comprises three distinct, though inseparable, aspects of the pneumatic metamorphosis—messianic, priestly, and angelic. For Abulafia, moreover, the matter of reception is critical to his understanding of the prophetic-messianic experience, as the enlightened mind, the soul unfettered from the chains of corporeality, receives the surfeit of the holy spirit, which is identified as the Active Intellect, the angelic Metatron, and as the wheel of letters that is the idealised Torah scroll. The experience of *unio mystica*, therefore, may be viewed phenomenologically in four ways: to cleave to the name, to be conjoined with the intellect, to be transformed into the demiurgical angel, and to be incorporated within the textual embodiment of the word of God.[68]

---

[66] Abulafia, *Ḥayyei ha-Olam ha-Ba*, p. 67.

[67] Idel, *The Mystical Experience*, pp. 116–19; Idel, *Messianic Mystics*, pp. 65–77, 303–6; Wolfson, *Language, Eros, Being*, p. 241.

[68] On the use of the expression 'word of God' by Abulafia to designate the Active Intellect, see Idel, *Language, Torah, and Hermeneutics*, p. 33. See also Wolfson, *Abraham Abulafia*, p. 141, and on the identification of the visionary and the Torah in the supreme state of ecstasy, see Scholem, *Major Trends*, p. 141.

'CHEMISTRY, THAT STARRY SCIENCE': EARLY MODERN CONJUNCTIONS OF ASTROLOGY AND ALCHEMY

Peter J. Forshaw

ABSTRACT: The Hermetic notion of a relationship between the heavens above and the earth below received its most authoritative statement in that ur-text of the alchemists, the *Tabula Smaragdina* or *Emerald Tablet*, a work that became accessible to Western readers with its translation into Latin during the twelfth-century Renaissance. This paper will discuss the relationship between astrology and alchemy in early modern Europe, with a particular focus on the writings of the celebrated medical reformer and occult philosopher Theophrastus Paracelsus of Hohenheim (1493–1541), significant followers such as Oswald Croll and Gerard Dorn, and the critical response of opponents like the alchemist Andreas Libavius to the Paracelsian conjunction of *astrologia* and *spagiria*.

> If you do not understand the use of the Cabalists and the old astronomers, you are not born by God for the Spagyric art, or chosen by Nature for Vulcan's work, or created to open your mouth about the Alchemical Art.[1]

So declares the iconoclastic Swiss medical theorist and hermetic philosopher Theophrastus Paracelsus of Hohenheim (1493–1541) early in the sixteenth century, emphasising the importance of both the Jewish mystical tradition of Kabbalah and astrology for his new Spagyric art of preparing chemical medicines, and for the alchemical preparation of the Philosophers' Stone. The relationship between astrology and alchemy is a somewhat contentious issue in the history of science, with historians of alchemy, in particular, either ignoring or tending to downplay the significance of astrology. Truth be told, scholars like William Newman, Anthony Grafton, and Joachim Telle are perfectly correct that, in the great majority of medieval and early

[1] Paracelsus, *De Tinctura Physicorum*, in *Aureoli Philippi Theophrasti Bombasts von Hohenheim Paracelsi Opera*, ed. Johann Huser, 2 vols. (Strasburg, 1603), Vol. 1, pp. 922–25, at p. 923.

modern alchemical treatises, there is little evidence of the authors having any specialist knowledge of astrology; indeed, it is summarily rejected by Geber's *Summa perfectionis*, the influential medieval work on practical alchemy, and by Robert Boyle, one of the leading figures in the Scientific Revolution.[2] Further reading, however, does suggest that astrology played its part in the theory and practice of at least a few significant individuals, including Paracelsus and some of his followers; the curious quote in the title comes from one such supporter of Paracelsian philosophy, the Englishman Thomas Moffet (1553–1604).[3] This essay considers the various ways knowledge of astrology contributed to their alchemical endeavours.

AS ABOVE, SO BELOW

The celestial science of astrology had, of course, been part of the higher division of the Seven Liberal Arts taught at universities since the Middle Ages, the Quadrivium, that included the mathematical sciences of arithmetic, geometry, music, and astronomy. In *De artibus liberalibus*, Robert Grosseteste (1175–1253), an important figure in the history of medieval natural philosophy, readily discussed the

[2] William R. Newman and Anthony Grafton, 'Introduction: the Problematic Status of Astrology and Alchemy in Premodern Europe,' in *Secrets of Nature: Astrology and Alchemy in Early Modern Europe*, ed. William R. Newman and Anthony Grafton (Cambridge, MA: MIT Press, 2001), pp. 1–37, at pp. 21–22; Joachim Telle, 'Astrologie et alchimie au XVIe siècle: à propos des poèmes didactiques astro-alchimiques de Christoph von Hirschenberg et de Basile Valentin', *Chrysopoeia*, Tome III, Fasc. 2 (April/June 1989): pp. 163–92; idem., 'Astrologie und Alchemie im 16. Jahrhundert. Zu den astroalchemischen Lehrdichtungen von Christoph von Hirschenberg und Basilius Valentinus', in *Die okkulten Wissenschaften in der Renaissance*, ed. August Buck (Wiesbaden: Harrassowitz, 1992), pp. 227–53. On Geber, see William R. Newman, *The Summa Perfectionis of Pseudo-Geber, A Critical Edition, Translation & Study* (Leiden: Brill, 1991). For Boyle, see Lawrence M. Principe, *The Aspiring Adept: Robert Boyle and his Alchemical Quest* (Princeton, NJ: Princeton University Press, 1998). For Boyle's rejection of horary astrology, see Robert Boyle, *Certain Physiological Essays* (London, 1669), pp. 56, 94, 102.

[3] Peter French, *John Dee: The World of an Elizabethan Magus* (1972; repr., London: Routledge, 2002), p. 127.

association of astrology with alchemy and medicine, despite his later misgivings about judicial astrology:

> Natural philosophy needs the assistance of *astronomia* more than that of the rest; for there are no, or few, works of ours or of nature, as for example the propagation of plants, the transmutation of minerals, the curing of sickness, which can be removed from the way of *astronomia*. For nature below effects nothing unless celestial power moves it and directs it from potency into act.[4]

This encouraged Jim Tester to argue that there were, apparently, 'three legitimate, even necessary, kinds of astrology: meteorological, alchemical and medical'.[5]

As part of their education, many university-trained physicians already had a knowledge of *iatromathematics* or medical astrology, and would have been familiar with the notion of preparing and applying *materia medica* dependent on the constellations. The classical authorities Hippocrates (c. 460–370 BC) and Galen (129–199/217 CE) both advised the physician to be an astrologer, to be aware of the most opportune astrological times to gather plants for remedies, when to prepare them, and when to administer them to patients. The same applied to phlebotomy, with a great deal of advice existing on when and when not to let blood. Doctors had tables of beneficial and critical days for disease and healing. One of the best known visual symbols of this relationship between the human being as the microcosm and the heavens was the *Melothesia* or Zodiac Man, so popular in medieval medical manuscripts and early modern texts, which assigned body parts to specific planets or signs.[6]

---

[4] Cited in Jim Tester, *A History of Western Astrology* (1987; repr., Woodbridge: The Boydell Press, 1999), p. 179.

[5] Ibid. For further consideration of Grosseteste's attitudes towards astrology and alchemy, in *De artibus* and *De generatione stellarum*, see James McEvoy, 'The Chronology of Robert Grosseteste's Writings on Nature and Natural Philosophy', *Speculum* 58, no. 3 (1983): pp. 614–55, at pp. 622–23. For a Latin edition of *De artibus liberalibus*, see Robert Grosseteste, *Die Philosophischen Werke des Robert Grosseteste, Bischofs von Lincoln*, ed. Ludwig Baur (Münster: Aschendorffsche Verlagsbuchhandlung, 1912), pp. 1–7, at p. 5.

[6] On the Melothesia, see Charles Clark, 'The Zodiac Man in Medieval Medical Astrology', *Journal of the Rocky Mountain Medieval and Renaissance Association* 3 (1982): pp. 13–38; Tamsyn Barton, *Ancient Astrology* (London: Routledge, 1994), pp. 189–90. On

The most representative alchemical statement of this relationship between the heavens above and the earth below is found in the *Tabula Smaragdina* or *Emerald Tablet* of the legendary Hermes Trismegistus, the alleged author of hundreds of treatises on astrology, magic and alchemy.[7] There we find the classic Hermetic formulation, 'What is below is like what is above; and what is above is like what is below'. This short enigmatic text was one of the first alchemical works to become available to Western readers in the twelfth century, when they were first beginning to enter Europe in Latin translation from Arabic sources.[8] The earliest translation was made by Hugo of Santalla, who was also the translator of Umar ibn al-Farrukhan al-Tabari's *Kitab al-Masa'il* ('Book of Questions'), an astrological work with an early reference to *al-kimiya* and dating from the eighth century, the earliest period of scientific prose writing in Arabic. Umar's book concerns astrological 'Interrogations', i.e. the calculation of a chart for the moment at which a particular question is asked (concerning, for example, medical, travel, or business-related issues). In the 79th chapter, 'On the knowledge of alchemy', he provides advice on how to judge whether or not a man knows the science of transmutation. Hugo's Latin translation was included in the *Liber trium iudicum*, a collection of three works on judicial astrology. This collection also contained another translation

---

medicine and astrology, see Anthony Grafton, and Nancy Siraisi, 'Between the Election and My Hopes: Girolamo Cardano and Medical Astrology', in Newman and Grafton, *Secrets of Nature*, pp. 69–131; Monica Azzolini, 'Reading Health in the Stars: Politics and Medical Astrology in Renaissance Milan', in *Horoscopes and Public Spheres: Essays on the History of Astrology*, ed. Günther Oestmann, H. Darrel Rutkin, and Kocku von Stuckrad (Berlin: Walter de Gruyter, 2005), pp. 183–206; Roger French, 'Astrology in Medical Practice', in *Practical Medicine from Salerno to the Black Death*, ed. Luis Garcia Ballester, Roger French, Jon Arrizabalaga, and Andrew Cunningham (Cambridge: Cambridge University Press, 1994), pp. 30–59.

7 See Hermes Trismegiste, *La Table d'Émeraude et sa Tradition alchimique*, ed. Didier Kahn (Paris: Les Belles Lettres, 1995); Antoine Faivre, *The Eternal Hermes: From Greek God to Alchemical Magus* (Grand Rapids: Phanes Press, 1995).

8 Peter Forshaw, 'Alchemical Exegesis: Fractious Distillations of the Essence of Hermes', in *Chymists and Chymistry: Studies in the History of Alchemy and Early Modern Chemistry*, ed. Lawrence M. Principe (Sagamore Beach: Science History Publications, 2007), pp. 25–38.

of an Arabic astrological work on interrogations that mentions alchemy, the *Liber Zahel de iudiciis* of the ninth-century Sahl ibn Bishr. Another of Sahl's works, translated into Latin as *De Electionibus,* on choosing the most astrologically propitious time to undertake any activity, also includes a section on alchemy.[9]

A further early Latin reference to the 'science of alchemy' can be found in the *Liber de naturis inferiorum et superiorum* of the scholastic philosopher Daniel of Morley (c. 1140–1210), who lists the 'scientia de alckimia' as one of the eight parts of astrology, explaining it to be knowledge of the 'transformation of metals into other kinds'.[10] Appearing not so long afterwards, the *De perfecto magisterio* provides the first known description of alchemy as 'Inferior Astronomy', with the statement that it is comparable to the *first,* 'Superior' Astronomy, i.e., astrology.[11] This was to become a fairly common description of the affiliation between the two arts throughout the middle ages and well into the early modern

---

[9] Charles Burnett, 'The Astrologer's Assay of the Alchemist: Early References to Alchemy in Arabic and Latin Texts', *Ambix* 39, Part 3 (1992): pp. 103–9. Burnett mentions two later Latin astrological works that draw material from these translations and mention alchemy: Leopold of Austria's *Compilacio de scientia astrorum* (late thirteenth century) and Roger of Hereford's *Liber de quatuor partibus astronomie* (late twelfth century). For further (brief) references to astrology in the context of Arabic alchemy, see Sami K. Hamarneh, 'Arabic-Islamic Alchemy—Three Intertwined Stages', *Ambix* 29, Part 2 (1982): pp. 74–87.

[10] On Daniel of Morley, see Gregor Maurach, 'Daniel von Morley, "Philosophia"', *Mittellateinisches Jahrbuch* 14 (1979): pp. 204–55, at p. 239: 'De dignitate eius [astronomiae] invenitur, quod illius partes, secundum quod dixerunt sapientes primi, octo sunt...scientia de alckimia, que est scientia de transformatione metallorum in alias species...' See also Barbara Obrist, *Les Débuts de l'Imagerie Alchimique (XIV*[e]*-XV*[e] *siècles)* (Paris: Éditions le Sycomore, 1982), pp. 42, 156; Robert P. Multhauf, *The Origins of Chemistry* (London: Oldbourne, 1966), p. 167 (although Multhauf mistakenly mentions *twelve* parts of astrology in Morley). Morley's list is almost identical to that found in the *De divisione Philosophie* of Dominicus Gundissalinus (fl. 1150), who includes the 'sciencia de alquimia' as one of the eight parts of natural science, glossing it as 'sciencia de conversione rerum in alias species'. See Dominicus Gundissalinus, *De Divisione Philosophiae,* ed. Ludwig Baur (Münster: Druck und Verlag der Aschendorffschen Buchhandlung, 1903), p. 20.

[11] Maurice P. Crosland, *Historical Studies in the Language of Chemistry* (London: Heinemann, 1962), p. 6.

period.[12]

The most obvious manifestation of this relationship was the series of traditional correspondences between the seven planets of the Ptolemaic system and the seven main metals of alchemy.[13] The association of metals and planets is mentioned by the pagan philosopher Celsus as early as the second century CE.[14] A passage attributed to the seventh-century Byzantine philosopher Stephanus of Alexandria provides what were to become the standard correspondences: Saturn-Lead; Jupiter-Tin; Mars-Iron; Sun-Gold; Venus-Copper; Mercury-Quicksilver; Moon-Silver.[15]

Moving forward in time to the early modern period, we discover that Oswald Croll (1560–1609), author of the *Basilica Chymica* (1608), the most popular seventeenth-century exposition of the alchemical philosophy of Paracelsus, continues to promote the relationship between 'Astrologia' and 'Chymiologia', explaining that

> ...we find out of ancient Records that Astrologers and Chymiologers were very near of kin; for the Caelestiall Astronomy is as it were the Parent and Mistresse of the inferiour, for as much as both have their own Heaven, their own Sun, their own Moon, their Planets, and their own proper Stars; yet so as that the astrology of superiour things hath to doe with the Chymiology of things inferiour.[16]

---

[12] Nicholas H. Clulee, '*Astonomia inferior*: Legacies of Johannes Trithemius and John Dee', in Newman and Grafton, *Secrets of Nature*, pp. 173–233, at p. 173.

[13] Tester, *A History of Western Astrology*, pp. 24–25.

[14] Celsus, quoted by Origen, in *Contra Celsum*, trans. Henry Chadwick (Cambridge: Cambridge University Press, 1953), p. 334, VI:22. See esp. n. 1 which provides other lists of correspondences, including Proclus, *In Timaeum* 1:43.5 and Olympiodorus, *In Aristotelis Meteorum* III.6.

[15] Marcellin Berthelot and C. E. Ruelle, *Collection des anciens alchimistes Grecs*, Vol. 1 (Paris, 1887), p. 84. For alternative early modern correspondences between planets and minerals, see David de Planis Campy, *Bouquet Composé des plus belles fleurs Chimiques* (Paris, 1629), pp. 65–66; Annibal Barlet, *La Theotechnie Ergocosmique* (Paris, 1653), p. 102.

[16] Oswaldus Crollius, *Philosophy Reformed & Improved in Four Profound Tractates* (London, 1657), p. 27. On Croll, see Owen Hannaway, *The Chemists & the Word: The Didactic Origins of Chemistry* (Baltimore: The Johns Hopkins University Press, 1975).

A similar belief in celestial influence on the generation of earthly metals can be found in the *Gloria Mundi* (1620):

> For nature makes the metals true bodies in the earth, some perfect, like gold and silver; others imperfect, like Venus, Mars, Saturn, and Jupiter, according to how planetary motion influences and works.[17]

Basil Valentine's widely read *Triumphwagen Antimonii* (1604) provides slightly more detail about the process:

> I find that all that is dug up from the bowels of the mountains is infused by the stars and celestial bodies, and derives its origin from a certain aqueous vapour, which, after being nourished by the stars for a long time, is reduced to a tangible shape by the elements.[18]

While none of this, so far, is evidence that alchemists actively made use of astrology in their laboratory practice, it is at least an initial indication of the presence of astrological thought in their world-view. In the *Novum Lumen Chemicum* (1604) of the Polish alchemist Michael Sendivogius (1566–1636), we discover that the Ptolemaic sequence of the planets provides insights into the possible order of transmutation of metals. We are informed that 'the vertues of the Planets doe not ascend but descend'. Mars is not made from Venus, but Venus from Mars, i.e., Copper is made from Iron, but not the other way round, the rationale behind this being that Venus is closer to the Earth in the Ptolemaic cosmos. Jupiter, we gather, is more readily transmuted into Mercury, i.e., Tin into Quicksilver, because the former is the second planet from the Firmament, the latter the second from the Earth. There is a particularly great correspondence we learn, between Saturn,

---

17 [Roberto Vallensis], *Gloria Mundi, sonsten Paradeiss-Taffel*, in *Deutsches Theatrum Chemicum*, ed. Friederich Roth-Scholtz, 3 vols. (1728–1733), Vol. 3, pp. 372–536, at p. 515.

18 Wayne Shumaker, *The Occult Sciences of the Renaissance: A Study in Intellectual Patterns* (1972; Berkeley: University of California Press, 1979), p. 177. On 'l'Astronomie inferieure des sept Metaux, & sur l'Harmonie de leurs systemes; ensemble des douze signes du Zodiac, & autres constellations du Ciel des Philosophes Hermetiques', see the end of 'L'Imprimeur au Lectuer', in Basile Valentin, *Revelation des Mysteres des Teintures Essentielles des Sept Metaux, & de leurs Vertus Medicinales* (Paris, 1678).

the first planet from 'heaven', and the Moon, the first from the Earth.[19]

In medieval and early modern alchemical works we find this association between the planets and metals influenced by theories drawn from the disciplines of medicine and agriculture, which respectively discuss the influence of the planets on stages of the gestation of the human foetus in the womb and the germination of seeds in the earth.[20] In two of the earliest alchemical works to appear in print, the *Liber Trium Verborum* of King Kalid and *De Alchimia* of Geber, we read:

> In Mercury are the works of the planets and their imaginations work in their parts and times, as in a foetus. Saturn works with its dryness in the first month of conception, causing congelation and constriction, Jupiter in the second month works with its benign heat and creates digestion; in the third month Mars takes over with its stronger heat and dryness to bring about division and disposition of the members and their parts, and so forth.[21]

This belief was still current in the seventeenth century when Johann Daniel Mylius (1583–1642) wrote of the development of the 'Philosophical Embryo' in his *Anatomia auri* (1628),[22] and the following year David de Planis Campy (1589–1644) included the very same information in his *Bouquet composé des plus belles fleurs Chimiques* (1629), explicitly stating that whoever bore this in mind for the generation of the

---

[19] Micheel Sandivogius [*sic*], *A New Light of Alchymie* (London, 1650), p. 26–27. Sendivogius explains that 'Chymists know how to change Iron into Copper without Gold, they know how also to make Quick-silver out of Tin: and there are some that make Silver out of Lead.' The same text can be found in the anonymous *Tractatus alter De Lapide Philosophico*, in Martin Ruland, *Lapidis Philosophici Vera Conficiendi Ratio* (Frankfurt, 1606), pp. 121–22.

[20] See the extended agricultural metaphor in Jodocus Greverus, *Secretum Nobilissimum et Verissimum*, in *Theatrum Chemicum, praecipuos selectorum auctorum tractatus de chemiae et lapidis philosophici antiquitate, veritate, iure, praesentia et operationibus*, 6 volumes (Ursel / Strasbourg, 1602–1661), Vol. 3 (1613), pp. 735–59.

[21] *Liber Trium Verborum Kalid Regis*, in *Theatrum chemicum*, Vol. 5 (1622), pp. 186–90, at p. 188.

[22] Johann Daniel Mylius, *Anatomia auri, sive Tyrocinium medico-chymicum* (Frankfurt, 1628), pp. 205–6.

philosophical work would reach the port when many others had suffered shipwreck.[23]

The notion that astrology had something 'to doe with the Chymiology' of terrestrial things was not restricted to the concerns of *Chyrosopoeia* or Gold-making. The late Middle Ages saw a great deal of interest in alchemical distillation. The fourteenth-century French monk Jean de Roquetaillade or John of Rupescissa (d. 1366) furthered the idea of an intimate relationship between upper and lower astronomy by naming the product of alchemical distillation, the Quintessence, the *Coelum* or 'Heaven'.[24] The relationship was further cemented by the famous fifteenth-century Italian astrologer, Marsilio Ficino (1433–1499), who associates the alchemical elixir with the cosmic Spirit of the World.[25] The 1512 edition of Hieronymus Brunschwig (c. 1450–1512), the first German edition of whose *Liber de arte distillandi* appeared with a German translation of Ficino's *Buch des Leben* (1505), includes images of various distillation processes taking place under the influence of specific signs of the Zodiac.[26]

---

[23] De Planis Campy, *Bouquet Composé*, pp. 20–21.

[24] See Michela Pereira, 'Heavens on Earth. From the *Tabula Smaragdina* to the Alchemical Fifth Essence', *Early Science and Medicine* 5, no.2, *Alchemy and Hermeticism* (2000): pp. 131–44, at p. 140. See also Leah DeVun, *Prophecy, Alchemy, and the End of Time: John of Rupescisssa in the Late Middle Ages* (New York: Columbia University Press, 2009), p. 130. See too her remark on p. 72 about Rupescissa's astrology being 'chiefly general and theoretical, rather than specific and operative'. In this, he fits into the mainstream of late-medieval alchemical authors on astrology.

[25] Marsilio Ficino, *Three Books on Life: A Critical Edition and Translation with Introduction and Notes*, by Carol V. Kaske and John R. Clark (Binghamton, NY: Medieval and Renaissance Texts and Studies and the Renaissance Society of America, 1989), III.3, p. 257; Sylvain Matton, 'Marsile Ficin et l'alchimie. Sa position, son influence', in *Alchimie et philosophie à la Renaissance. Actes du colloque international de Tours (4-7 Déc. 1991)*, ed. Jean-Claude Margolin and Sylvain Matton (Paris: Vrin, 1993), pp. 123–92.

[26] Hieronymus Brunschwig, *Liber de arte Distillandi de Compositis* (Strasburg, 1512), ff. 144r, 145r, 146v, 149r, 154v. For the edition published with Ficino's *Buch des Leben,* see Hieronymus Brunschwig, *Medicinarius. Das buch der Gesuntheit. Liber de arte distillandi Simplicia et Composita* (Strasburg, 1505).

Fig. 7.1: Hieronymus Brunschwig, Liber de arte distillandi *(1512), sig. 154v (detail)*

'AGAINST THE DISEASES'

The sixteenth century was to see a major change of direction in alchemical endeavour, at least among followers of the iconoclast Paracelsus, who declared in his *Buch Paragranum* (1531), that

> It is not that one should be saying, 'Alchemy makes gold or it makes silver.' Here is the real purpose: to make *arcana* and to direct them against the diseases.[27]

Paracelsus is the first known writer of his period to have called himself an Iatrochemist, one combining both medicine and chemistry.[28] For this he is often considered the father of *Chymiatria*, or Chemical Medicine, introducing the homeopathic use of poisonous minerals and metals,

[27] Paracelsus (Theophrastus Bombastus von Hohenheim, 1493–1541), *Essential Theoretical Writings*, ed. and trans. Andrew Weeks (Leiden: Brill, 2008), p. 221.

[28] James R. Partington, *A History of Chemistry*, 4 vols. (London: Macmillan & Co., 1961–1970), Vol. 2, p. 135.

prepared by his Spagyric art, as part of his astro-alchemical pharmacopoeia.[29]

In addition to the well-known 'As above, so below' maxim of the *Emerald Tablet*, for the Paracelsians there was also the related action of the internal and external stars. Paracelsus took the medieval medical idea of correspondences between the astral firmament and parts of the human organism, and argued for the existence of a corresponding firmament within man.[30] This was an 'inner astronomy' with correspondences between the *astra* and the seats of disease, as well as astral concordances in the preparations of remedies.[31] Disease arose, however, not from an imbalance of bodily humours, as had been the belief since antiquity, but from contagion by external poisons having their ultimate origins in the stars. A Paracelsian physician had to be aware not only of how man's *physical* body derived its nourishment from the fruits of the earth, but how his *sidereal* body did from the stars.[32] The combined knowledge of philosophy and astrology, 'that is the nature of diseases and their remedies, with all their combinations and conjunctions', was essential for the practice of a Paracelsian doctor. The true physician had to know about the stars in order to understand the causes of sickness, and about alchemy in order to prepare the *arcana*.[33]

In *De Sigillis Planetarum*, the seventh book of *Archidoxis Magica* (c. 1570), Paracelsus declares the importance of combining astrological knowledge of the heavens with natural magical knowledge of the earth:

None can deny that the superior stars and influences of

[29] Telle, 'Astrologie et alchimie au XVIe siècle', p. 164.

[30] See, for example, *Paragranum*, in Paracelsus, *Essential Theoretical Writings*, p. 217. See also J. R. R. Christie, 'The Paracelsian Body', in *Paracelsus: The Man and his Reputation, his Ideas and their Transformation*, ed. Ole Peter Grell (Leiden: Brill, 1998), pp. 269–91.

[31] Walter Pagel, *The Smiling Spleen: Paracelsianism in Storm and Stress*, (Basel: S. Karger, 1984), pp. 15, 149–50.

[32] See Dane T. Daniel, 'Invisible Wombs: Rethinking Paracelsus's Concept of Body and Matter,' *Ambix* 53, no. 2 (July 2006): pp. 129–42.

[33] Paracelsus, *Essential Theoretical Writings*, pp. 215–17: 'The physician must decidedly base his knowledge on the stars. He must define medicine in accordance with the stars, recognizing that the *astra* are both above and below. And since medicine can do nothing without the heavens, it must be guided by the heavens'.

> heaven have very great weight in transient and mortal affairs...And it is even possible for man himself to bring these into a certain Medium, so that they may effectively operate, whether this Medium be a Metal, a Stone, an Image, or something of the sort.[34]

The significance of alchemy is implicit in the statement emphasising the importance of metals:

> But this is most important of all: to know that the seven planets have greater force in nothing than they possess in their proper metals.[35]

In the fourth book, *On the Transmutation of Metals and their Times*, the reader is advised to consult a table indicating the most appropriate times to transmute one metal into another. The example provided is somewhat perverse, for we learn that the best time to turn gold into silver is in the hour of the Moon when the Moon is in the 6th degree of Cancer.[36]

It is not only the substances that should be considered in relation to the stars, but also the very laboratory vessels that contain them. In a mid seventeenth-century work on *Chymical Transmutation* (1657), claiming to be an English translation of Paracelsus, we find advice on constructing a Spagyrical vessel, the 'Matrix' or 'Philosophers' Egg':

> Since therefore our Matter is our Radix and Foundation both of the white and red [Stone], our Vessel necessarily ought to be made after this manner, that the matter therein may be ruled by the Celestial bodies; for the Celestial influences, and the invisible impressions of the Stars, are chiefly necessary for this work; otherwise it is impossible to attain to the excellent Oriental, Persian, Chaldean, and Egyptian Stone, by any means;...[Anaxagoras] did very much make known our Vessel to the Cabalists, and that according to the true Geometrical measure and proportion; and how it ought to be built of a certain Quadrature in a Circle, whereby the Spirits and soul of our matter being separated from their body, may

---

[34] *Liber Septimus Archidoxis Magicae, De Sigillis Planetarum,* in Paracelsus, *Operum Medico-Chimicorum sive Paradoxorum, Tomus Genuinus Undecimus* (Frankfurt, 1605), pp. 154–55.

[35] Ibid., p. 155. For more, see Nicholas Goodrick-Clarke, *Paracelsus: Essential Readings* (Berkeley: North Atlantic Books, 1999), p. 192ff.

[36] *Liber Quartus Archidoxis Magicae, De Transmutatione Metallorum & Tempore eorum,* in Paracelsus, *Opera Medico-Chimica,* pp. 137–38.

> be elevated in the altitude of their Heaven.[37]

The same material can also be found two years later in another translation, *Paracelsus, His Aurora, & Treasure of the Philosophers* (1659),[38] edited by the Belgian Hermetic Philosopher Gerard Dorn (1530–1584), a major proponent of 'spagiric Astronomy', who passes on similar advice in one of his own works on metallic transmutation:

> Our vessel must be such, that in it matter can be influenced by the heavenly bodies. For the invisible celestial influences and the impressions of the stars are necessary to the work.[39]

Although he disapproves of such experimental combinations of astrology and alchemy, the Saxon alchemist Andreas Libavius (1560–1616) reports that the Paracelsian Heinrich Khunrath (1560–1605) made use of glass vessels for the specific purpose of employing celestial light and heat, in imitation of God's 'Fiat Lux' at the moment of creation.[40] This may well have been the case, for in *De Igne Magorum Philosophorumque* (1608), Khunrath writes of harnessing the powers of the heavens for the physical-magical animation or vivification of the inferior elemental fire of a lamp or coals by means of the fire or heat of the Sun,[41] by directing its rays with a Beryl crystal, a steel mirror, or round flask full of cold water, called an 'incensoria'.[42] Given Khunrath's interest in natural magic, we are left with the distinct sense that these rays may have been conceived as having more than just a

---

[37] Paracelsus, *Of The Chymical Transmutation of Metals & Minerals*, trans. R. Turner (London, 1657), p. 20, Chapter V, 'Of the second Spagyrick Instrument, which is the Matrix or Philosophers Egge'.

[38] Paracelsus, *His Aurora, & Treasure of the Philosophers* (London, 1659), Chapter XVIII, 'Of the Instruments and Philosophical Vessel', pp. 56–57.

[39] Gerard Dorn, *Congeries Paracelsicae chemiae de transmutationibus metallorum*, in *Theatrum Chemicum*, Vol. 1 (2nd ed., 1659), pp. 491–568, at p. 506.

[40] Andreas Libavius, *Examen Philosophiae Novae, quae veteri abrogandae opponitur* (Frankfurt, 1615), p. 144. On Libavius, see Bruce T. Moran, *Andreas Libavius and the Transformation of Alchemy: Separating Chemical Cultures with Polemical Fire* (Sagamore Beach, MA: Science History Publications, 2007).

[41] Heinrich Khunrath, *De Igne Magorum Philosophorumque secreto externo et visibili* (Strasburg, 1608), pp. 27, 31.

[42] Ibid., pp. 77, 28–29, 32.

calorific value, but rather imbued the alchemical substances with the virtues of specific planets and signs.

THE REGIMEN OF FIRE

The vast majority of alchemical authors, however, never exhibit more interest in astrology than a readiness to employ the signs of the zodiac as an analogy for their 'Regimen of Fire,' comparing the degrees of heat in the athanor to the increasingly seasonally warm zodiac signs of Aries, Taurus, Gemini, Cancer, and Leo, with the implication that, just as the Sun causes all things to mature and ripen in the world as it progresses through the twelve signs, so too the gradual increase of heat in the athanor will eventually produce the Philosophers' Stone.[43] Thus, at the most rudimentary level of understanding, the three astrological Fire signs were taken to symbolise degrees of heat in the alchemical furnace. [Pseudo]-Geber's *De Alchimia* describes the three degrees as 'weak' (*debilis*), 'strong' (*firmus*), and 'perfect' (*perfectus*), being respectively when the Sun enters Aries and is in its exaltation, when it enters its own domicile of Leo, and thirdly when it enters Sagittarius, with the recommendation, 'Understand, then, because with these grades the work of alchemy is completed'.[44] In *Anatomia Auri*, Mylius qualifies this information by explaining that the first stage is called weak because the heat is weak and the stage is the 'order of water'; the second stage in Leo is more fervid on account of the greater heat and is the 'order of air'; the third degree has nothing to do with burning heat, but is the 'order of dessicating air,' because at this stage the removal of heat leads to rest and true tranquility.[45]

It is not uncommon to find the advice: 'The beginning of this work is in the entrance of the Sun into Aries', or, 'One should distill this Stone with the Sun in Aries and Taurus'.[46]

---

[43] Ibid., pp. 47–48, 61–62.

[44] Geber, *De Alchimia libri tres* (Strasburg, 1531), Caput xcvii. On the astrological theory of exaltation, see William Lilly, *Christian Astrology* (London, 1647), pp. 102, 104, 'A Table of the Essentiall Dignities of the Planets according to Ptolomy'.

[45] Mylius, *Anatomia auri*, pp. 205–6. My thanks to Lawrence Principe for his advice on the sequence of baths.

[46] Giovanni Bracesco, *De alchemia dialogi duo* (Lyon, 1548), p. 117. See also Guido de Montanor, *Scala Philosophorum*, in Manget, *Bibliotheca*

For some this simply meant that, as above, one should start the operation with a moderate heat; for others, it was a direct reference to the astrological importance of the Vernal Equinox, when the Sun begins a new zodiacal cycle.[47] According to Limojon de Saint-Didier's *Le Triomphe hermétique* (1699), Denys Zachaire advises that the operation should begin at Easter,[48] while the Cosmopolite (Michael Sendivogius) says it should be when all living things appear animated with a new fire, when one sees rams, bulles and young bergers 'clearly expressing in this spiritual allegory the three months of Spring by the three celestial signs that correspond to them, Aries, Taurus, & Gemini'.[49] Some admit that other times are possible, but the result won't be as good.[50] Khunrath writes of physico-magically using a mirror on the correct 'day and hour of the Sun' (i.e. Sunday at sunrise) when the sun is entering the sign of Aries. This certainly sounds as though he considered it astrologically propitious to kindle a new fire in his laboratory at the start of Spring, one which he perhaps kept perpetually burning throughout the year.[51] Discussing the Paracelsian Jacques Gohory (1520–1576), Didier Kahn notes his recommendation to heed the advice of Raymond Lull's declaration that when 'our Sun is exalted in the sign of Aries, at that moment our work should begin because it's the first degree of sublimation', presumably when the sap is rising.[52]

Chemica Curiosa, Vol. 2, p. 137, and *Consilium Coniugii, seu De Massa Solis & Lunae*, in idem., 250.

[47] Egidius de Vadis, *Dialogus inter Naturam & Filium Philosophiae*, in *Theatrum Chemicum*, Vol. 2 (1602), p. 106.

[48] Louis Figuier, *L'Alchimie et les Alchimistes: Essai historique et critique sur la Philosophie hermétique* (Paris, 1854), p. 151. See also Antoine-Joseph Pernety, *Dictionnaire mytho-hermétique* (Paris, 1758), p. 486.

[49] Limojon de Saint-Didier, *Le Triomphe hermétique, ou La Pierre Philosophale victorieuse* (Paris, 1699), p. 91.

[50] Christian Friedrich Sendimir von Siebenstern, *Chymischer Mondenschein* (Frankfurt, 1760), p. 30.

[51] Khunrath, *De Igne Magorum*, p. 21. Khunrath gives the date in the old Julian calendar as 12th March.

[52] Didier Kahn, 'Alchimie et littérature à Paris en des temps de trouble: Le Discours d'Autheur incertain sur la pierre des philosophes (1590)', *Réforme, Humanisme, Renaissance* 41, no. 1 (1995): pp. 75–122, at p. 100. Louis XIII's physician David de Planis Campy even suggests that it should also be when the constellation of Pegasus is rising on the morning of the equinox, presumably because of the myth that the winged horse sprang from the blood of

One authority, [Johannes Grassaeus or Chortalassaeus], author of the *Cabala Chemica*, tells us that

> every thing has its Astronomy or its star in itself to which you must be attentive, when for example it is near its good ascendant, & its greatest planet (from which seed it descends) is exalted: for then any herb (or whatever else it may be) is most effective in its virtue, then it must be collected and put to use...for when the conjunction and exaltation of any planet is observed, their genera (be they metals, minerals, herbs, gems and stones) under heaven are united with them, then the rays of stars enter into these bodies, which augment and exalt their virtues to such an extent, that through them not only are the ills in men cured spiritually (as with touch of a sigil hanging from neck) but also admirable, nigh on impossible things can be achieved with natural magic.[53]

Doubtless having in mind the common notion of alchemy being the transmutation of lead to gold, the alchemical author and musician Giles Duwes (d. 1535) offers an interesting astrological argument for the significance of Aries in his *Dialogus inter Naturam & Filium Philosophiae*: 'With this they say that the upper Sun is exalted in the sign of Aries, which is the Fall of Saturn and the House of Mars; hence the work on the lower sun [gold] is exalted in the house of Mars, i.e., in Aries, which is your oven in which coldness, the property of Saturn, is suppressed, and heat exalted, which is the cause of the exaltation of your sun.'[54]

It is also relatively common to find allusions to the exaltations of the Sun in Aries and the Moon in Taurus. The *Mutus Liber* (1677) includes several engravings with a Ram on the left, the Sun above, and a Bull on the right, beneath the Moon. John Dee hints at these two exaltations in his *Monas Hieroglyphica* (1564), when discussing the symbolism of his hieroglyph: 'Accordingly we propose for the

---

Medusa, whose very look could turn things to stone. See David de Planis Campy, *Traite des playes faites par des mousquetades* (Paris, 1623), p. 59.

[53] Johannes Grasseus [Johann Grasshoff], *Physica Naturalis Rotunda Visionis Chemicae Cabalisticae,* in *Theatrum Chemicum,* Vol. 6 (1661), pp. 343–81, at pp. 351–52.

[54] Egidius de Vadis, *Dialogus inter Naturam & Filium Philosophiae,* in Manget, *Bibliotheca chemica curiosa,* 2 vols. (Geneva, 1702), Vol. 2, pp. 326–35, at p. 330.

consideration of philosophers the labours of Sun and Moon around the earth; in particular, how the Moon, while the Sun's splendour is in Aries, receives a new dignity of light in the next adjoining sign (namely Taurus) and is exalted above her innate powers....how the Sun...in the very house of Mars (namely in our Aries) it is said to be triumphant in its exaltation'.[55] In his *Opus Medico-Chymicum* (1618), Johann Daniel Mylius develops this theme: 'without the spirits and souls of Mars and Venus it is not possible for either the Sun or Moon to be exalted to surpassing perfection, which Philosophers who cultivate astrology can perfectly well understand, since the Sun and Moon have their [respective] exaltations in Aries, domicile of Mars and Taurus, domicile of Venus'.[56] He adds, 'For which reason he who desires to be proficient in the secret arts and Chymia, ought well to consider celestial bodies, their qualities and nature, their location; unless he does this whatever he undertakes will be done in vain'.[57]

THE PROGRESS OF THE SUN

Moving from considerations of the fiery triplicity and the first two zodiac signs to the simplest instance of a zodiacal progression, the Bohemian physician Daniel Stolcius's *Viridarium Chymicum* (1624) recommends a sequence through the four signs of Aries, Cancer, Libra and Capricorn:

> Phœbus runs through the zodiac in a year, and recreates all seeds with his rays. Hence learn the four degrees of our toil...they are the Ram, the Crab, the Scales, and Capricorn.[58]

There are several ways of interpreting this: Phœbus, that is, the Sun, passes through the four 'cardinal' signs of the Zodiac, representing 1) a circulation through the elemental

---

[55] C. H. Josten, 'A Translation of John Dee's "Monas Hieroglyphica" (Antwerp, 1564), With an Introduction and Annotations', *Ambix* 12, no. 2/3 (1964): pp. 83–222, at p. 167.
[56] Johann Daniel Mylius, *Opus Medico-Chymicum* (Frankfurt, 1618), p. 75.
[57] Ibid., 80.
[58] Daniel Stolcius, *Viridarium Chymicum* in John Read, *Prelude to Chemistry: an Outline of Alchemy, its Literature and Relationships* (London: G. Bell & Sons, 1936; repr., Cambridge, MA: MIT Press, 1966), p. 264.

sequence fire, water, air and final coalescence in earth, i.e., the Philosophers' Stone; 2) a temporal process stretching from vernal equinox to hibernal solstice; or 3) a thermal progress indicated by seasonal variation, Spring to Winter.[59] This passage calls to mind the seventh treatise of the *Splendor Solis* manuscript, attributed to Salomon Trismosin, complete with its own beautifully illustrated pictorial sequence of seven alchemical operations, each governed by a different planet in the heavens:

> I have observed a bird which the philosophers call Orsan. It flies when it is in Aries, Cancer, Libra or Capricorn, and you can obtain it in perpetuity from rich minerals and precious mountain stones.[60]

Other writers go further and consider the complete sequence of zodiac signs the ideal template for the progress of the alchemical work. David de Planis Campy provides a table associating each of the twelve zodiac signs with a different alchemical process, commencing with Aries.[61] Another, somewhat later French alchemist, Antoine-Joseph Pernety (1716-1796) includes the same sequence in his *Dictionnaire Mytho-Hermetique* (1758).[62] Elias Ashmole's edition of George Ripley's *The Compound of Alchymie, Conteining twelve Gates* in the *Theatrum Chemicum Britannicum* (1652) includes a circular figure calling to mind the nested spheres of the Ptolemaic cosmos, with the description: 'Our heaven this Figure called is/ Our table also of the lower Astronomy'. Ripley's Wheel includes the twelve glyphs for the signs of the zodiac, separated into their four elemental groups in subordinate wheels at the four angles of the

---

[59] On cardinal signs, see Lilly, *Christian Astrology*, p. 88.

[60] Salomon Trismosin, *Splendor Solis*, trans. Joscelyn Godwin, (Grand Rapids, MI: Phanes Press, 1991), p. 72. For reproductions of these images, see Alexander Roob, *The Hermetic Museum: Alchemy & Mysticism* (Cologne: Taschen, 2003). On the golden bird (*Oyseau doré*) and its astrological travels, see also *Le Livre des Figures Hieroglifiques de Nicolas Flamel Escrivain*, in *Philosophie Naturelle de Trois Anciens Philosophes Renommez Artephius, Flamel, & Synesius, Traitant de l'Art occulte, & de la Transmutation metallique* (Paris, 1682), pp. 45–88, at p. 63. A similar passage can be found in Stanton J. Linden, *The Alchemy Reader: From Hermes Trismegistus to Isaac Newton* (Cambridge: Cambridge University Press, 2003), p. 133.

[61] De Planis Campy, *Bouquet Composé*, pp. 992–93.

[62] Pernety, *Dictionnaire Mytho-Hermetique*, p. 99.

diagram. The very fact that there are 'Twelve' Gates encourages speculation that the processes may be related to the zodiac sequence. Jean Jacques Manget's edition of Ripley in the *Bibliotheca Chemica Curiosa* (1702) also includes a table that associates Zodiac signs with particular stages of the alchemical work.[63]

Zodiac Sign	Alchemical Process (David De Planis Campy) (Ripley in Ashmole)		(Ripley in Manget)
Aries	Calcination	Calcination	Calcination
Taurus	Congelation	Solution	Coagulation
Gemini	Fixation	Separation	
Cancer	Dissolution	Conjunction	Solution
Leo	Digestion	Putrefaction	Trans-mutation
Virgo	Distillation	Congelation	Sublimation
Libra	Sublimation	Cibation	Distillation
Scorpio	Separation	Sublimation	
Sagittarius	Inceration	Fermentation	
Capricorn	Fermentation	Exaltation	Putrefaction
Aquarius	Multiplication	Multiplication	
Pisces	Projection	Projection	

Table 1: Comparing correspondences between zodiac signs and alchemical processes

Although no explanations are provided for these correspondences, there is a certain internal logic to some of the attributions. Looking at De Planis Campy's associations, it would make astrological sense, for example, to start Calcination (burning) under the first fire sign, Aries, just as Digestion under the second fire sign, Leo, and Inceration (melting) under the third, Sagittarius. Congelation, the process of making something solid, likewise seems appropriate for the first earth sign, Taurus. Dissolution under the first water sign, Cancer, also sounds reasonable. Some processes, however, are slightly more puzzling: while

[63] Elias Ashmole, ed., *Theatrum Chemicum Britannicum* (London, 1652), p. 117. For the table, see Georgius Ripleus, *Liber Duodecim Portarum*, in Manget, *Bibliotheca Chemica Curiosa*, 2 volumes (Geneva, 1702), Vol. 2, pp. 275–85, at p. 275.

Sublimation makes sense under the second air sign, Libra, why would Fixation take place under Gemini, when surely a more volatile process would be expected for the first air sign? There, at least, Ripley's 'Separation' might make sense if it indicated the separation of spirit from matter, although there is very little agreement elsewhere in the two sequences, save at the beginning and end.

De Planis Campy provides the one instance, so far, of a writer mocking practitioners for attempting to prepare herbal spagyric remedies under the wrong constellation: 'If some abstractor of quintessence were so ill-practiced in his Art, that he wanted to extract the waters of Hermes under [the sign of] Libra, he would find his water greatly diminished in virtue and humor'—the reason being that the virtues in the leaves and herbs are generally at their most potent between the beginning of Taurus and Cancer (i.e., May and June), but are already fading by the end of September/October when the sun is in Libra.[64] We should not forget that, since Ptolemy's *Tetrabiblos*, the practice of astrology has included an interest in meteorology; a similar consideration of seasonal variation can be found in what sounds like sensible advice for the time to collect plants and prepare and administer remedies:

> Regarding the times to work chymistry, we mean the season, month, and day. In the season we observe what plants are accompanied by the virtues that we require, which are most often Spring and Autumn. With the months pay attention to the beginning, middle and end. The same observation holds for the day, regarding the collection, preparation, and administration of chemical medicines...Furthermore, if one wishes to prepare the remedies, humidity augmented alters the quality that we wish to extract. And nevertheless we see that in order to have a quantity of the oil of Sulphur, that which is distilled in the bell?, one should extract it on a rainy and very damp day: we observe the same with the extraction

---

[64] De Planis Campy, *Bouquet Composé*, p. 255. It is possible that by the phrase 'abstracteur de quinte-essence', De Planis Campy is making a humorous reference to Rabelais' *La vie inestimable du Grand Gargantua, pere de Pantagruel, iadis composée par l'abstracteur de quinte essence* (1535). See also De Planis Campy, *L'Hydre Morbifique exterminée par l'Hercule Chimique* (Paris, 1629), pp. 496–99 for a discussion of the preparation of Spagyric remedies in an astrological context.

> of the oil of tartar *per deliquium*.[65] And if you observe the rains, don't neglect the winds, in so far as these are caused and excited by the planets, as well as the signs, they shouldn't be neglected no less than their causes; and why would a knowledge of their dispositions and qualities be useless when we find those of the planets and signs so useful and necessary?[66]

Everything discussed so far, however, has all been an intimation of astro-alchemical relationships rather than developed discussions. For those we owe thanks to the editorial work of the antiquary Elias Ashmole (1617–1692), friend of the famous English astrologer William Lilly (1602–1681).[67] Under the anagrammatical pseudonym James Hasolle, Ashmole published Arthur Dee's *Fasciculus Chemicus: Or Chymical Collections. Expressing the Ingress, Progress, and Egress of the Secret Hermetick Science, out of the choisest and most Famous Authors* (1650). No astrologer familiar with Ashmole's birthdate (23 May 1617) would have had difficulty identifying his horoscope, covering the face of the authorial bust in the book's frontispiece, nor would they have missed the astrological connotations in the title's use of 'Ingress, Progress, and Egress'.[68]

The publication included an astro-alchemical work, the *Arcanum: Or, The grand Secret of Hermetick Philosophy*. Ashmole introduces it as 'The Work of a concealed Author. Pene nos unda Tagi', which is an anagram of the French alchemist Jean d'Espagnet (1564–1637).[69] There we are

---

[65] See John Henry, 'Boyle and Cosmical Qualities', in *Robert Boyle Reconsidered*, ed. Michael Hunter (Cambridge: Cambridge University Press, 1994), pp. 119–38, at p. 120: 'Common tartar which stays dry in the air and is only dissolved in water with difficulty is changed by moderate calcining, so that it spontaneously turns into that liquor which chemists call tartar *per deliquium*'.

[66] De Planis Campy, *Bouquet Composé*, p. 244.

[67] On Lilly and Ashmole, see Patrick Curry, *Prophecy and Power: Astrology in Early Modern England* (Oxford: Polity Press, 1989), pp. 28–40 and *passim*. See also Bruce Janacek, 'A Virtuoso's History: Antiquarianism and the Transmission of Knowledge in the Alchemical Studies of Elias Ashmole,' *Journal of the History of Ideas* 69, no. 3 (2008): pp. 395–417.

[68] On Ashmole's birthdate, 23 May 1617, see C. H. Josten, 'Elias Ashmole, F.R.S. (1617–1692)', *Notes and Records of the Royal Society of London* 15 (July, 1960): pp. 221–30.

[69] James Hasolle, *Fasciculus Chemicus: Or Chymical Collections. Expressing The Ingress, Progress, and Egress, of the Secret Hermetick*

provided with a 'Signifer of Philosophers with the Houses of the Planets', that is a chart with the astrological signs and their rulers.[70] We are also provided with 'The Interpretation of the Philosophers Scheme', showing evident familiarity with alchemical symbolism, especially the colour changes of Black Primal Matter to White Lunar Stone and then finally the Red Solar Stone:

> Philosophers in handling their Philosophical work, begin their yeare in Winter, to wit, the Sun being in *Capricorne*, which is the former House of *Saturne*...In the Second place the other House of *Saturn* is found in *Aquarius*, at which time *Saturne*, i.e., the Blacknesse of the Dominary work begins after the 45 or 50 day. *Sol* coming into *Pisces* the work is black, blacker than black, and the head of the Crow begins to appear.
>
> The third month being ended, and *Sol* entring into *Aries*, the sublimation or separation of the Elements begins. Those which follow unto *Cancer* make the Worke White. *Cancer* addeth the greatest whitenesse and splendour, and doth perfectly fill up all the dayes of the Stone or white Sulphur, or the Lunar worke of Sulphur....In *Leo* the Regal Mansion of the Sun, the Solar work begins, which in *Libra* is terminated into a Rubie-Stone, or perfect Sulphur.
>
> The two Signes *Scorpius* and *Sagitarius* which remaine, are indebted to the compleating of the Elixir. And thus the Philosophers admirable young taketh its beginning in the Reigne of *Saturne*, and its end and perfection in the Dominion of *Jupiter*.[71]

Note, however, that in this scheme the Work begins in the sign of Capricorn. For the alchemist there would be some sense in this as Capricorn's ruling planet is Saturn, which equates with Lead in alchemy, one of the candidates for the

---

Science, out of the choisest and most Famous Authors (London, 1650), 156 ff.

[70] *Fasciculus Chemicus*, p. 266 [mispaginated as p. 252].

[71] *Fasciculus Chemicus*, p. 267–68, 'The Times of the Stone'. See [Jean d'Espagnet], *L'Ouvrage secret de la Philosophie d'Hermez*, in *La Philosophie Naturelle restablie en sa Pureté* (Paris, 1651), p. 376, 'Le Zodiaque des Philosophes, avec les Maisons des Planettes', followed by the explanation 'Temps de la Pierre'.

materia prima of the Great Work.[72] Cancer, ruled by the Moon, makes sense as a symbol for the perfection of the White Stone, and Leo for the commencement of the work on the Red Stone doesn't seem too remiss.

A few more ambitious souls even suggested that attention to the Fixed Stars would provide indicators of when certain operations should take place. Writing in the early seventeenth century, Wolfgang Hildebrand, for example, suggests that Aldebaran at 6° Gemini is a good point to start sublimation, Achoraye at 23° Taurus is a good time for Coagulation, and Alnaya at 1° 58′ Capricorn is a good time for Putrefaction.[73] Leonard Thurneisser (c. 1530–1596) encourages similar reflections in his *Megale Chymia, vel Magna Alchymia* (1583),[74] and Pierre-Jean Fabre (1588–1658) devotes a great amount of space in one of his works, *Opus Pan-Chymici* (1651), to alchemical interpretations of the classical fables connected with the constellations. Arcturus, we learn, represents 'our white sulphur'; Cygnus is simply 'sulphur', and Perseus is 'mercury of the Chymists'.[75]

THE PROGRESS OF THE MOON

For some, however, it was not the progress of the sun, but of the moon that should be heeded, as was generally the case in medical astrology. Not long after alchemy's arrival in the West, Constantine of Pisa's thirteenth-century *Book of the Secrets of Alchemy* similarly concerns itself with the effects of good and bad lunations, and explicitly declares the necessity of some astrological knowledge:

---

[72] See, for example, Johann Joachim Becher, *Chymischer Glücks Hafen* (Halle, 1726), p. 493, who relates Capricorn and its time of the year to the process of putrefaction, often seen as the starting point of the whole alchemical reduction of matter to its primal state. Paracelsus makes no reference to Capricorn, but instead to the Raven's Head, which Becher is presumably taking as a synonym. See *Aurora Thesaurusque Philosophorum* (Basel, 1577), p. 48.

[73] Wolffgang Hildebrand, *Ein new außerlesen Planeten-Buch* (Erfurt, 1613), Sig [K4^r]ff. On Hildebrand, see Joachim Telle, 'Die Magia naturalis Wolfgang Hildebrands,' *Sudhoffs Archiv* 60, no. 2 (1976): pp. 105–22.

[74] Leonhart Thurneysser zum Thurn, *Megale Chymia, vel Magna Alchymia* (Berlin, 1583), pp. 113–14, 'Von Qualitet/ Natur und Art der 28. Mansionen'.

[75] On Fabre, see Bernard Joly, *La Rationalité de l'Alchimie au XVIIe siècle* (Paris: Vrin, 1992).

> It is as impossible to lick heaven with one's tongue as it is impossible to enter upon the practice of alchemy other than through the congealing of mercury, of which many are ignorant and which cannot be taught reliably except through the motion of the upper bodies, especially the orbit of the moon.[76]

Probably the best-known alchemical image illustrating the notion of the generation of the metals under the influence of the constellations is found in the fifteenth-century *Aurora Consurgens*.[77] There we see a personification of the moon sitting in the centre of the Zodiac, afterbirth flowing from between her legs. While in Ripley's zodiac scheme the sun in Leo is associated with the process of transmutation of base metal to gold, and in D'Espagnet's sequence it marks the start of the Red Work, in this image we have the maternal moon giving birth to, presumably, the Philosophers' Stone. As Barbara Obrist points out, the moon is holding an astrolabe pointing in the direction of the sign of Leo.[78]

---

[76] Barbara Obrist, 'Visualization in Medieval Alchemy', *HYLE: International Journal for Philosophy of Chemistry* 9, no. 2 (2003): 131–70. See Constantine of Pisa, *The Book of the Secrets of Alchemy*, ed. Barbara Obrist (Leiden: Brill, 1990), pp. 236–47, 264–66.

[77] See the image in Obrist, *Les Débuts de l'imagerie alchimique*, Illustration 53, and in Mino Gabriele, *Alchimia e Iconologia* (Udine: Forum, 1997), Illustration 63.

[78] Obrist, *Les Débuts de l'imagerie alchimique*, p. 224.

Fig. 7.2: Aurora consurgens, *Leiden Vossianus Ms. Chym F. 29*

SIGNS AND SUBSTANCES

Although up to now we have discussed the Constellations predominantly as temporal markers for the stages of the alchemical process, this is not the case with all such references to the zodiac signs. What may look like a consideration for the timing of an operation can also be a veiled hint to use a substance. In Sendivogius' *Novum lumen chemicum*, for instance, the reader is told that 'the work is to be begun in Aries'. William Newman and Lawrence Principe, in their research into the alchemist George Starkey and his interpretation of Sendivogius, have replicated some of Starkey's experiments and come to the conclusion that Starkey understands that by 'Aries', the work should begin with Iron, i.e., the metal associated with the ruler of Aries, Mars.[79] The Constellations are quite often related to particular substances, although here, as with their correspondences with alchemical processes, while there are

[79] William R. Newman and Lawrence M. Principe, *Alchemy Tried in the Fire: Starkey, Boyle, and the Fate of Helmontian Chymistry* (Chicago: University of Chicago Press, 2002), p. 193.

some overlaps, there are also many variations. Ripley, for example, associates many of the signs with different kinds of salts: warlike Aries with Saltpetre, a main component in gunpowder; pacific Libra with Sal Armoniac and its more agrarian use as a fertiliser; while Taurus is linked with Vitriol, possibly due to Venus being the planetary ruler of Taurus and the related metal, copper, being an ingredient in Vitriol.[80]

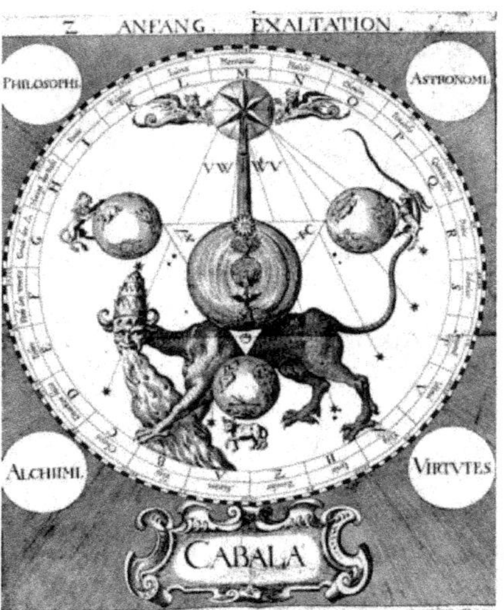

Fig. 7.3: Stephan Michelspacher, Cabala, Spiegel der Kunst und Natur *(1616), Fig. 2. Beginning: Exaltation.*

A somewhat more developed set of relations can be found in Stephan Michelspacher's *Cabala, Spiegel der Kunst und Natur* (1616), which includes some slightly puzzling engravings in which the author displays a readiness to combine several occult sciences, including astrology and alchemy, but giving us little explanation of his intentions. One engraving entitled 'Beginning: Exaltation' includes both alchemy and astrology (together with Philosophy and Virtues) in a list of four practices ultimately deriving from Paracelsus's four pillars of medicine in the *Paragranum*. Arrayed in a circle, we find alchemical substances associated with letters of the alphabet, and inside the circle an

[80] True, saltpetre or potassium nitrate can also be used as a fertiliser.

apocalyptic-looking beast surrounded by curious groupings of astrological symbols in relation to three spheres:

1) Leo beside a sphere containing Scorpio, Cancer, and Aries, all being related to the fiery principle of Sulphur (thus the planetary relationships are Double Mars, the Sun, the Moon; the elements of the signs are Double Water, Double Fire);

2) Capricorn beside a sphere containing Pisces, Sagittarius, and Aquarius (Double Jupiter and Double Saturn, the elements being Earth, Water, Fire, and Air), all related to the watery alchemical principle of Mercury;

3) Taurus beneath a sphere containing Libra, Gemini, and Virgo (Double Mercury and Double Venus, the elements of the signs being Double Earth, Double Air), all related to the third Paracelsian principle Salt (here Vitriol).

Fig. 7.4: Stephan Michelspacher, Cabala, Spiegel der Kunst und Natur *(1616), Fig. 3. Middle: Conjunction.*

In Michelspacher's next engraving, entitled 'Middle: Conjunction', we see a hill with figures representing the seven planets/metals: Venus, Mars, Sol in ascending order on the left, Saturn, Jupiter, Luna ascending in that order on the right, with Mercury at the summit. Within the hill stands what looks like a classical temple containing an alchemical furnace, with seven steps leading up to the entrance, each step named after a different alchemical process. Encircling this image is the zodiac, but with the signs grouped according to their planetary rulerships and following the sequence of the figures on the hill.

Zodiac Sign (Ruler)	Alchemical Substance
Taurus (Venus)	Mercury
Libra (Venus)	Verdigris
Scorpio (Mars)	Vitriol
Aries (Mars)	Sulphur
Leo (Sun)	Antimony
Virgo (Mercury)	Sal Ammoniac
Gemini (Mercury)	Cinnabar
Cancer (Moon)	Orpiment
Sagittarius (Jupiter)	Salt
Pisces (Jupiter)	Saltpetre
Capricorn (Saturn)	Tartar
Aquarius (Saturn)	Alum

Table 2: Michelspacher's correspondences between Signs and Substances

Again, it is not too difficult to figure out some of these attributions: as Verdigris is the natural patina of copper, its association with Venus and Libra isn't much of a puzzle. The same can be said for linking Mars and Scorpio with the ferrous sulphate version of Vitriol. As in his previous engraving, Michelspacher associates Aries with fiery Sulphur. It is interesting to see that he links Leo and the Sun with Antimony, the very substance alchemists used to purify gold of other metal alloys. De Planis Campy has very different ideas, linking Aries with Antimony; Leo with Gold; Scorpio with Sulphur; Taurus with Asphalt or Bitumen; Cancer with Sal Armoniac; and Libra with Roman Vitriol.[81]

[81] De Planis Campy, *Bouquet Composé*, p. 992.

Perhaps these practitioners really did see a genuine correspondence between the celestial planets and signs and their terrestrial substances, but the great variation in attributions suggests that individual alchemists were assigning *Decknamen* or 'cover' names to their substances as a way of concealing, or hinting at, their identity. In his *Zwölf Schlüssel* (1602), Basil Valentine seems to be doing just that in a passage when he groups the 'bestial or quadruped' signs of the zodiac on one side and the 'manly or humane' on the other:

> Let me tell you allegorically that you must put into the heavenly Balance the Ram, Bull, Cancer, Scorpion, and Goat. In the other scale of the Balance you must place the Twins, the Archer, the Water–bearer, and the Virgin. Then let the Lion jump into the Virgin's lap, which will cause the other scale to kick the beam. Thereupon, let the signs of the Zodiac enter into opposition to the Pleiads, and when all the colours of the world have shewn themselves, let there be a conjunction and union between the greatest and the smallest, and the smallest and the greatest.[82]

We do have at least one instance of a work on alchemy, the *Clavis Sapientiae*, included in Benedictus Figulus's *Thesaurinella Olympica* (1682), advising the calculation of a sick person's nativity (birth chart) and then relating that to minerals necessary for a cure:

> Knowing the star ruling in his nativity, let us make a figure of that man of the mineral that is of the part of that planet in the hour, and let the star contrary in nature be aspecting the figure, and let it be with it in that sign in which that star is in fortune…And if the ruling star is not known...let us make seven figures of the seven stones that are of the nature of the seven Planets, and they are these: of the nature of Saturn, Arsenic, of the nature of Jupiter, Loadstone, of the nature of Mars, Ruby, of the nature of the Sun, Alum, of Venus, of Mercury, and of the Moon, Crystal…[83]

---

[82] Basil Valentine, *Zwölf Schlüssel, in Fratris Basilii Valentini Benedictiner Ordens Chymische Schriften alle/ so viel derer verhanden* (Hamburg, 1700), p. 24–74, at p. 63. English translation from John Read, *From Alchemy to Chemistry* (London: G. Bell & Sons, 1957; repr., NY: Dover Publications, 1995), p. 57.

[83] *Sapientissimi Arabum Philosophi, Alphonsi, Regis Castellae, &c. Clavis Sapientiae,* in Benedictus Figulus, *Thesaurinella Olympica Aurea Tripartita* (Frankfurt, 1682), pp. 351–52. Yes, this passage is slightly

To make matters worse, more than one alchemist suggests using the astrological symbols as alphabets for encoding alchemical secrets.[84] One misguided but imaginative soul, Jean de Bonneau, in his *Abrege de l'Astronomie Inferieure* (1644), even argues that the astrologers 'usurped' their symbols for the planets and constellations from the Hermetic Philosophers! Ptolemy, we learn, borrowed 'all the ornaments of his doctrine from Chymistry'.[85] Sometimes speculation on the astrological symbols leads to more intriguing results. Perhaps inspired by the notion that alchemical Electrum is a combination of all the seven metals, John Dee 'discovered' his symbol of the Hieroglyphic Monad, which is, as it were, a conjunction of all seven astrological symbols for the planets. Aware of Dee's concern with astronomy in *Propaedeumata aphoristica* (1558) and the alchemical concerns in the *Monas Hieroglyphica* (1564), Nicholas Clulee argues that Dee's work contributed to the development of a new conception of 'Inferior astronomy' in the Renaissance, indeed, the unification of alchemy with astrology and magic.[86] Somewhat surprisingly, one of the major opponents of Dee and the Paracelsians, Andreas Libavius (1555–1616), actually saw this as a promising theoretical development in that his compound or composite master hieroglyph could be further analysed and synthesised to generate other hieroglyphs in order to create some symbolic coherence in the way alchemical substances were symbolically depicted.[87]

---

garbled. It is not much better in the German versions in Artefius, *Clavis maioris sapientiae* (Strasburg, 1699), pp. 47–48 or Artephius, *Clavis Sapientiae* (Leipzig, 1736), p. 38.

[84] De Planis Campy, *Bouquet Composée,* pp. 994–95; Andreas Libavius, *Rerum Chymicarum Epistolica Forma ad Philosophos et Medicos qvosdam in Germania, Liber Tertius* (Frankfurt, 1599), p. 348.

[85] [Jean de Bonneau] I. D. B., *Abrégé de l'Astronomie Inferieure, Expliquant le Systeme des Planetes; les douze signes du Zodiac & autres Constellations du Ciel Hermetique. Avec un Essay de l'Astronomie Naturelle, contre les Systemes de Ptolomée, Copernic, & Ticho Brahé* (Paris, 1644), sig biii[r], 8–9.

[86] Nicholas H. Clulee, 'The *Monas hieroglyphica* and the Alchemical Thread of John Dee's Career', *Ambix* 52, no. 3 (2005): pp. 197–215, at p. 211.

[87] Libavius, *Rerum Chymicarum,* Vol. 3, pp. 345–48. See Peter J. Forshaw, 'The Early Alchemical Reception of John Dee's *Monas Hieroglyphica*', *Ambix* 52, no. 3 (2005): pp. 247–69.

THE COMBINATIONS OF STELLAR INFLUENCES

From combinations of astro-alchemical hieroglyphs, let us turn to the consideration of planetary combinations in the heavens, as already hinted at in Michelspacher's 'Middle: Conjunction' engraving, for there is some evidence of slightly more sophisticated considerations of the relevance of the conjunctions, oppositions, trines and squares of planetary aspects to the alchemical enterprise.

In his *Cabala Chemica* (1658), Johann Grasshoff (c. 1560–1623) is a rare case of an author who tries to give some explanation why an understanding of celestial forces may be useful for the alchemist:

> The reason why this operation must be done at the point of the conjunction of the planets and stars is worthy of observation: Any matter seeks new form, i.e., whichever of the planets comes into conjunction, this celestial, spiritual and material matter desires to enter into all living forms, and on account of its attractive nature or property the celestial fire...as it were attracts life to itself in an instant, and conjoins with it, to such an extent that in this way spirit and soul mix and unite with one another. Earth...likewise in an instant draws that body to itself, to such an extent that these two conjunctions are made in the same point in time, whence all vegetables, animals & minerals draw their origin...If truly you were familiar with this conjunction beforehand, by means of astronomy, and the metals and vegetables which have the influence, nature and complexion of those stars, by your contact you could conveniently bring it about that they would then enter not into the earth but into bodies of such a kind (because like rejoices in like), on which occasion such wonderful things would be accomplished through this miracle of nature. Any Philosopher or Chymist ought to grasp something else here too.[88]

Grasshoff goes on to informs his reader that different fruits are produced when one planet enters into a conjunction or a trine aspect with another planet: when the exalted Sun is conjunct with Mars (i.e., when they are together in Aries), a greater but also darker red carbuncle is produced than otherwise. If exalted Saturn (in Libra) is conjunct with the

[88] Johannes Grasseus [Johann Grasshoff], *Physica Naturalis Rotunda Visionis Chemicae Cabalisticae* in *Theatrum Chemicum*, Vol. 6, pp. 343–81, at p. 351.

Moon when the heavens are clear and bright, you'll obtain beautiful, clear, transparent/white and pure crystals; if the heavens are dark and cloudy and the conjunction is with Saturn and Mars, you'll get quite other crystals. Quicksilver owes its origin to the planet Mercury and, depending on which planet it conjoins with, various kinds of quicksilver arise, and it transmutes itself into the nature of whichever planet it conjoins, according to the sign of the Zodiac in which the ruler is placed.

De Planis Campy similarly relates that practitioners have argued for the importance of observing planetary aspects:

> Almost all the chemical philosophers have assured us that not all times are suitable for beginning our work. That's why they desire that we observe the influence and conjunction of certain stars, like the conjunction of the sun with the moon, or with mercury. Certain wish to have us pay attention to the waxing of the moon, and others to its waning.[89]

Immediately after having spoken of conjunctions in a temporal, astrological sense, however, De Planis Campy then turns to the use of conjunction in a physical alchemical sense, adding: 'By this advice one should immediately understand by the conjunction of the sun with the moon or with mercury, the second and third regimens of the work, i.e., Cibation and Fermentation, for that concerns the conjunction of Gold with the universal dissolvant, called Moon by similitude'. Does the advice of the 'Chemical philosophers' genuinely concern observation of heavenly phenomena and the alignment of alchemical processes with those celestial aspects, for instance the 'conjunction' of Gold with the universal dissolvant at the moment when the Sun conjuncts the Moon (during an eclipse), or should the advice be taken purely symbolically? Given the proliferation of alternative

---

[89] David de Planis Campy, *L'Ouverture de L'Escolle de Philosophie Transmutatoire Métallique* (Paris, 1633), pp. 164–65. De Planis Campy also refers the reader to the *Turba Philosophorum*. There, for example, in the 68th dictum, Attamus relates that 'the second work is performed from the tenth day of the month of September to the tenth [degree?] of Libra'. Turba Philosophorum, in Theatrum Chemicum, Vol. 5, p. 48: 'Secundum autem opus fit a decimo die mensis septem. ad 10. librae'. See also in the same volume, *Platonis Quartorum cum commento Hebuhabes Hamed*, p. 125, where we read the advice to attempt alchemical Solutio after astrological Conjunctio, and Coagulatio after Oppositio.

interpretations of texts like the *Emerald Tablet*, I imagine different practitioners interpreted this advice on different levels.

Choosing the Moment

The best example to be found of an alchemist displaying familiarity with astrology exists in a long alchemical poem, *The Ordinall of Alchimy* (1477) by the English alchemist Thomas Norton, who writes:

> The *fift Concord* is knowne well of Clerks,
> Betweene the *Sphere of Heaven* and our *Suttill Werks*.
> Nothing in Erth hath more Simplicitie,
> Than th'elements of our *Stone* woll be,
> Wherefore thei being in warke of Generation,
> Have most Obedience to Constellation:
> Whereof Concord most kindly and convenient
> Is a direct and firie *Ascendent*,
> Being signe common for this Operation,
> For the multitude of their Iteration:
> Fortune your *Ascendent* with his *Lord* also,
> Keeping th'aspect of *Shrewes* them fro;
> And if thei must let, or needely infect,
> Cause them to looke with a *Trine* aspect.
> For the *White warke* make fortunate the *Moon*,
> For the Lord of the *Fourth house* likewise be it done;
> For this is *Thesaurum absconditum* of olde Clerks;
> Soe of the *Sixt house* for *Servants* of the Werks;
> Save all them well from greate impediments,
> As it is in Picture, or like the same intents.
> Unlesse then your *Nativity* pretend infection,
> In contrariety to this Election...[90]

Ashmole is clearly extremely interested in what Norton has to say about the use of astrology for alchemy, and he provides quite extensive 'Annotations and Discourses' at the

---

[90] Thomas Norton, *The Ordinall of Alchimy*, in Elias Ashmole, ed., *Theatrum Chemicum Britannicum* (London, 1652), pp. 99–100. The horoscopes to be discussed can be found on p. 91. On Norton, see Antonio Clericuzio, 'Norton', in *Alchemie: Lexikon einer hermetischen Wissenschaft*, ed. Claus Priesner and Karin Figala (Munich: Verlag C.H. Beck, 1998), pp. 258–59. See also Richard Kieckhefer, *Magic in the Middle Ages* (1989; repr., Cambridge: Cambridge University Press, 2003), p. 135ff.

end of his collection.[91] Concerning Norton's rhyme on 'Generation' and 'Constellation,' he explains:

> Here our Author refers to the Rules of Astrologie for Electing a time wherein to begin the Philosophicall worke, and that plainly appears by the following lines, in which he chalkes out an Election fitly relating to the Businesse.[92]

As should be apparent from the examples of Sahl ibn Bishr, Norton is following the centuries-old practice of catarchic or horary astrology for choosing the most opportune time for an activity.[93] Grosseteste had emphasised the necessity of such a practice in his *De artibus liberalibus*: 'Truly, in the preparation of the stone, by which the transmutation of metals is achieved, the election of times is no less necessary...This substance isn't prepared at just any hour, but when the Sun is in exaltation in Libra [*sic*], away from any aspect of the malefics'.[94] It is evident that this practice was not universally accepted by alchemists from a comment by Basil Valentine in *His Last Will and Testament* (1657):

> Those who think themselves to be the wisest doe say, that it is a vanitie to observe mathematically the stars above, and to order any work after seasonable dayes and houres, it is something said, but not so well grounded. But this is most certain, that if you work according to common course, otherwise than we do, following onely your own fancies, then is your labour in vaine.[95]

Ashmole then turns to Norton's 'Rules for framing an Election' and what is the most intriguing aspect of Norton's work: four alchemical horoscopes, again rather perplexing in their style. Ashmole, too, is at first bemused by the non-

---

[91] Ashmole, *Theatrum Chemicum Britannicum*, p. 437ff.

[92] Ibid., p. 450.

[93] For an early modern reference, see Georgius Raguseius, *Epistolarum mathematicarum seu De Divinatione, Libri Duo* (Paris, 1623), p. 195, *Electio temporis opportuni in Chymicae artis operibus qualis esse debeat, secundum Astrologos*.

[94] Grosseteste, *Die Philosophischen Werke des Robert Grosseteste*, p. 6: 'Haec substantia non qualibetcumque hora praeparatur, sed cum fuerit sol in exaltatione librae ab aspectu malorum'. Perhaps there is scribal error here, for the Sun is exalted in Aries and Saturn in Libra.

[95] Basilius Valentinus, *Letztes Testament* (Strassburg, 1651), pp. 27–28. English translation from *His Last Will and Testament* (1657), p. 10.

standard symbols for the planets, but provides a helpful explanation:

> It is also worthy of our Observation to see how the Author continues his Vailes and Shadows, as in other parts of the Mistery, so likewise in the very Figures of some of the Planets, for he does not exhibite them under the Characters commonly now (or then) used, but Hierogliphically in Figures agreeable to their Natures...For Saturn is pointed out by a Spade, Jupiter by a Miter, Mars by an Arrow, Venus by a beautifull Face, Mercury by the figure (in those daies) usually stamped upon the Reverse of our English Coyne: Onely the Sun and the Moon are left us in that fashion the Auncients bestowed upon them.[96]

Fig. 7.5: Thomas Norton, The Ordinall of Alchimy *(1477), in Elias Ashmole (ed.),* Theatrum Chemicum Britannicum *(1652), p. 91.*

[96] Ashmole, *Theatrum Chemicum Britannicum*, p. 452.

Texts in the centre of Norton's two lower charts inform us that the top left horoscope concerns itself with 1) 'Division, Separation and Correction', i.e., the initial stages of alchemy; the top right with 2) 'Conjunction of the Elements and of things produced by the elements'; the bottom left 3) 'Conforms to the Secret Operation of Nature', while the bottom right 4) 'Perfects and brings all things to their term', i.e., results in the perfect conjunction of materials and produces the Philosophers' Stone.[97]

Ashmole makes it obvious that he is in a quandary about how to interpret the four charts. He knows that it is impossible for them to be 'the elected times for the Authors owne Operations...but are rather fained and invented', and he is fully aware that

> some Planets...are not placed in that exact order (for houses and signes) as Astronomicall Rules direct, and the Doctrine of Astrologie requireth.

For example,

> you have Venus placed in every Figure so remote from the Sun, that Astronomers must count it absurd, since she is never above 48 degr. Elongated from him; and yet in the third Figure she comes not within the compasse of a Sextile Aspect, nay in the second she is almost in Opposition to him.

However, he has complete faith in Norton, who 'manifests himselfe a learned Astrologian', and has refrained from making any alterations to the charts. He admits to having 'taken some paines to Calculate the places of the Planets for severall years about the Authors time, but cannot finde the three Superiors and place of the Sun to be in those Signes wherein he has posited them'.[98]

---

[97] Norton, *Ordinall*, p. 91: 1) 'Figura superior dividit, Seperat, et Corrigit'; 2) 'Superior figura est Elementorum conjunctionis atque Elementatorum'; 3) 'Haec operationem Naturae Secretam comfortat'; 4) 'Inferior haec perficit et terminat omnia'.

[98] I tried the same by checking ephemerides for 1470s, especially for the conjunction of Sun, Moon, and Mercury in Leo in the fourth chart, which so obviously symbolises the production of the Philosophers' Stone. Such a conjunction did not occur in 1477, the alleged year of the *Ordinall*'s composition, but there was one on 1–2 August 1475, and another in 19–20 July 1479. Admittedly, none of

Be this as it may, there is still sufficient material of interest in these 'fained and invented' charts to illustrate or educate potential astro-alchemists. Although Grafton and Newman dismiss 'what seems at first a serious commitment to catarchic astrology' as 'mere window dressing, for Norton's elections are manifestly unworkable',[99] perhaps there is another way of considering Norton's line in the extract above, 'As it is in Picture, or like the same intents'. In her article on the 'Uses of Pictures' in Leonhard Fuchs' *De historia stirpium* ('On the History of Plants'), Sachiko Kusukawa discusses Fuchs' declaration that each picture taken from a plant was rendered *absolutissima* by including its roots, stems, leaves, flowers, seeds, and fruits. This sense of 'completeness' manifested in two ways: some pictures were diachronic, showing different stages of a plant, such as the flowers and fruits, on the same bush at the same time. Other pictures depict one plant containing variations, such as a rose, showing *rosa canina* and *rosa sylvestris*. Such a picture was not, then, 'natural', but an *Idealbild*.[100]

Could it be that Norton's alchemical horoscopes should be viewed in a similar light? Perhaps the perfect chart for creating the Philosopher's Stone would be as rare as the heavenly configurations for the birth of Christ? If such were the case, then a pragmatic alchemist would have to make do with the best celestial positions available, hoping to find a time when as many propitious signs and houses, as many benign aspects as possible, could enhance the success of his work. Looked at in this way, then, the near opposition (176°) between Venus in Pisces and the Sun in Virgo in the second chart should not be seen as a blunder, but instead as indicating that either of these chart placements is beneficial for the alchemist busy conjoining the elements. For example, Venus is in the fourth house in this (and two other) charts, the house of *Thesaurum absconditum*, as Norton says, or

---

the other planets matched their chart positions. This data can be found in the online ephemerides at Astrodienst: <www.astro.com/swisseph/swepha_e.htm>.

[99] Newman and Grafton, *Secrets of Nature*, p. 21.

[100] Sachiko Kusukawa, 'The Uses of Pictures in the Formation of Learned Knowledge: The Cases of Leonhard Fuchs and Andreas Vesalius,' in *Transmitting Knowledge: Words, Images, and Instruments in Early Modern Europe*, ed. Sachiko Kusukawa and Ian Maclean (Oxford: Oxford University Press, 2006), pp. 73–96, at pp. 77–78, 80.

'Treasure-trove', as Lilly writes in his *Christian Astrologer*.[101] Lilly advises:

> If the Question be concerning Treasure absolutely, without knowledge whose or what it was, *viz*, whether there be any in the place or ground suspected or no; observe in the Figure whether Jupiter or Venus or the Moon's North Node be in the fourth house, there's then probability of Treasure being there...

True, we're talking about the Philosophers' Stone, but Norton implies the same rules apply. If we then check which sign Venus is in, we see that it is Pisces, its exaltation, another excellent reason to indicate such a placement. Although such an approach to reading Norton's charts may not be entirely satisfactory, it does at least seem a more constructive approach than 'the evident conclusion that Norton was an astrological incompetent'.[102]

Ashmole was certainly not the only reader of the *Ordinall*, and indeed, in his vast manuscript collection he possessed the notes of another occult philosopher familiar with Norton's work: *The Astrologicalle Judgmentes of Phisick and other Questions* (1606) of Simon Forman.[103] The Elizabethan astrologer-physician echoes Norton's advice concerning the choice of ascendant:

> When thou firste goest about to make the philosophers stone...take heed thou put Sagittarius in ascendent which is a direct signe ascendent common apt to receyve all natures.[104]

Forman provides further advice and explanations of this kind, but goes far beyond Norton in his advice on what to check for before commencing the Work. Let the prospective alchemist beware, for instance, if the Lord of the eighth house is in the fourth, for 'evill sprits will spoill yt in the end

---

[101] Lilly, *Christian Astrology*, pp. 202, 215.

[102] Newman and Grafton, *Secrets of Nature*, p. 21.

[103] Oxford, Bodleian Library, Ms. Ashmole 389: *The Astrologicalle Judgmentes of Phisick and other Questions writen by Simon Forman Dr of Astronomy and Phisick 1606—Lambeth the 16 of September in which is comprised his experience for 20 yeares before And many yeares after*, f. 636, *De lapide philosophico*. Lauren Kassell, *Medicine & Magic in Elizabethan London: Simon Forman, Astrologer, Alchemist, & Physician* (Oxford: Oxford University Press, 2005).

[104] Forman, *Astrologicalle Judgmentes*, f. 636: *De lapide philosophico*.

or cause the losse or destruction therof'. If, however, the Lord of the Ascendant is in the eleventh, 'thou shallt obtain favour and help of the angells', and if the Lord of the Ascendant is in conjunction with the Lord of the eleventh in the ninth house, you are fortunate, for 'the good angells of god shall showe thee in dreams and revelations helpe othe[r] knowledg[e] therof'. If the Lord of the ninth house is poorly aspected, your books will be unreliable, and if this is the case with the Lord of the sixth, you'd better look for new servants, for those you have are 'evill & unfortunat'.[105] Forman also provides advice that may explain, for example, Norton's conjunctions of Saturn and Jupiter in the second and third charts: 'Yf youe worke the work of [Saturn] then fortunat [Saturn] and dominus 4[e] [the Lord of the fourth house]'.[106]

Now let us return to the final piece of advice in the extract from Norton's poem:

> Unlesse then your *Nativity* pretend infection,
> In contrariety to this Election...[107]

In line with Ptolemy's sixth aphorism in the *Centiloquium*,[108] Ashmole explains this by emphasising that it is necessary not only to draw up a chart for the alchemical work, but also to consult the birth chart of the alchemist, so that

> By an apt position of Heaven, and fortifying the Planets and Houses in the Nativity of the Operator, and making them agree with the thing signified; the Impreßion made by that Influence, will abundantly augment the Operation.[109]

Unfortunately, we don't have a copy of Norton's nativity, but we do have that of an important figure in the history of alchemy, George Starkey (1628–1665), alias Eirenaeus Philalethes, none other than the person who initiated Robert

---

[105] Ibid., ff. 638–39.
[106] Ibid., f. 636.
[107] Norton, *The Ordinall of Alchimy*, pp. 99–100.
[108] Ashmole, *Theatrum Chemicum Britannicum*, p. 451: 'Though an Election of a Day or houre be well made, yet will it prove of little advantage unlesse sutably constituted to the scheame of the Nativity, because else it cannot divert that evill which in the Nativity the Planets threatned.' See Ptolemy, *Centum Sententiae, quod Centiloquium dicunt*, in *Claudii Ptolemaei Omnia, quae extant, Opera* (Basel, 1541), pp. 500–4, at p. 500.
[109] Ashmole, *Theatrum Chemicum Britannicum*, p. 450.

Boyle, father of English chemistry, into the secrets of alchemy. Starkey's chart can be found in John Gadbury's *Collectio Geniturarum: or, A Collection of Nativities*, published ten years after Ashmole's *Theatrum Chemicum Britannicum* in 1662. It is not a particularly extensive interpretation, but Gadbury does write:

> Dr George Starkey, born anno 1628, June 9. 11h. 33.m. A.M. Sub Lat. 52.d. This our Native, is not onely an accomplished Scholar and Physitian, but a very excellent Chymist; as his several Books on that subject already published, &c. do sufficiently prove. To denote which Astrologically, he hath Virgo, the most ingenious Signe of the whole Zodiack, ascending the Horoscope, and Mercury Lord thereof, is in Gemini...[110]

We also have the nativity of Gadbury's friend, the self-proclaimed Rosicrucian John Heydon (1629–1667):

> John Heydon, Chemist and Astrologer, married widow of Nicholas Culpeper; We have here the geniture of a person of an uncommon desire for searching into the occult mysteries of nature, and for obtaining of knowledge of all the curious arts. This propensity is very aptly described by the great strength of Mercury, lord of the ascendant [Gemini], in his exaltation [Virgo], in sextile to Jupiter, and in trine aspect of the Sun, whereby the mental endowments of the native far excelled the common lot of mankind.[111]

So, it would appear that anyone wishing to work with the alchemists' Philosophical Mercury should have the planet

[110] John Gadbury, *Collectio Geniturarum: Or, A Collection of Nativities* (London, 1662), p. 130. See also John Partridge, *Defectio Geniturarum: Being an Essay toward the Reviving and Proving the True Old Principles of Astrology* (London, 1697), p. 310. On Starkey, see William R. Newman, *Gehennical Fire: The Lives of George Starkey, An American Alchemist in the Scientific Revolution* (Chicago: University of Chicago Press, 2003) and William R. Newman and Lawrence M. Principe, *Alchemy Tried in the Fire: Starkey, Boyle, and the Fate of Helmontian Chymistry* (Chicago: University of Chicago Press, 2002).

[111] Ebenezer Sibley, *A new and complete illustration of the celestial science of Astrology* (London, 1822), p. 880; p. 872 for Heydon's Geniture. On Heydon, see Donald R. Dickson, *The Tessera of Antilia: Utopian Brotherhoods & Secret Societies in the Early Seventeenth Century* (Leiden: Brill, 1998), p. 215f.

Mercury prominent in his nativity.[112] Both Starkey and Heydon have it double-strength, the former with Virgo Ascendant and Mercury in Gemini, the latter with Gemini Ascendant and Mercury in Virgo. The differing ascendants may explain the essentially practical laboratory approach of Starkey and the more speculative approach of Heydon. Heydon's rather idiosyncratic mixture of alchemy, astrology, and geomancy can be found in *Elhavareuna or the English Physitians Tutor in the Astrobolismes of Mettals* (1665), which includes a variant of one of the laboratory pictures from Norton's *Ordinall* along with his collage-like renditions of alchemical engravings and 'astromantic' charts for various elixirs.[113]

This essay began with one of the first Arabic astrological texts to mention alchemy, in a book on 'Interrogations'. Let our final example of the relationship between the two arts be an early modern interrogation in William Lilly's *Christian Astrology* (1647). A client asked Lilly to draw up a chart to discover 'Whether he should obtain the Philosopher's Stone? or, that Elixar by which such wonders are performed?' Although the rising sign was Virgo, as it was in Starkey's nativity, here the Lord of the ascendant, Mercury, is conjunct Saturn and square Mars. As the ascendant and its Lord signify the querent, and Mercury has bad aspects to the two malefic planets in astrology, Lilly's advice was 'to decline his further progresse upon that subject', partly due to errors in his choice of materials and composition, partly due to indications of unreliable assistants, but mainly because 'where Mercury is corrupted, there the fancy or imaginative part is imbecill'.[114] As I am beginning to feel a certain sympathy with Lilly's querent, perhaps this is a good place to bring this essay to its conclusion.

---

[112] See Lilly, *Christian Astrology*, p. 627: 'Of Mercury when he is Lord of the Profession'; p. 628 'if [Mars] have any testimony with [Sol] and [Mercury], he inclines to Chimistry'.

[113] John Heydon, *Elhavareuna or the English Physitians Tutor in the Astrobolismes of Mettals* (London, 1665), p. 158, shows a laboratory scene based on Norton's *Ordinall* (Ashmole, *TCB*, p. 102). See, for example, p. 166, the 'figure' (chart) for the Elixir of Iron.

[114] Lilly, *Christian Astrology*, p. 442–44. See also p. 429: 'If one shall profit by his Knowledge, be in what kind it will; Chymistry, Chyrurgery, &c or if he be perfect'.

CONCLUSION

Even a focus on Paracelsian writings can bring up conflicting approaches. Iconoclast that he was, Paracelsus rejected much traditional astrology in favour of his own 'astrosophy'.[115] Joachim Telle rightly highlights contrasting Paracelsian representations of 'Paracelsus,' citing the pseudepigraphical *Coelum philosophorum*, which sounds much more like pseudo-Geber in its insistence that 'There is no use in calculating or in knowing how the 12 signs of the zodiac and the seven planets move and act. There's no need to worry about knowing at what time, on what day or at what hour such or such planet is good or bad: for that neither adds or removes anything, neither helps nor prevents anything in the natural art of alchemy'.[116]

As I hope has been shown, there is quite a lot of evidence to suggest that more than a few alchemists also took account of astrology, considering it just as natural to relate the heavens to their artificial laboratory microcosm as physicians did to the human microcosm. Different individuals placed more or less emphasis on how important a familiarity with astrology was to their handling of metals and minerals, their choice of vessels, or the timing of their processes. This being said, there does not appear to have been a great unanimity in either their theories or practices. At times the associations and correspondences make perfect sense; at others there is only a sense of mystification. I would prefer to assume that this is due to my own ignorance rather than to their sheer flights of fancy.

---

[115] Walter Pagel, *Paracelsus: An Introduction to Philosophical Medicine in the Era of the Renaissance* (Basel: S. Karger, 1958), pp. 65ff, 72.

[116] Telle, 'Astrologie et alchimie au XVIe siècle', p. 191. Telle also provides instances from works accepted as genuine, such as *De mineralibus*, where Paracelsus rejects traditional astrology. Similar discriminations can be observed in well-known followers, including Croll, *Philosophy Reformed & Improved*, p. 30, and Dorn, *Congeries Paracelsicae*, pp. 557–58.

PART THREE:

Celestial Symbolism in the Contemporary World

KATHARINE MALTWOOD, H. P. BLAVATSKY AND THE ORIGINS OF THE GLASTONBURY LANDSCAPE ZODIAC

Anthony Thorley

ABSTRACT: A Landscape Zodiac is a representation of the zodiacal signs and symbolism found in the texture and morphology of rural and urban landscape which broadly mirrors the constellations of the ecliptic. Are these mysteries of the British landscape ancient constructions or a modern enthusiasm? Are they simply effusions of a heated creative imagination and psychological projection, or do they have a more secure status in some dimension of imaginal reality? This paper outlines the origins and history of the discovery of the Glastonbury Zodiac, the first of over forty that have been described in different parts of the UK in the last seventy years. The Glastonbury Zodiac was allegedly first described by the artist Katharine Maltwood (1878–1961) in the 1920s, but this paper will trace an earlier origin reaching back into Theosophical writings of Madame Blavatsky which strongly suggests that, not withstanding their rich inherent symbolism of the astrological signs as evidenced by correlated place-names, myths and legends, and compelling synchronicities, landscape zodiacs are a modern phenomenon with no certain documentary or historical evidence for their existence before the 1880s.

Over the last eighty years the subject of landscape zodiacs has attracted minimal attention from the Academy and it is only recently through my own research that have I been able to systematically examine this phenomenon in a more academic context. In this account, I intend to explore some of the issues which arise from my research into the origins of the concept of the landscape zodiac and the specific role of Katharine Maltwood as the first person to describe a zodiac in the English landscape.

But first, it is necessary to ask what *is* a landscape zodiac, or more intuitively perhaps, remembering that the zodiac is usually in the sky, what on *earth* is a landscape zodiac? Unsurprisingly, there is no formal definition available and I suspect that clarifying a definition will preoccupy me in my unfolding research, but here is a preliminary attempt.

> A landscape zodiac is a representation of the zodiacal signs and symbolism found in the texture and morphology of rural and urban landscape which broadly mirrors the zodiac in the constellations of the ecliptic.

I might follow that definition with an equally broad fact sheet about Landscape Zodiacs in order to have some idea of the scope of the field from the outset. The most well known landscape zodiac is at Glastonbury in Somerset, England.[1] It is approximately circular and is about eleven miles in diameter, while some of its zodiacal figures or effigies are several miles across. The boundary of each zodiacal figure is demarcated by natural features (rivers, cliffs, contours of hills and boundaries of woods) and man-made or altered features (roads, paths and field boundaries), and the zodiacal constellations as seen on the planisphere of the northern sky may broadly correlate with the zodiacal effigies on the ground. Over forty similar zodiacs, each with their own characteristics, have been described in the United Kingdom, and there are others in Europe and elsewhere.[2] However, whilst advocates of landscape zodiacs offer no consensus or clear explanation for their origins or construction, they commonly describe the following characteristics: individual zodiacal effigies have place-names often developed over many centuries which, it is claimed, strongly correlate with the name or the characteristics of the effigy on which they are found; similarly, there are rich elements of myth, legend and history, including contemporary events, which correlate with specific zodiacal signs; personal experiences relating to landscape zodiacs include synchronicities, perceptual distortions in time, divinatory experiences and a sense of the numinous more commonly found in descriptions of sacred place.[3]

The landscape zodiac phenomenon has attracted the interest of New Age spirituality and Earth Mysteries enthusiasts, but has been almost totally neglected by the

[1] M. Caine, *The Glastonbury Zodiac: Key to the Mysteries of Britain* (Kingston: Mary Caine, 1978).

[2] P. Heselton, 'A Provisional List of Terrestrial Zodiacs in Britain', *Terrestrial Zodiacs Journal* 1(1989): pp. 30–31.

[3] For a useful discussion of these characteristics see Yuri Leitch, ed., *Signs and Secrets of the Glastonbury Zodiac: An Anthology from the Maltwood Moot* (Glastonbury: Aeon Publications, 2013).

Academy[4].

Although there is no clear explanation for the origins of landscape zodiacs, in the sense that there is no clear indication as to who or what might be behind their construction, a clearer picture however, emerges from various accounts of their recognition or so-called discovery. The most classic material regarding recognition or discovery is to be found in the case of the Glastonbury Zodiac. This was first discovered in the last years of the First World War (around 1917) by a remarkable and intriguing Englishwoman, Katharine Emma Maltwood (1878–1961).[5]

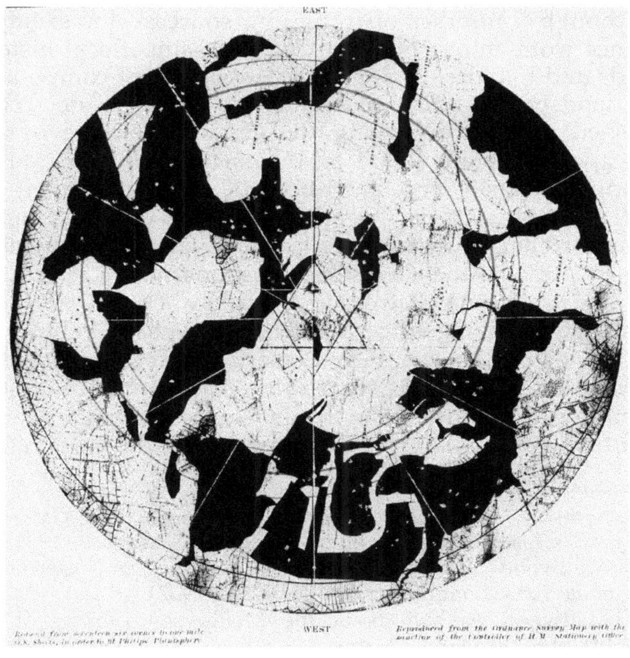

Fig. 8.1: Katharine Maltwood's 1935 Glastonbury Zodiac from A Guide to Glastonbury's Temple of the Stars.

[4] See for example, A. Ivakhiv, *Claiming Sacred Ground: Pilgrims and Politics at Glastonbury and Sedona* (Bloomington: Indiana University Press, 2001), pp.112–13; P. Rahtz and L. Watts, *Glastonbury: Myth and Archaeology* (Stroud: Tempus, 2003), p. 63.

[5] R. A. Brown, *Katharine Emma Maltwood, Artist: 1878–1961* (Victoria: Sono Nis Press, 1981).

As a recognised artist, sculptor and mystical scholar, Katharine Maltwood made no public statement about her discovery of the Glastonbury Zodiac before 1935, when she anonymously published *A Guide to Glastonbury's Temple of the Stars*.[6] Two years later, using the name K. E. Maltwood, she broke her anonymity and in 1937 published the *Air View Supplement* to her *Guide*.[7] My own research into the Maltwood Archive has established that Maltwood first discovered the Zodiac almost twenty years before in 1917, but in this present account I want to focus on the origins of the basic idea of a landscape zodiac and what range of factors influenced Katharine Maltwood in her discovery.[8]

Although a number of intriguing sources—for example, evidence from stories of West Country saints, local history, legends and folklore—point to the possibility of components of a landscape zodiac in Somerset, there is no actual published account regarding the certain existence of any landscape zodiac anywhere on the earth before 1888, when Madame Blavatsky first described the phenomenon in *The Secret Doctrine*.[9] Thus, in the absence of any earlier descriptions before 1888, the concept of a landscape zodiac as a whole entity has to be considered a modern idea.[10]

In 1877, Helena Petrovna Blavatsky (1831–1891), the indefatigable driving force and primary energy behind the founding of the Theosophical Society in 1875, presented to the world *Isis Unveiled*, her two-volume account of the role of ancient mystery traditions in the development—past, present and future—of mankind.[11] In this vast work, there are only a

---

[6] Anonymous (Katharine Emma Maltwood), *A Guide to Glastonbury's Temple of the Stars* (London: John M. Watkins, 1935).

[7] K. E. Maltwood, *Air View Supplement to a Guide to Glastonbury's Temple of the Stars* (London: John M. Watkins, 1937).

[8] The Maltwood Archive is at the McPherson Library of the University of Victoria, B.C., Canada.

[9] H. P. Blavatsky, *The Secret Doctrine: The Synthesis of Science, Religion and Philosophy* (1888; repr. Pasadena, California: Theosophical University Press, 1974).

[10] The concept of the landscape zodiac may be a modern idea, but many of the zodiacs themselves have convincing features which suggest centuries of coherent development before their modern recognition. The basis for this accumulating coherence and evidence for an apparent early presence of zodiacal features in the landscape are issues being addressed by the author in his current postgraduate research.

[11] H. P. Blavatsky, *Isis Unveiled: A Master-key of the Mysteries of*

few references to astrology and the zodiac, and Blavatsky tends to take a rather understated and reserved position on these subjects, as if she privately acknowledges their universal importance but is also holding the most inner mysteries and significance of these areas close to her chest: the enduring esoteric!

In her wide-ranging and relentless text, Blavatsky pursues a complex model of the twelve signs of the zodiac, heavily influenced by the 1870 treatise on the Rosicrucians by Hargreave Jennings, noting his statement that regarding the zodiac: 'the whole theory requires a key of explanation to render it intelligible; which key is only darkly referred to as possible, but refused absolutely, by these extraordinary men, as not possible to be disclosed'.[12] To which Blavatsky tantalisingly notes: 'The said key must be turned *seven* times before the whole system is divulged. We will give it but *one* turn, and thereby allow the profane one glimpse into the mystery. Happy he, who understands the whole!'[13]

The 'one glimpse' involves an exploration of an esoteric representation of the zodiac known as Ezekiel's Wheel, in which the first six signs, Aries to Virgo, represent the ascending line of the macrocosm of creation into increasing spirituality. Libra represents the balancing sign of Man, and the remaining five signs, Scorpio to Pisces, represent the descending line of the microcosm into increasing terrestrial materiality.[14] The tension between the forces of ascending spirituality and descending and more grounded materiality are crossed at the pivotal centre-point of man, represented as Libra the balance, and 'in the subjective as well as in the objective worlds, they are the two powers which through their eternal conflict keep the universe of spirit and matter in harmony'.[15] Thus

> the *balance* is there, ever sensitive at the intersection point. It regulates the actions of the two combatants, and the combined effort of both, causes planets and 'living souls' to pursue a double diagonal line in their revolution through

---

Ancient and Modern Science and Theology (1877; repr. New York: J. W. Bouton, 1887).

[12] H. Jennings, *The Rosicrucians: Their Rites and Mysteries* (1870; repr., London: George Routledge, 1887), p. 72.

[13] Blavatsky, *Isis Unveiled,* Vol. 2, p. 461.

[14] Ibid., p. 463.

[15] Ibid.

> Zodiac and Life; and thus preserving strict harmony, in visible and invisible heaven and earth, the forced unity of the two reconciles spirit and matter. [16]

In *Isis Unveiled*, this is Blavatsky's strongest and most developed statement as to how, through astrological and zodiacal forces both spiritual and heavenly, material and earthly matters are in close and inseparable relationship.

Another aspect of Blavatsky's glimpse into the ancient function of the zodiac is to acknowledge the key role of cycles of transformation in the esoteric teaching of ancient peoples. She describes the zodiac as: 'This stupendous conception, the ancients synthesized for the instruction of the common people, into a single pictorial design—the Zodiac, or celestial belt'. [17]

It is also most significant that Blavatsky identified this notion of ancient teaching not only in the heavens but in addition at a specific locality on the earth: part of ancient greater Asiatic India, from where Blavatsky believed all the wisdom teachings originated—seen by her literally as the 'cradle of humanity'. She wrote:

> Tradition says [that] there was a vast inland sea, which extended over Middle Asia, north of the proud Himalayan range, and its western prolongation. An island, which for its unparalleled beauty had no rival in the world, was inhabited by the last remnant of the race which preceded ours. This race could live with equal ease in water, air, or fire, for it had an unlimited control over the elements. These were the 'Sons of God'; not those who saw the daughters of men, but the real *Elohim*....It was they who imparted Nature's most weird secrets to men, and revealed to them the ineffable, and now *lost* 'word'. [18]

According to Blavatsky, this 'word' has travelled all around the earth and still resonates in the hearts of privileged men, and that 'the hierophants of all the Sacerdotal Colleges were aware of the existence of this island, but the "word" was only known to the...chief lord of every college, and was passed on to his successor only at the moment of death'.[19]

And where is that great inland sea today? Blavatsky tells

---

[16] Ibid.

[17] Ibid., p. 456.

[18] Blavatsky, *Isis Unveiled,* Vol.1, p. 589.

[19] Ibid., p. 590.

us that it is now where the mysterious Gobi desert lies:

> Around no other locality, not even Peru, hangs so many traditions as around the Gobi Desert....Beneath the surface are said to lie such wealth in gold, jewels, statuary, arms, utensils and all that indicates civilization....Occasionally some of the hidden treasures are uncovered, but not a native dare touch them, for the whole district is under the ban of a mighty spell...[a] prehistoric people awaiting the day when the revolution of cyclic periods shall again cause their story to be known for the instruction of mankind.[20]

So here in *Isis Unveiled* Blavatsky has set out two potent foci, two dynamic theatres of ancient wisdom, each enabling the passing of esoteric teachings from the Sons of God to the hierophants and chiefs of the mystery schools, and so on to the common man. One is the zodiac, the source of eternal balance and the mystical conduit between the spiritual and the material, seen as heaven and earth, and the means of instruction of the mysteries to the common man. The other is a beautiful island in a vast inland sea now washed over and hidden by the sands of the Gobi Desert: the location of the *Elohim,* and a place of great wisdom teaching which spread through the hierophants to all the man's great civilisations on earth.

We have here the principle of the dynamic zodiac, and we also have the principle of a specific locality, a landscape of wisdom teaching. Only the final stage, a formal connection to provide a landscape zodiac, is lacking. Thus, in *Isis Unveiled* Blavatsky has set out all the pieces of the jigsaw which will be taken up in her next book, *The Secret Doctrine* (1888), and laid out as the first explicit description of the landscape zodiac.

Let us see how this transition takes place. In her preface to *The Secret Doctrine,* now appearing some ten years after *Isis Unveiled,* Blavatsky tells us that her new work was originally meant only to be a revision of the older text. But the material had so expanded and the treatment so developed that only about twenty of the pages of the original had survived into the new book.[21] And what is the great change that has taken place between these two vast accounts? She gives us a hint in her Introduction, as she clarifies the nature of the 'Secret

[20] Ibid., p. 589.
[21] Ibid., p. vii.

Archaic Doctrine' when she returns to earmark the zodiacal theme.

> Speaking of the keys to the Zodiacal mysteries as being almost lost to the world, it was remarked by the writer in *Isis Unveiled* some ten years ago that: 'The said key must be turned *seven* times before the whole system is divulged. We will give it but *one* turn, and thereby allow the profane one glimpse into the mystery. Happy he, who understands the whole!' The same may be said of the whole Esoteric system. One turn of the key, and no more, was given in *Isis*. Much more is explained in these volumes.[22]

What we find is that in *The Secret Doctrine* the significance of astrology and the zodiac is a much more prominent and confidently presented theme. Indeed it is a major issue, for as Blavatsky tells us:

> The Secret Doctrine teaches that every event of universal importance, such as geological cataclysms at the end of one race and the beginning of a new one, involving a great change in mankind, spiritual, moral and physical—is pre-cogitated and preconcerted, so to say, in the sidereal regions of our planetary system. Astrology is built wholly upon this mystic and intimate connection between the heavenly bodies and mankind; and it is one of the great secrets of Initiation and Occult mysteries.[23]

The gradual move between her two great works to increasingly emphasise the role of astrology is further evidenced in an article in the 1881 Theosophist magazine. Here Blavatsky sets out a strong and thoughtful case for both astrology and numerology, a theme further consolidated in the *Secret Doctrine* as a whole chapter: 'The Zodiac and Antiquity'.[24]

She also returns in *The Secret Doctrine* to the tradition of the inland sea and the beautiful island, going over the exact paragraphs which appeared in *Isis Unveiled*.[25] But now, in a key section on page 502, the connection between sky and

---

[22] Ibid., p. xxxviii.

[23] Blavatsky, *Secret Doctrine,* Vol. 2, p. 500.

[24] H. P. Blavatsky, 'Stars and Numbers', *The Theosophist,* June, 1881 at http://www.blavatsky.net/...StarsAndNumbers.html (accessed 25th May 2010); Blavatsky, *Secret Doctrine,* Vol. 1. pp. 647-668.

[25] Blavatsky, *Secret Doctrine,* Vol. 2, p. 220.

land, zodiac and landscape, is at last made. First she invokes the traditional maxim of Hermetic teaching:

> As above so below. Sidereal phenomena, and the behaviour of the celestial bodies in the heavens, were taken as a model, and the plan was carried out below, on earth. Thus, space, in its abstract sense, was called 'the realm of divine knowledge', and by the *Chaldees* or Initiates *Ab Soo*, the habitat (or Father, *i.e.*, the source) of knowledge, because it is in space that dwell the intelligent Powers which *invisibly* rule the Universe.
>
> In the same manner and on the plan of the Zodiac in the *upper* Ocean or the heavens, a certain realm on Earth, an inland sea, was consecrated and called 'the Abyss of Learning'; twelve centres on it in the shape of twelve small islands representing the Zodiacal signs...and were the abodes of twelve Hierophants and masters of wisdom. This 'sea of knowledge' or learning remained for ages there, where now stretches the Shamo or Gobi desert.
>
> It existed until the last great glacial period, when a local cataclysm, which swept the waters south and west and so formed the present great desolate desert, left only a certain oasis, with a lake and one island in the middle of it, as a relic of the *Zodiacal Ring* on Earth.[26]

The 'Zodiacal Ring on Earth' was the source of the sea of knowledge transmitted to mankind, but the *primary* source for the Ring itself was the celestial zodiac, the 'realm of divine knowledge' and those intelligent powers which invisibly rule the universe. Blavatsky is establishing that landscape zodiacs are very special places on Earth, places where great mysteries and universal truths are studied and imparted to all mankind. Blavatsky next identifies that one of the places on Earth where there are other zodiacs is in Britain. She describes Egyptian priest initiates journeying in a north-westerly direction through the Straits of Gibraltar and then overland to Great Britain. This journey predates the separation from the continent of Britain into an island.[27] She writes that their purpose was 'supervising the building of *menhirs* and dolmens, of colossal Zodiacs in stone, and places of sepulchre to serve as receptacles for the ashes of generations to come'.[28]

---

[26] Ibid., pp. 502–3.
[27] Ibid., p. 750.
[28] Ibid.

And who made these cyclopean structures, asks Blavatsky, such as Stonehenge and Carnac? Not the druids, she observes, as they were only the heirs to the cyclopean lore left to them by generations of mighty builders, magicians both good and bad and indeed the giants of old.[29] And if the reader is left in any doubt, Blavatsky tells us that: 'In Cornwall and in ancient Britain the traditions of these giants are, on the other hand excessively common; they are said to live down to the time of King Arthur'.[30] Here now we have in Blavatsky's writings of 1888 all the components of the landscape zodiac at Glastonbury taken up by Katharine Maltwood by 1917: zodiacs in Britain, giants and King Arthur.

So now let us leave Madame Blavatsky's rich contribution of zodiacal lore and turn to examine Katharine Maltwood's quest into the Glastonbury Zodiac and see if we can develop a narrative of discovery. What is clear about Maltwood's own first discoveries of effigies in the land around Glastonbury is that initially they were totally unrelated to Blavatsky's material and unrelated to the idea of a zodiac in the landscape. This is a very fundamental and important point in understanding the origins of the Glastonbury Zodiac.

Maltwood was particularly interested in Arthurian romance, and she tells us that in the Welsh story of *Owein and Lunet* there is a lion which kills a man-eating giant.[31] Indeed, Maltwood admits that the lion, as a heraldic and mythic beast, is 'an integral part of Arthurian romance'.[32]

At some stage, possibly before the First World War, Maltwood obtained a copy of Sebastian Evans' new translation of the thirteenth-century Arthurian romance: the *High History of the Holy Graal*, first published in 1898.[33] She comments that in the *High History*, the lion is several times represented as being killed by Arthur's knights, 'but like the giants, it lives to fight another day'.[34] Potent symbols in her imagination, lions and giants were emerging as insistent images in her creative world.

---

[29] Ibid., p. 754.
[30] Ibid.
[31] Maltwood, *A Guide*, p. 16.
[32] Ibid.
[33] Evans, S., *A High History of the Holy Graal* (London: J. M. Dent, 1898).
[34] Maltwood, *A Guide*, p. 16.

Evans' *High History* seems to have made a big impression on Maltwood. In the last paragraphs of the book, it is strongly inferred that the Latin of the early thirteenth-century manuscript was 'taken in the Isle of Avalon, in a holy house of religion that standeth at the head of the Moors Adventurous, where King Arthur and Queen Guenievre lie'.[35] This holy house is unequivocally Glastonbury Abbey, and the implication to be drawn may be that the author of the romance was a Glastonbury monk. The *High History* had a direct link with Somerset, and as she read the convoluted and obscure stories in the romance of Arthur and his knights going about their adventures, Maltwood came to realise that the landscape on which they were taking place was the local landscape of the Somerset Levels, much of it around Glastonbury itself. How she came to this decision or opinion we do not know,[36] but it seems to have led her to begin to study the one-inch Ordnance Survey maps and possibly to make visits to the Somerset countryside to see for herself.[37] She explains that

> [b]y noting the places within this area where the Knights appear to meet one another, and knowing the time it would take to ride there from the last encounter, it is possible to gradually make a complete itinerary and map of the Quest disregarding, of course, fabulous distances which obviously refer to the starry sky.[38]

She must have pored over maps (there are over one hundred Ordnance Survey maps in the Maltwood Archive)[39], always mindful of the presence in the legends of giants and lions.

---

[35] Ibid., p. 4.

[36] A useful reality-check on this process has been provided by Paul Weston. See Paul Weston, *Mysterium Artorius: Arthurian Grail Glastonbury Studies, An Introduction and Invocation* (Glastonbury: Avalon Aeon, 2007), pp. 106–9.

[37] As a sculptress and artist, Katharine Maltwood had a strongly-developed sensation function coupled with her interest in mystical symbolism and so would have had a natural eye to detect form and meaning in landscape morphology and topography. This must have assisted the likelihood of her detecting effigies with symbolic meaning. It is noteworthy how many other people who discover landscape zodiacs are artists or designers, e.g. Mary Caine for the Kingston Zodiac, Nigel Ayres for the Bodmin Zodiac etc.

[38] Maltwood, *A Guide*, p. 8.

[39] Brown, *Katharine Emma Maltwood, Artist*, p. 62, note 52.

Then one day, with lions particularly in mind, she tells us '[i]t was whilst pondering on the characteristics of this fabulous creature, with Bartholomew's map, Sheet 34, open, that the Cary River was seen to take the outline of a lion, with the ancient capital of Somerset "a load of nine hundred rocks between the two paws"'.[40]

This was a pivotal moment in Maltwood's life, almost a kind of epiphany, and she never forgot it. Writing much later in 1944 in her book *The Enchantments of Britain* she returns to that magical moment.

> I shall never forget my utter amazement when the truth dawned on me, that the outline of a lion was drawn by the curves of the Cary river below the old capital town of Somerset. So that was the origin of the legendary lion that I had been questing! A nature effigy, and a god of sun-worshippers! Leo of the Zodiac....Obviously, if the lion was a nature effigy then the dragon, griffon, and the giants etc., must be likewise; this was the most thrilling moment of my discovery.[41]

So we learn from *The Enchantments of Britain* that Maltwood is clearly certain that the lion of Arthurian adventure is also Leo of the Zodiac. But how did she make that connection, or rather make the transition from finding effigies linked to the Arthurian romance, such as lions and giants, and realise that they were signs of a zodiac outlined on the ground?

In her published material, she never clearly tells us, but the story seems to have been known by that third lady of landscape zodiac studies (after Maltwood and Mary Caine) and master of so much local lore about the Glastonbury Zodiac, the late Elizabeth Leader (1908–1998). Writing in 1969, Leader tells us that after Maltwood had found a lion and a giant, 'A friend suggested to her that she might have chanced upon one of the great zodiacs or star patterns that had been rumoured as existing. These were supposed to be in the Andes, Britain and in the Gobi Desert'.[42] This appears to be an unmistakeable reference to Madame Blavatsky's contribution in *The Secret Doctrine*.

---

[40] Maltwood, *A Guide*, p. 16.

[41] K. E. Maltwood, *The Enchantments of Britain* (Victoria: Victoria Publishing Company, 1944), p. 81.

[42] E. Leader, 'The Somerset Zodiac: Myths and Legends', p. 8, in *Glastonbury: A Study in Patterns*, ed. M. Williams (London: R.I.L.K.O., 1969).

In an unpublished manuscript entitled 'King Solomon's Table', dating probably from the late 1940s, Maltwood confirms this story.

> It was when I was making a map...that I found the lion effigy, outlined by the Cary River and considerable earthworks. Had it not been for a chance remark by a London friend, that it might be 'Leo' of the Zodiac, I should have never followed up the quest; but once on the trail there was no looking back, and the circle of nature effigies were gradually discovered.[43]

We have to presume that the London friend was the same person who told her about the Blavatsky descriptions of great zodiacs in the landscape of Britain, and again presume that in following up the quest, sometime around 1917, Maltwood read for herself Blavatsky's writings on the zodiac.[44] We know that she was interested in Theosophical ideas, had Theosophical friends and acquaintances, and later published in Theosophical Journals, but never actually became a Theosophist herself.[45] We can only suspect that it was about this time (probably around 1917) that she wrote down her reaction to Blavatsky's quotation from page 502, showing that she was genuinely intrigued and impressed.

However, it was not until 1944 in *The Enchantments of Britain* that Maltwood first sets out in public her reaction to Blavatsky's idea of the zodiac on the land. Here she comments:

> But the interesting fact remains, that this time Blavatsky traces records of the building of a British Zodiac, as coming from Egypt, whereas the 'Zodiacal Ring on Earth' which she

---

[43] K. E. Maltwood, Draft Article, 'King Solomon's Table', undated. Maltwood Archive, McPherson Library, University of Victoria, B.C., Canada.

[44] Sometime in 1977 the astrologer and author, Judy Hall, was in conversation with the mystic and esotericist, Christine Hartley (1897–1985). Hartley was a pupil of Dion Fortune (Violet Firth) (1890–1946) and knew others in Glastonbury esoteric circles during the 1930s. Hartley told Judy Hall that the London friend (still unidentified at present) 'put her [Maltwood] in the right direction', the implication being that Maltwood had no sense or awareness of a zodiacal basis to the effigies before this important encounter. (Judy Hall, 2012. Pers. Comm.).

[45] Brown, *Katharine Emma Maltwood, Artist,* p. 37.

> describes on Page 502, with twelve little islands set in an inland sea, hails from Asia. The latter exactly pictures the 'Temple of the Stars', in England, for the signs are set in what are called on the map, the Sea Moors. The low-lying land has now been more or less drained but in Winter time the little hills appear as islands emerging from the flooded Sea Moors. Probably the 'Initiates' came via Egypt, to lay out this Zodiac.[46]

Clearly Maltwood is commandeering the Blavatsky Gobi Zodiac, with its twelve little islands, and finding that it rests very comfortably in the Somerset countryside. She also tellingly comments on the role of giants, quoting from a Blavatsky section of the *Secret Doctrine* entitled 'Are Giants a Fiction?'

> Of Giants who were 'in the earth in those days' of old, the Bible alone has spoken to the wise men of the West, the Zodiac being the solitary witness called upon to corroborate the statement in the persons of Atlas and Orion, whose mighty shoulders are said to support the world.[47]

This passage is most significant because Maltwood first identified her own giant of the Glastonbury Zodiac as Orion, and even though this becomes the zodiacal sign of Gemini, she maintained a key role for this figure, always referring to it as a giant and how it mystically related to the deeper mysteries of the whole Zodiac.[48]

But to return to Maltwood's first reaction to Blavatsky's writings on the earthly zodiac, we can examine a small square of handwritten paper which I found among her writings in the Maltwood Archive in Canada.[49] The paper is undated, but for some reason has been kept over the years, as if for Maltwood herself it carried a special importance. There is little doubt that it represents her first reaction to Blavatsky's exposition. In the following rough note, she writes:

---

[46] Maltwood, *Enchantments*, p. 32.

[47] Ibid.

[48] Maltwood, *A Guide*, pp. 103–9.

[49] The author would like to thank the staff of the McPherson Library, University of Victoria, B.C., Canada for their enthusiasm and patient assistance when he worked on the Maltwood Archive in May 2007.

> Blavatsky says p. 502 in Vol II The Secret Doctrine 'As above so below. Sidereal phenomena & the behaviour of the celestial bodies in the heavens, were taken as a model, & the plan was carried out below on earth. In the same manner'... [text then copied exactly as quoted above].... This 'sea of knowledge' or learning remained for ages there, (where now stretches the Shamo or Gobi desert). ? Where did she get this from? Was it the Tiamat legend about Bel and Merodach. It is not in the B[ritish] M[useum] translation I have.[50]

Maltwood is clearly intrigued as to the origins of Blavatsky's source in order to propose a land-based zodiac and cannot match it from her own considerable knowledge and material. Note how she has underlined the key phrase 'and the plan was carried out below on earth', and writes with double question marks: '? Where did she get this from?'

We may well echo Maltwood's basic question and ask ourselves: from where *did* Madame Blavatsky get this? Without any precedent that I am aware of in mystical or historical literature, the zodiac on Earth in the landscape seems to be a completely original idea, possibly not deriving even from Blavatsky's own creative thinking or alleged wisdom teaching sources but simply arising as a visionary insight or through some form of private channelling from a spirit or guide.[51] At present we have to say that the ultimate source of Blavatsky's inspiration, beyond her own creative stream, is unknown.

So what are we to conclude from this brief examination of the creative minds of these two formidably talented women? Blavatsky seems to have first proposed the idea of a land-based zodiac reflecting the twelve signs around 1888, but there is no description or suggestion that her twelve islands in the Gobi Desert were recognisably shaped in the form of specific zodiacal signs: the lion, the bull, the crab, etc. They are simply featureless islands. However, around 1917, whilst independently pursuing specific images of the Arthurian romance of the *High History of the Holy Graal*, Maltwood finds the precise shapes of a lion, a giant and other creatures in the texture and maps of the Somerset countryside. Then, after encountering Blavatsky's twelve islands through a chance meeting with a London friend, Maltwood goes on to dress

---

[50] K. E. Maltwood, Undated Paper, Maltwood Archive, McPherson Library, University of Victoria, B.C., Canada.

[51] H. P. Blavatsky was a well-known psychic and medium.

her effigies with symbolic form amalgamating Arthurian and zodiacal images into a common, united mythos and so creating the first actually identified landscape zodiac, or as she herself always termed it, 'Glastonbury's Temple of the Stars'.

In this interweaving of creative process between these two extraordinary women, the idea of the landscape zodiac had materialised from visionary speculation into actual geographical recognition and the stage was now set for almost a hundred years of further discovery, profound development of role and function, and many new accounts of distinctive experience and rich description.[52]

[52] See for example a wide compilation in Leitch, ed., *Signs and Secrets of the Glastonbury Zodiac: An Anthology from the Maltwood Moot*.

READING THE FUTURE IN THE LANDSCAPE: ZANADROANDRENA LAND, CENTRAL EAST MADAGASCAR

Christel Mattheeuws

ABSTRACT: In the article I show that one can compare reading the landscape of Zanadroandrena land with reading a horoscope because the people make it like a horoscope, 'locating an individual in a unique position within their world while at the same time providing a personal link to the cosmos by a ritual iteration of the moment of birth or origin of the world' (Brady, 2011). The Zanadroandrena do not only practice astrology but they live it and live in it. But, one can only see how deeply astrology is embedded in their life, creations and history if one learns to look with astrological eyes. Astrological images of the land are not ready to be grasped with the naked eye but are yet invisible grids of the landscape that appear only with an astrological vision. This astrological vision is linked with the vision of deep history. The deeper one digs into the past, the more apparent the future becomes. Reading the future can only be done with the past in front of you. In Zanadroandrena land the *imago mundi* is infinitely present in all the Zanadroandrena creations, never finished being read.

In the heart of Bezanozano, about 170 km northeast of the capital Antananarivo, Zanadroandrena land stretches out over 20 km^2 along the river Mangoro. Marshy valleys and small to moderate rivers etch the highlands of the area forming the foundations of wet rice fields and dry prairies, manioc fields and human settlements. The Zanadroandrena are part of the Zafinadriamamy, descendants of Andriamamilazabe, who migrated from the area of Ankazondandy in North Merina and their communal tomb to settle in West Bezanozano. It is in the life paths of these immigrants that villages, tombs, fields and other human places have appeared, disappeared and changed in the landscape. They emerged as the result of fruitful relationships with the land that 'gives' and of which the dead had become part by building new tombs. Astrological politics used in the foundation of places during the rituals of blessing weave the invisible texture of the land that people inhabit. The astrological paths that form the texture can be read as the formative gestures of this land, not only indicating the movements of rooting and growth, but also their direction and intention. Growing in the Malagasy language (*maniry*) means literally, searching for the future, seeking to become (something else). The study of astrological practice reveals the texture of the world as a relational meshwork. Nothing happens randomly since whatever moves also pulls and pushes. The art of living is about learning what to do in such a world in order to make one's movements and actions life-generative. By following the life-paths of people that have directions, we can learn what is life-generative in their view and in their actions, and we can also learn all the possibilities, eventualities and choices that are part of the perpetual search for good directions, so that whatever is pulled or pushed into motion in one's movements end up in a tightening of the meshwork. In this article I will elaborate this theme and I will show how we can read people's future in the landscape by explaining into which direction their past and present creations of tombs, villages, ritual centres and rice fields pull and push people's history. It is like reading a horoscope, the astrologer's map, as an *imago mundi* locating an individual in a unique position within their world while at the same time providing a personal link to the cosmos by a ritual iteration of the moment of birth or origin of the world. This constant activity of returning to the moment of origin places the map in circular time, which Eliade defined as sacred time, 'indefinitely recoverable, indefinitely repeatable'.[1] I argue that in the Zanadroandrena case one can read the landscape as a horoscope only because the Zandroandrena has made it like a horoscope. In Zandroandrena land, there is a very thin line between the sacred and the profane. It only depends on how far the Zandroandrena community live their life according cosmo*eco*logical principles.

---

[1] Mircea Eliade, *The Sacred and the Profane: The Nature of Religion*, trans. and ed. Willard R. Trask (London: Harcourt, 1987), p. 69, in B. Brady, 'The Horoscope as an Imago Mundi: Rethinking the Nature of the Astrologer's Map', in *Astrologies: Plurality and Diversity*, ed. Nicholas Campion and Liz Greene (Ceredigion, Wales: Sophia Centre Press), pp. 47–61.

LEARNING TO READ THE ZANADROANDRENA SACRED LANDSCAPE

On 7 December 1999 I arrived in the field with hardly more than a box of pencils and a few empty notebooks as my anthropological survival kit to gather information for my doctoral dissertation. I did not, however, consider my arrival as a blank starting point. I had myself and the village people along whose life paths I was planning to walk. These paths, I hoped, would reveal some points of interest which I would then study in more detail. Curiously, nothing specific in people's life caught my special attention. For almost two years I participated in various planned and unforeseen events or activities as they happened to take place. In so doing, I became rather puzzled by the particular movements, developments, directions, stagnations and growths of my own work and started to wonder by what exactly it was moved and shaped. Back in Europe in September 2001 I identified my research as bearing testimony to the Zanadroandrena life into which I had been pitched and which I wanted to keep central to the dissertation. My focus on astrological practices in my ethnography came only after one year of going through my field notes again and again. Finally, only in the process of reading and writing anthropology to define what astrology is and does, I became fully aware that my 'eye-with-ness' of Zanadroandrena life had made me into a different kind of anthropologist. My eyes had become Zanadroandrena eyes that in their vision of Zanadroandrena land recognised the kind of world described by Ingold.[2] In his disagreement with the view that human societies appear on the surface of a biologically preformed world, Ingold wanted to show that 'if "the perceiver" is always moving, learning and becoming, then the same must surely also go for the world he or she perceives and inhabits'.[3] He has been in the first place preoccupied with the form-giving processes shaping the

---

[2] Christel Mattheeuws, 'Towards an Anthropology in Life: The Astrological Architecture of Zanadroandrena Land in West Bezanozano, Madagascar' (PhD Thesis, University of Aberdeen, 2008).

[3] Tim Ingold, *The Perception of the Environment: Essays on Livelihood, Dwelling and Skill* (London: Routledge, 2000), pp. 166–68.

world in which humans also participate, and of which my ethnography became an example.

In order to understand my experiences of transformation as 'a perceiver' one can turn Ingold's argument around and say that *'perceivers' in a relational world that is constantly moving and changing must also change, transform and become.* This is the argument of Goethe (1749–1832), who dedicated his life not only to writing poetry but also to studying nature. Goethe said of his own approach to nature: 'that which has been formed is immediately transformed again, and if we would attain in some measure a living comprehension of nature, we must ourselves remain as mobile and plastic as the example nature presents to us'.[4] He has formulated a specific research methodology that studies nature from within, through the process of developing new organs of perception. Doing research along Goethean lines means treading a path of conscious development, of believing that we are the most adequate instrument to know the world on the one hand, and always wanting to make it a better instrument on the other. Since the researcher develops and refines his or her capacities of seeing in the research process, we can only learn Goethean methodology by doing it.[5] My own research path started with my arrival in the village without study topics and questions, yet with myself and my intention to work along the people's life-paths. This starting point alone already carries the core of the Goethean method. It is empirical as it stays all the time with the phenomena we study, even, as we shall see, in the process of understanding, reading and writing. The researcher studies with a 'receptive will', an openness toward the phenomena we want to know better.[6] We must constantly remain aware of our temptations to formulate preconceived opinions and conceptions that we tend to impose on the phenomena rather than letting the

---

[4] Margaret Colquhoun, 'An Exploration in the Use of Goethean Science as a Methodology for Landscape Assessment: the Pishwanton Project', *Agriculture, Ecosystems, and Environment* 63 (1997): p. 149.

[5] Craig Holdredge, 'Doing Goethean Science', *Janus Head* 8, no. 1 (2005): pp. 27–52. Jochen Bockemühl, 'Transformations in the Foliage Leaves of the Higher Plants', in *Goethe's Way of Science: A Phenomenology of Nature*, ed. D.Seamon and A. Zajonc (New York: State University of New York Press, 1998), pp. 115–28.

[6] Henri Bortoft, *The Wholeness of Nature: Goethe's Way of Science* (Hudson, NY: Lindisfarne Press, 1996), p. 242.

phenomena speak for themselves. The whole learning process is a conversation with nature (including people), demanding that we listen first before we question or answer. We don't go to nature, but let nature come to us and let it gradually reveal itself.[7]

The process starts with 'exact sense perception' and 'exact sensorial imagination or visualisation'.[8] The first sees (handles, tastes, hears, smells) the phenomena as accurately as possible in the here and now and as substantial form. This detailed observation should be carried out in as many manifestations and circumstances as possible over time and in different places. The second way of perceiving is actively remembering these observations. Goethean scientists state that, in order to experience more vividly what is observed, to get more connected with the phenomena and seeing the becoming or relationships, we must go through the process of visualisation of the seen. This has been my experience when I went through the observations in my notes in order to write them out in very detailed accounts. The most vital moments were those in which I felt my work was growing and gaining direction in the movements and relations that this process of visualisation revealed. This kind of experience gave me the feeling that I could go on with my observations anew and afresh. Many anthropologists in the field follow this method of engaged, twofold seeing. However, errors and biases often creep in when anthropologists return home and begin to render as rather passive the field experiences of observing and learning what others do, in contrast to the active thinking and exchange that goes on with colleagues. Ethnography is often an account of how other people become enclosed in culture, while anthropology, reflecting on this, looks for explanations beyond the observed facts.[9]

The Goethean method shows that perception and visualisation are very active and demands a serious involvement of the observer in the field. But in order to let the phenomena reveal themselves as meaningful, instead of looking for explanations beyond them, we must become

---

[7] Holdredge, 'Doing Goethean Science', pp. 27–52.

[8] Isis Brook, 'Goethean Science as a Way to Read Landscape', *Landscape Research* 3, no. 1 (1998): pp. 51–69. Colquhoun, 'An Exploration in the Use of Goethean Science', pp. 145–57.

[9] See also Tim Ingold, 'Ethnography is not Anthropology', *Proceedings of the British Academy* 154 (2008): pp. 69–92.

rather passive in a third step, which Goethe called 'knowing in beholding'. In my case, since I had become aware of the relational developments of my research with occurrences during fieldwork, I made every possible effort to bring the field along with me when I returned to Europe. Due to the physical distance, this was only possible in the shape of my notes and in my memories. This intensified the active sensorial imagination, making the connection with the field so strong that it started to live in me and gradually revealed itself, directing the movements of my thoughts. Although I had already become aware in the field of something that moved my research in specific directions, it is only away from the field that I have understood it as having been grasped by the movements of Zanadroandrena life that shaped Zanadroandrena land. This kind of consciousness appears as what Colquhoun calls 'inspiration', which can give the experience a spiritual dimension. In Hoffmann's words, I had become conscious of the formative gestures in Zanadroandrena land. A gesture is not merely a movement, but an indication of intention or direction.[10] Several months later I would relate the formative gestures with astrology, an idea that popped up in 'a sudden intuitive moment'. In a moment, I saw the fullness of my research, and I also saw the astrological gestures in every single place or activity, as well as in the complete land. According to Goethe, I had finally seen the 'primal phenomenon', the 'theory' of Zanadroandrena land. This marked the beginning of my active reading and writing period, during which I became aware that my eyes had changed. Goethe explains that 'every new object, clearly seen, opens up a new organ of perception in us'. According to Goethe, the eye owes its existence to the light. The eye sees light because it has developed in it.[11] Likewise, I could see astrology because my eyes had been developed in an astrological attitude to life, making me intimately entwined with Zanadroandrena land. Struck by the similarities I found in Ingold's studies concerning form-giving processes, I would identify the astrological attitude to life later, as a continuous generation of life with specific

---

[10] Nigel Hoffmann, 'The Unity of Science and Art: Goethean Phenomenology as a New Ecological Discipline', in D.Seamon and A. Zajonc, eds., *Goethe's Way of Science*, p. 134.

[11] Jeremy Naydler, ed., *Goethe on Science: A Selection of Goethe's Writings* (Edinburgh: Floris Books, 1996).

directions (gestures, characters) shaping the land and all its *in*habitants. In the finest reading of the happenings in Zanadroandrena land, I would learn that only the destiny that was given to the land as a whole and to the Zanadroandrena community was the 'primal gesture'. It was in the light of this destiny that the appearances of everything else in the land should be understood, as the whole that is reflected in all its parts, giving Zanadroandrena land its particular character in comparison to other people's land. The Zanadroandrena call this destiny the spirit or knot of their land from which their customs and habits are derived.

ZANADROANDRENA LAND AS AN *IMAGO MUNDI*

The spirit or knot of Zanadroandrena land lie at its very heart in the middle of the vital river Bemahia and is called Ambatomiskana, Where the Rock Obstructs (nr. 1 on fig. 9.1). This stone is the house of the *kalanoro* Kelilavavolo, the Small Girl with the Long Hair.[12] One of the ancestors of the Zanadroandrena clan, Rainivao, is said to have disappeared there for three days dwelling in the caves beneath the water learning all the secrets of the land that Kelilavavolo protects. Until today, the Zanadroandrena fetch rainwater from a small hole on top of the stone when they go to their ritual centre in the month of March on their birth destiny Alasaty. Although people don't say it with so many words, I believe that Zanadroandrena land in all directions is basically a manifestation linked to the generative forces of this place. It is the only place that has no destiny because it is the axis mundi of Zanadroandrena land, the knot where all the destinies of the land join. It is the centre where people go to adjust their destiny in times of severe crisis. And it is also the centre from which their cosmo*eco*logical perception of the land and the creation of their ancestral village in particular have to be understood. The Zandroandrena perceive their land in the shape of a quadrant formed by the four cardinal directions surrounded by twelve destinies equivalent to our Zodiac. Their dwelling place where you find the villages, tombs, ritual centres, elevated stones and the source of their main rice field system has emerged in the southeast, the destiny Alasaty (Lion). This destiny was given by their Great

---

[12] A *kalanoro* is an invisible female creature.

Ancestor and astrologer Ralaisikidy when he erected the ritual centre of Zanadroandrena land in Voara.

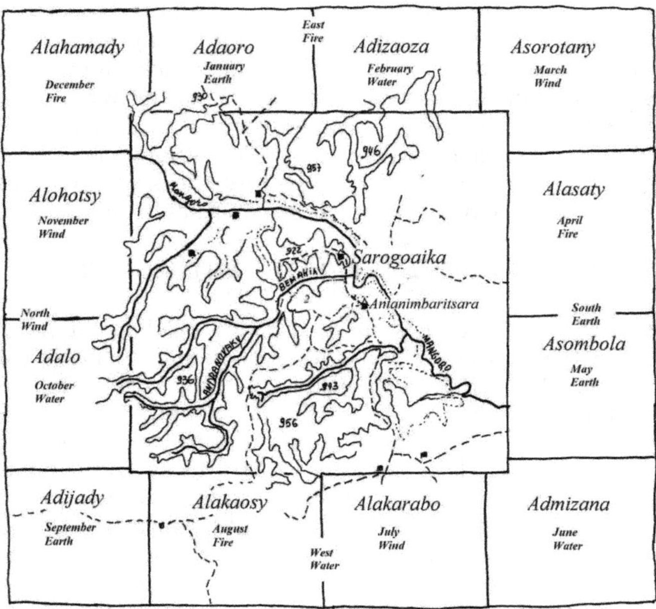

Fig. 9.1: An astrological view of Zanadroandrena land

Voara was the first but yet abandoned settlement of the Zanadroandrena. When I visited the Zanadroandrena in March 2003 the central wooden post of this centre had collapsed. The healer in charge of the wellbeing of this place told his family that they could renew the post on the first of April of the same year when the new moon brought the destiny of Alasaty. 'If they did not, it would take them about thirty years before they could plan another renewal', he said. The year 2003 was namely a year out of the ordinary because in that year the sun, moon and stars coincided. The Zanadroandrena follow an astrological moon calendar without intercalary months to bring the calendar in synchronisation with the solar year. This means that the astrological months brought by new moon drift backward yearly with eleven to twelve days away from their starting point to reach a new beginning only in about thirty-three years. The Zanadroandrena use the new moon that brings

Alasaty at the end of March or the beginning of April as reference for a new era to start, which happened in 2003. They do not recall the date of the foundation of the ritual centre, however, but I estimate it to have been in the 1870s. Based on computer simulation, on Saturday 21 March 1874 new moon brought the destiny Alasaty. If my estimation is correct then the ritual centre, carrying the destiny Alasaty, was elevated on that day in the generative Alasaty forces of the *axis mundi* Ambatomisakana.

March/April[13] is the time when a cold wind pulls the forces down, giving the meaning of Alasaty as the simmering fire and of the ritual centre called *jiro*, 'a light in the darkness'.

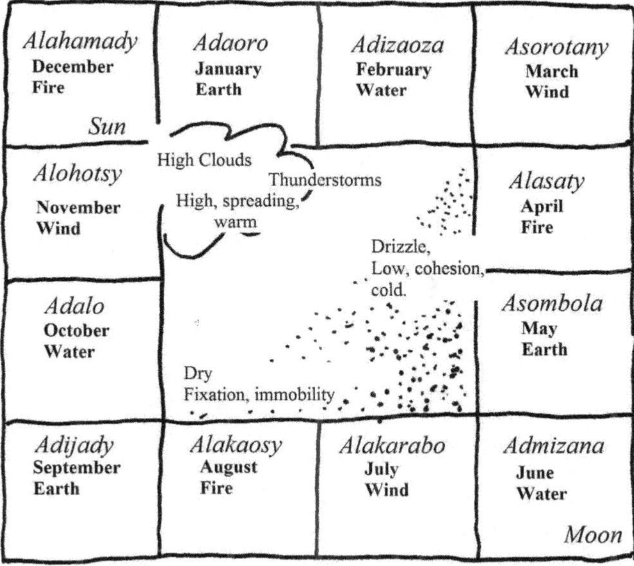

Fig. 9.2: Destinies and weather.

I consider the origin of Zanadroandrena land in West Bezanozano in the destiny Alasaty as the particular way the Zanadroandrena participate in the world's becoming comparable to one's birth chart beautifully described by

[13] March is the beginning of autumn in the southern hemisphere.

Brady as 'an encyclopaedia of the self, written at birth but never finished being read'.[14] The life paths of the Zanadroandrena would look differently if the clan was given a different destiny. The Zanadroandrena do not like a hierarchical society like the Merina kingdom in the seventeenth and eighteenth century where the kings and queens reigned in the destiny Alahamady, 'the sun in the zenith high above the earth'. They look for more horizontal, equalizing and incorporating sensations, more moderate kinds of hierarchies based on the system of precedence.

They resist the introduction of ideologies supporting an external elitist minority by stimulating their autonomy and fetching forces from close to the earth, by distributing sacrifices among their extended family members, land and ancestors, by fostering the relationships with spirits and creatures dwelling in their land, and by fixing the realisations of their dead in their land. To achieve this they wait until the sky is hanging low above the earth like the morning fog of autumn or the low compact clouds bringing drizzling rain during wintertime (see fig. 9.2). Asorotany, the mother of Alasaty, is the leading destiny of the earth, 'the horizon where day and night appear and disappear', 'the source of everything growing on earth', 'the owner of the *jiny*'.[15] Asorotany is the destiny where earth and sky, sun and moon encounter each other on dark moon and when the moon will gradually take over the leadership from the summer sun. The encounter between heaven and earth must turn the primordial dead or killing nature of the earth into 'life-giving forces', *hasina*. The name of the ritual centre 'light in the darkness' refers to what becomes visible in the shade or darkness of the heaven-earth encounter, namely the caring forces dwelling close to the earth like *Vazimba*,[16] spirits of deceased healers, and the spirits of the ancestral medicines revealing matters about the well-being of Zanadroandrena land and its people through the possession of a Zanadroandrena healer during the rituals at the centre in March (fig. 9.3).

---

[14] Brady, 'The Horoscope as an Imago Mundi', p. 59.

[15] *Jiny* are forces in the realm of death. It is often used in contrast with life-giving forces, *hasina*.

[16] Vazimba are spirits of people from the past who dwelled in the region long before Zanadroandrena settlement.

Fig. 9.3: The ritual centre (jiro) *of Voara*

Fig. 9.4: Ritual of Blessing

Fig. 9.5: Ritual of Blessing

It is in the light of the first settlement of Voara with its ritual centre that we have to understand the life paths of the Zanadroandrena community in which villages, tombs, fields and other human places have appeared, disappeared and changed in the landscape pulling and pushing the history of the community. Every start of an activity or building must be blessed. The astrological day on which this happens gives the birth destiny of the activity or building. The act of blessing calls all the life-giving beings (*hasina*) such as Andriamanitra (God), Zanahary (His Children), the Vazimba and the Ancestors breathing upon the event through blowing rum in the direction of the sacred northeast Alahamady, 'the position of the morning sun in wintertime' by the three eldest people present to make it alive. While the horoscope is an image of the sky showing the relative positions of the sun and other stars, the moon and other planets, the quadrangular shape of Zanadroandrena land or the setting of every ritual of blessing represent an image of the land drawn by the relative positions and actions of the people present, construction works, artefacts, animals, and even 'empty spaces' for the invisible forces (figs. 9.4 and 9.5). Although every image shows a different constellation according to the kind of event and place, they are all *whole*, carrying the full range of forces needed to get a fruitful result, not only of the ritual but of the whole life path of activity or building started. This life path, having a destiny, will depend on the relational movement and changes of the meshwork created during the blessing as well as on past and new elements that will be encountered. Fruitful and life-giving relationships are built with all the features of the world, which grow, help and protect each other. A dangerous attack such as sorcery, thunder, flood, or robbery that can be stopped in traditional ways by using the magic of ancestral medicines confirms that 'the place where the Zanadroandrena live is still sacred' as people say in these circumstances. I interpret this as meaning that the land and the community are still developed along with cosmo*eco*logical principles unlocking *hasina*.

Plaiting the Past into the Future

Human settlement and world (trans)formation are a commitment towards the land and its inhabitants. This is embodied in the Adaoro path in the village that people use when they go out in the open highlands searching for new challenges. Adaoro is the destiny of 'a bull with horns directed forward always ready to fight', expressing the efforts and struggles to fulfil one's life. Life is not something given or happening of itself. One must gain it during life time, tightening the meshwork of relations as the fundament for further generations. Becoming an ancestor within the community is the result of successes accomplished during one's life. Someone who failed will be forgotten. Someone who achieved something will be remembered after death and will be requested to come during rituals of blessing. In an agricultural community as the one of the Zanadroandrena, achievements come in the first place from the good relationships with the land. A tomb is a place where humans become part of the landscape, their dissolving bodies adding to the fertility of the land and their spirit joining the blessing *hasina*. Yet, the living community is first grown by the land through their fruitful relationship with it, before they ever built a tomb in a new settlement. In other words, a tomb is in the first place the result of the land that gives, and not the point of departure for ownership of the land. Building a tomb is part of the process of belonging to the land, and of the continually growing and changing embodiment of the forces of that land.

Customs of the ancestors, traditions (*fomban'drazana*) are, in their most original form, the maintained, grown and renewed paths of ancestors, which they have walked during their lifetimes. All the greatest ancestors that are remembered by the living lie at the basis or start of something. They continue to live on as creatures that transcend the lifetimes of the living in the directions of both the past and the future. They are not however permanent; their temporality is that of bones, some kinds of hardwood and stones. Those bodies also have their lifetimes and degenerate, slowly but surely. They can be conditionally renewed, but that depends on the living, their affinity with the land and the beings inhabiting the land, and also on the strength and will to survive of the ancestors themselves. The footsteps of other ancestors, of people less remarkable in their lifetimes, are much less permanent and might be part of less permanent traditions. Following the traditions means, for the living, to follow the footsteps of the (extra-ordinary) ancestors, to live and move in their footsteps, and according to each person's capacity and rhythm, to acquire the skills to grow into the ancestor's realizations and also maintaining and growing them. By growing into the ancestor's realizations, relations already established in an engagement with the world continue, but can also be developed and deepened. Moreover new relationships can evolve through encounters in and with the world. Living human beings walk the paths of, and with, the ancestors, seeking to make their land and community more stable in its destiny. The architecture of the land, established by the ancestors, works on and changes at the same time in and with the people who follow the traditions. The land of the present Zanadroandrena is the land established by Ralaisikidy in an engagement with the world, and yet is already a metamorphosis of that land. The Zanadroandrena astrological architecture of the world is a continuous process that generates past and future in the present as a centre of generative and morphogenetic processes. The *fomban'drazana* is the poiēsis of Zanadroandrena life, the particular way by which the Zanadroandrena, alive or as ancestors, are and move in the world wherein their land appears as an endless creative unfolding of a set of relations in which subjects appear and take on their particular forms, each in relation with the other.

Life is carried on in upward-and-downward movements. Like in a plaiting work where 'male and female strings'[17] are twisted into all kind of patterns and shapes according to the directions into which the strands are interwoven. Or like the relational travels of the winds, clouds, planets and stars. It also moves back and forth taking up what happened 'before' (*aloha*) along with what will happen 'after' (*aoriana*). The Zanadroandrena share with other Malagasy the idea that the future comes from behind with the past in front of you.[18] Water is considered to be the best vehicle of *hasina* since it overcomes all obstacles in upward and downward, and back and forward movements, gaining force during its course. Ambatomisakana, 'Where the Rock Obstructs', is the place where the living body of marshes is transformed into the life-giving body of rice fields. The water used in circumcision rituals for the young boys to become fully member of the community is Adaoro water (flood), fetched from the place

---

[17] The 'male and female strings' are the un-doubled leaves or stalks of the reeds that are fetched in the marshes.

[18] Oyvind Dahl, 'When the Future Comes from Behind: Malagasy and other Time Concepts and some Consequences for Communication', *International Journal for Intercultural Relations* 19, no. 2 (1995): pp. 197–209.

where the vital river Bemahia flows into the main river Mangoro. This spot is a place where the water is turning around for a while before it continues its journey. Making a *fotitra*, practically means to give an astrological destiny to a place considered as both an origin and a source, like the ritual centre giving life to the villages and Ambatomisakana giving life to the rice fields. The destiny of a *fotitra* is always double: one destiny has to make strong, and the other has to make mobile. In other words, a good *fotitra* always gives both, strong entanglements and new offshoots. It makes life possible as a basis for growing wealth like children, rice and cattle.

There can be also ruptures and breaches in someone's path however. The present settlement of the Zanadroandrena is an example. The Zafimamy migrations were caused by a social and political crisis in Imerina. An astrological reading of the history of Zanadroandrena settlement and growth in the present land shows that the present community has been born of 'a bare hill' (destiny Alakaosy/Adijady/Adalo in the west, spring time in August, September and October) making their place 'a forested hill' (Asorotany/Alasaty in the southeast). According to the local history of people from Amoridrano, a settlement 20km southwest of Zanadro-andrena land, Ralaisikidy and other members of the extended family would have lived there for a while before continuing their migration in an eastward direction towards different places. 'Once they had left, they never came back to get some ancestors to take along with them to their new settlement', informants told me. Although I gathered some evidence that the family members who arrived in Voara kept contact with Imerina in the beginning, they have eventually left this past behind.

The west is the directional place of the dead earth like 'the fertile soil of the lowlands' (Alakarabo), 'the hard soil of the highlands, wood, stones, bones, the grains of rice' (Alakaosy), or the 'immobile rotten water of a pool' (Adalo). Spring is the driest and hottest season of the year without wind and rain, causing the dead of the highland vegetation making the dry soil appear. Adijady is the destiny of division: separating a new year from the old one, present from past, like the umbilical cord that is cut after the birth of a newborn, the dead that are separated from the living during funerals, or the rice germ that dies after it germinated.

The rupture created in the west has to be overcome so that bodies can remain alive and the past is not separated from the present. People solve the first problem by bringing bodies into a field of nurture and care where people, the land, the ancestors and even the divine principle maintain and grow each other. For example, during the month of October or the beginning of November (Adalo, Alohotsy), Zanadroandrena women go to the grave of the chief *Vazimba* to offer a cock; this ritual is called *tranga taona*, 'the appearance of the year'. While the ritual at the *jiro* in March is meant to close the earth (to stop the urge of vegetative life) and to request the protection of the well-being of people, rice and cattle, the ritual in October is meant to open the earth and to pray for the health of the children and of caring and nurturing agents like mothers, healers and *Vazimba*. The second problem is solved by bringing the past along with the present through the rituals of the *famadihana*, a second burial of the dead so they can become ancestors, and are preferentially held in the months of August, September, and October. This ritual is in particular important when moving corpses from an old tomb to a new one. This did not happen in the new Zanadroandrena settlement. They decided to live a new future in the area that became eventually Zanadroandrena land.

READING THE FUTURE IN ZANADROANDRENA LAND: THE ROLE OF TOMBS

During my stay in 2000 and 2001 among the Zanadroandrena I became aware of something significant going on, whatever it was, because of so many important things happening. On 11 March 2000 (Alasaty of the month Adizaoza),[19] people went for the first time since the healer Ramariavelo died in 1992 to the ritual centre in Voara with a new healer. The same year, on 6 October a delegation of women went for the first time after the death of Ramariavelo to the grave of the Vazimba chef to pray for the health of children and nurturing agents. On 5 May 2000 (Asombola in the month Alasaty) a long process of making two elevated stones for Ranampy

[19] Every astrological moon brings the 12 destinies along, starting with the destiny of the month, and each destiny lasting for 2 or 3 days till the next new moon will bring the following month's destiny.

and his brother started, followed by his famadihana on 28–29 September (Adijady in the month Adijady), on 3 September 2000 (Adalo/Alohotsy [?] in the month Alakoasy) the fundaments of a new village were laid and last but not least, on 23 April 2001 (Alasaty in the month Alasaty) a small delegation from the village Sarogaika went for the first time in public to the ritual centre Analabemazava across the Mangoro. During a short visit in March 2003 I would find the confirmation in the collapse and renewal of the ritual centre that the Zanadroandrena follow a 33 year cycle, the time the astrological moon needs to get back to its place of origin. Therefore, I believe that the events I observed in 2000 and 2001 were part of the strengthening, renewing or rejuvenation of Zanadroandrena rooting and growing forces, in particular of two lineages, the one of Ranampy and the one of Ratsimanosy. I will highlight this through an elaboration of the politics of tomb building. In order to do so, I must enter the oral history of the Zanadraondrena clan which is a complex matter. I will stay as simple as possible and keep the names of villages and people reduced as much as the narrative allows me to.

We have already seen that the first tomb was built by Ralaisikidy at the fringes of his settlement in the destiny Adalo. This tomb was built as a result of the fruitful relationship with the land and its inhabitants that was given the destiny Alasaty, the counterpart of Adalo. Alasaty, 'the simmering fire', and Adalo, 'the Vazimba water or source', form the axis along which the Zanadroandrena forces evolve in relation with their land, reinforced by the rituals at the jiro and at the grave of the Vazimba chef. When two new tombs were built, this grave has been emptied of all the bodies, except the bodies of the female Kialabobo and her daughter,[20] and also the gun of Ralaisikidy to keep the grave protective of the community.

[20] I was not able to trace the identity of Kiala Bobo. It could be a niece of Ralaisikidy.

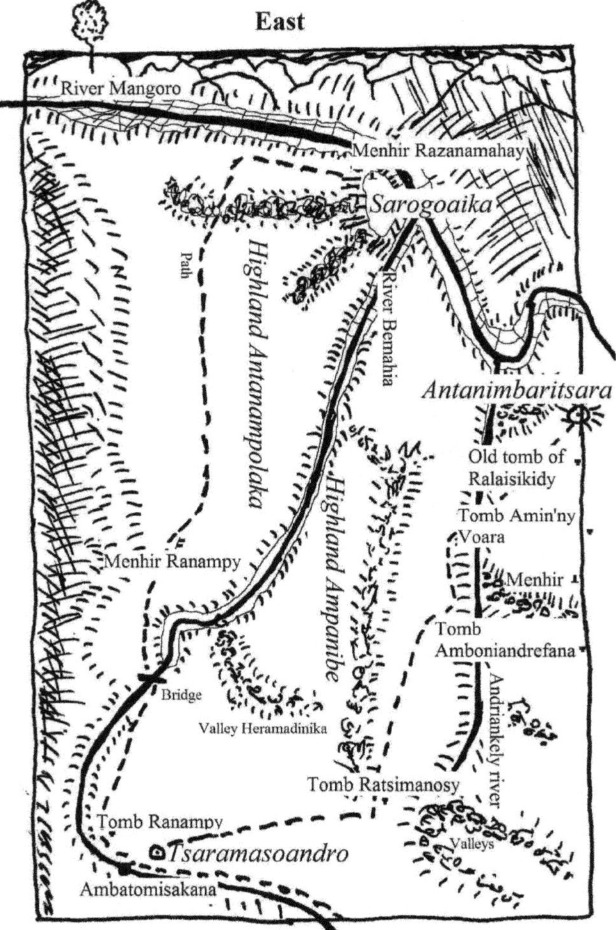

Fig. 9.6: The Southeast of Zanadroandrena land

The bodies of this first grave have been buried in the new tomb Amin'ny Voara, 'On Voara' also called the grave of Ralaisikidy. The second new tomb Ambony Andrefana, 'Up in the West' is called the grave of Rainitanety who was

Ralaisikidy's son.[21] The building of both tombs commenced almost simultaneously around 1921.

The new tomb of Ralaisikidy was built very close to his first tomb on the territory of his village Voara in the destiny Admizana. In this tomb all the Zanadroandrena members are allowed to be buried but in practice this is not the case. Above the entrance is an inscription giving a number of names of people buried in the tomb in two columns: the one on the left refers to the corpses of the 'descendants of a male' buried in the north part of the tomb and the one on the right points to the south part where the 'descendants of a female' are buried. I will not go here into detail because the ancestors of this tomb and the people of the oldest living village Antanimbaritsara connected most closely to it are not directly involved in the reading I will give.

The tomb Ambony Andrefana is built in the destiny Alakarabo on the other side of the Andriankely valley separating Voara from this place. Although the construction was started almost simultaneously with the new tomb of Ralaisikidy, one can say that it is born of the tomb at Voara, reducing the lineages who are allowed to be buried in it. The tomb has been built in two stages over a period of 15 years. It was initiated by Rainivao who is the founding father of the Zanadroandrena lineages living presently in Sarogoaika. After his death the work was taken up by his son and a cousin. It is called the tomb of Rainitanety because it is the grave of 'descendants of females' who do remember Rainitanety as their 'Great Ancestor'. More concrete, the descendants of the daughter and sister of his son Randrianaina I can be buried in Ambony Andrefana. The first person mentioned in the left hand column of the inscriptions above the entrance is Rafarilangita, the father of

[21] In reality, Rainitanety had chosen to be buried alone to show his discontent towards his father who wanted to bury his second slave wife in his own tomb. This has never happened but Rainitanety had commented: *'I prefer to be buried alone than risking to be mixed with filthy forces, including the wish that had polluted the family tomb'*. When people call the second new tomb built in the vicinity of Rainitanety's grave 'the tomb of Rainitanety', it is understood as the tomb of those who consider him as the Great Ancestor of their lineage. Three rice field plots near the axis mundi Ambatomisakana reveal which lineages consider Ralaisikidy, his son Rainitanety or his grandson Andrianaina I as their respective Great Ancestor; See Mattheeuws, 'Towards an Anthropology in Life', pp. 50–57.

Rainivao who was originally a 'descendant of a female' who should have been buried in Voara South. But he became the leading ancestor of the people buried in Ambony Andrefana North as the husband of Randrianaina I's daughter, making them 'descendants of a male'. Rainivao has in this way been able to consolidate his power and autonomy since 'descendants of males' are considered to be rooted more strongly in the land than 'descendants of females' who are the veins through which the Zanadroandrena sacred forces flow (they give birth to the Zanadroandrena healers). However, according to the present perception of the people not all the descendants of Rainivao are 'descendants of a male'. The lineage splits into the 'descendants of a male' led by his son Randriamahay and the 'descendants of a female' led by his daughter Razakabity. The members of her lineage are never considered as being as much rooted in Sarogoaika as the descendants of Randriamahay. Yet, economically and politically, they are far stronger especially through her son Ranampy who is considered as founder of the present village Sarogoaika.[22] He has followed his grandfather's tactic and strengthened the influence of his lineage by building a new grave. He moved Razakabity from Ambony Andrefana to this grave and made her the Great Ancestor. The only line represented in Ambony Andrefana North is the lineage of Randriamahay. Ambony Andrefana South is reserved for the descendants of Randrianaina I's sister. They form the 'descendants of a female' of the tomb. But the south part of the tomb has lost its strongest lineage to a new lineage tomb, the tomb of Ratsimanosy.

Off-branching should not be seen as a separation. For example during the *famadihana* of Ranampy, the lineage members of Razanamahay were included among the 'children of the grave' as being the descendants of the indivisible unity of 'a brother and a sister', respectively Razakabity and Randriamahay. They supported the organising lineage with new shrouds for Razakabity. If a new grave is built, people will always pay attention to ensure that no corpse of a common ancestor of the old grave and new grave will be moved to the new grave.

[22] In the time of Rainivao there were some settlements in Sarogoaika that never developed into a permanent village however.

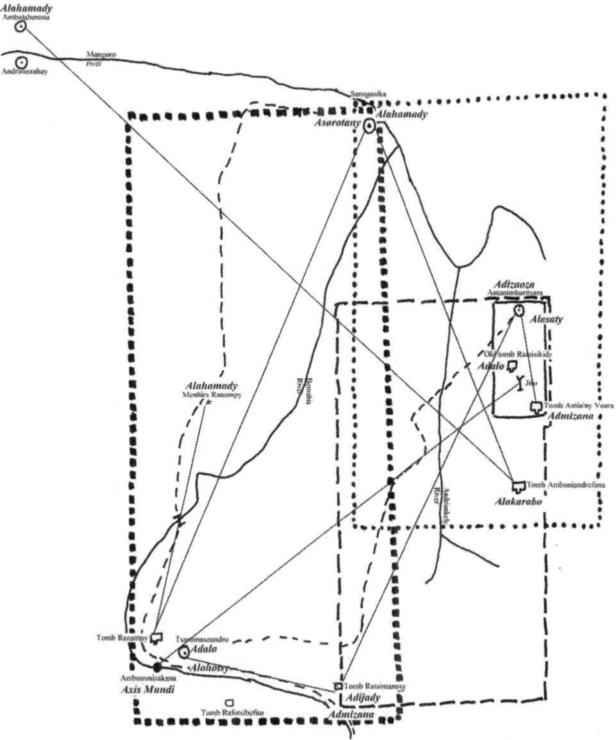

Fig. 9.7: Astrological representation of the relationship between tombs and villages in the southeast of Zanadroandrena land.

In the case of Ranampy it was unthinkable that Rainivao should have been removed to his grave. The 'parent' who stays in the old tomb keeps the members of the old and new grave together. However, the astrological reading of the relationships between tombs and villages only makes sense if the different tombs are seen in their own right revealing different invisible images of the earth as drafted in the representation of figure 9.7. The destinies of the villages and tombs are given in bolt. The arrows show the relevant relations between villages and tombs. For example, the village of Sarogoaika carries the destiny Alahamady in relation to the tomb Ambony Andrefana, and the destiny Asorotany in relation to the tomb of Ranampy. The tomb of

Ratsimanosy is said to carry the destiny Adijady. But this is only in relation to Antanimbaritsara. It becomes Admizana in relation to the new village Tsaramasoandro where most of Ratsimanosy's sons live.
In my view, the shifts of destinies give new meaning and directions to the future generations involved. Based on this and on the significant happenings I have experienced during my visits, I believe that Ranampy as a descendant of Rainivao and Ratsimanosy as a descendant of Randrianaina I's sister become the most recent leaders of Zanadroandrena history. Their offspring might take in it new directions, without denying that more rooted and older lineages also have their role to play. While Zanadroandrena history started in Voara as being the southeast corner of the southeast part of Zanadroandrena land seen from the axis mundi Ambatomisakana, there could be a shift towards the northwest corner of the same area (nr. 2 on fig. 9.1). My reading of the meshwork of this corner is as follows:

> *Antampon'Ambemahia is an example of a place that is getting little by little its own particular history, having started as the manioc field of Razafindrakoto and* ***Ratsimanosy****. 'The village of* ***Rainivao****' is situated northwest of Antampon'Ambemahia. A long plot over the northern length of the highland is now a manioc field of the youngest son of* ***Ranampy****, who inherited it from Razafindrakoto. Northwest of that plot is the lineage grave of Ranampy and his brother, which watches out over the Bemahia River. Southwest of Antampon'Ambemahia, on its highest point, is the empty grave of* ***Rafotsibefito*** *(Antampon'dRafotsibefito),*[23] *looking out over her land of origin Sahamanana, where Rafotsibefito is buried now in the Zafindrenimorona tomb Amboniala (near Morarano Gare). Southeast of the temporary grave of Rafotsibefito is the lineage grave of Ratsimanosy (Antampon'Andriakely), which watches out over the valley of the Andriakely. This river is very closely linked to the village of Antanimbaritsara: it is considered as their river, and not the river of the people of Sarogoaika. Except for the field cultivated by the descendants of Rainivao, the other two thirds of Antampon'Ambemahia has been a manioc field of*

[23] Rafotisbefito was the wife of Rainivao from the Zafindrenimorona clan.

Ratsimanosy. During his life, he had expressed the wish that his descendants should establish a new village on his field. This happened in 2000 on a lower part, northeast of his tomb. This is the village Tsaramasoandro (Nice Sun), so called because of the place being immersed in the nice sunrise and sunset [see fig. 9.8]. Antampon'Ambemahia is to me the origin of the younger life-giving forces that flow through the Zanadroandrena body.

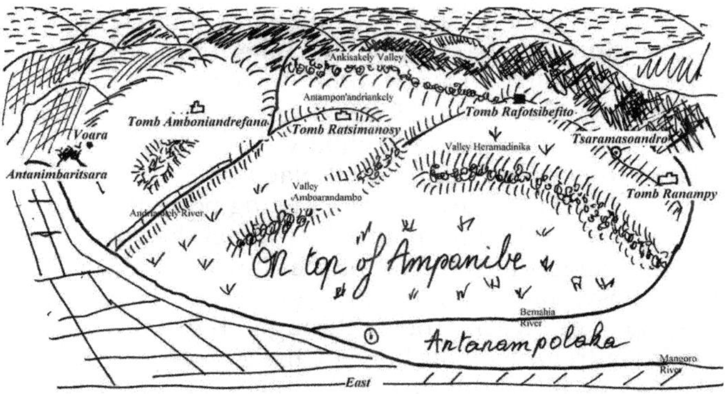

Fig. 9.8: Highland view of Antampon'amBemahia

Rainivao and Ratsimanosy, who have both lived together for a while in 'the village of Rainivao', lie both at the origin of respectively the new villages Sarogoaika and Tsaramasoandro born of the village Antanimbaritsara. Both people were born of 'descendants of female', yet have themselves become Zanadroandrena young male forces by initiating tomb building. While Ratsimanosy is remembered as the direct origin of Tsaramasoandro, Rainivao is remembered particularly through his grandson Ranampy. Ranampy is considered to be the first settler of present Sarogoaika and is also the father of the first Zanadroandrena healer born of that village. The backgrounds of the two people are similar. Both are Zanadroandrena through their mother and consolidated their rooting and realisations by building a tomb. But the politics of doing this are different.

Ratsimanosy chose as a young adult to leave his fraternal village of origin Andranovola and to stay in Zanadroandrena land where he had the right to be through his mother and grandmother. He married Rafako, a sister of Razanamahay who gave birth to Rakotondrasoa, the present healer of the *jiro* of Voara. Then he got remarried to Rasoanantoandro, daughter of Rasoavololona who is buried in Voara, who gave birth to many children. The building of a new grave consolidated his shift from his ancestral village and made Antanimbaritsara his new village of origin by bringing the male ancestors from Andranovola. All his children stay in Zanadroandrena land where they have recently built a new village on a place that he had indicated during his life. My question is, will his future descendants still be taken into consideration as the vehicle of inheritance of 'Zanadroandrena sacrality' that is only 'grown' into bodies of 'descendants of a female'? I am curious to see what happens.

Ranampy was still a little child when his mother returned to Zanadroandrena land, her place of origin, after her husband had passed away. Yet, as being born of another clan since children follow normally their father, he was often said to be less rooted in the land of the Zanadroandrena. Unlike Ratsimanosy, however, he did not try to counteract this perception of weak roots through the marriage of a Zanadroandrena woman born of the Zanadroandrena land, and most of his children left the place searching for other challenges. Only his youngest son stayed in Zanadroandrena land. Relations are yet very strong between the emigrant children and their land of origin. Unlike Ratsimanosy, Ranampy has also not put a male ancestor as head of his tomb, but rather Razakabity, his mother. His history and the history of his descendants accentuate their being 'descendants of a female' in many ways. This is, in my view, important for the flow of the recent off branching of the 'Zanadroandrena sacrality' giving Sarogoaika the opportunity to develop a magico-religious autonomy with respect to its father village Antanimbaritsara. Two sons of Ranampy are the first healers of Sarogoaika, and connected to a ritual centre erected by the daughter of Rainivao at the other side of the Mangoro on Zafindrenimorona territory.[24] Thus, the choice to consolidate his being 'descendant of a female' is maybe not only in this lineage's own interest but of the interest of the well-being of all the inhabitants of Sarogoaika.

---

[24] This ritual centre is shared by healers of the Zanadroandrena clan of Sarogoaika and the Zafindrenimorona clan of Ambodifany who are related as cousins.

CONCLUSION

In the introduction I have compared reading the landscape of Zanadroandrena land with reading a horoscope, locating an individual in a unique position within their world while at the same time providing a personal link to the cosmos by a ritual iteration of the moment of birth or origin of the world. The article became an argument that one can read Zanadroandrena land as a horoscope only because they make it like a horoscope. The Zanadroandrena do not only practice astrology but they live it and live in it. Zanadroandrena astrology is their cosmo*eco*logy understood as a poeisis of life, the continuous generation of Zanadroandrena life and creations. One can only see how deeply astrology is embedded in their life, creations and history if one learns to look with astrological eyes. Astrological images of the land are not ready to be grasped with the naked eye but are yet invisible grids of the landscape that appear only with an astrological vision. This astrological vision is linked with the vision of deep history. The deeper one digs into the past, the more apparent the future becomes. Reading the future can only be done with the past in front of you. In Zanadroandrena land, the *imago mundi* is infinitely present in all the Zanadroandrena creations, never finished being read.

THE CELESTIAL IMAGINARY IN WEIMAR CINEMA

Jennifer Zahrt

ABSTRACT: This article explores celestial symbolism in two films of German silent cinema: Paul Wegener's 1920 *Der Golem, wie er in die Welt kam* (*Golem, How He Came into the World*) and F. W. Murnau's *Faust*.[1] These films emerge out of and respond to the increasing popular appeal of astrology and the occult in Germany at the turn of the twentieth century. Both films are set during the astronomical revolution, that is, in early modern Europe, when the distinctions between astronomy, astrology, and the occult were not yet clearly established. As such, these two films depict figures who access magical powers via reading the sky as symbol—making esoteric knowledge exoteric by its projection on the screen. Through a close reading of scenes from these two films, I'll show how Wegener and Murnau used fiction to foreground a contemporary re-engagement with ways of knowing that hinge on the observation of the sky. In addition, the rich technical production history of these films will be explored to further establish a better understanding of the structural affinities between content and its creation in Weimar Cinema.

Weimar Germany was a hotbed of exploration into astrological and occult themes.[2] By 1920, when actor and director Paul Wegener released his third film about the legendary Golem, Theosophical publishing houses had been printing material about astrology, Theosophy, and the occult

---

[1] Paul Wegener, *Der Golem, wie er in die Welt kam*, (1920), and F. W. Murnau, *Faust, eine deutsche Volksage*, UfA-Studios (1926), respectively. All images in this article are taken from public domain copies of these films.

[2] For example, see Ellic Howe, *Astrology: A Recent History Including the Untold Story of Its Role in World War II* (New York: Walker and Company, 1967), and James Webb, *The Occult Establishment* (La Salle, IL: Open Court Publishing Co., 1976). In the context of film, see Tom Gunning, 'Der frühe Film und das Okkulte', in *Okkultismus und Avantgarde: von Munch bis Mondrian 1900–1915* (Frankfurt: edition tertium, 1995), pp. 558–61.

in German for nearly thirteen years. Interest in these topics only intensified during World War I.[3] Wegener's first two Golem films, which appeared in 1914 and 1917 respectively, were set in Wilhelmine Germany, but after the war he chose to remake this story as a seventeenth century tale. The third film's early modern setting—during a pivotal moment for astronomical progress and the practice of astrology—allows Wegener to create a cinematic experience where the audience is explicitly exposed to and asked to participate in negotiations between myth, symbol, word, and moving image. Six years later, director F. W. Murnau released a highly successful film version of the Faust legend. While his story bases itself mostly on Goethe's rendition, the beginning of the film—the part that concerns us here—deviates significantly from Goethe's classic text, in part I argue, as a response to the occult revival, which by that point was in full swing in Germany. The differing ways these films depict the sky points not only to the advances in cinematic technology in terms of the production of the film images themselves, but also to a cultural shift in Weimar Germany's appreciation of astrology and the occult.

GOLEM

In Berlin, around 8 P.M. on the evening October 29, 1920, just behind the clouds, if there were any, a waning gibbous moon rose in the east and the constellation Cygnus, the swan, hovered overhead. Inside the theatre, the UfA Palast am Zoo, *Der Golem: Wie er in die Welt kam* was about to make its world premiere. The film opens with an establishing shot of the starry night sky above the jagged lines of the buildings in the early modern Jewish ghetto. The sky and the buildings are visible in equal ratio, with the tallest building appearing to nearly touch a constellation of stars shining brighter than all the others, a possible reference to Cygnus, clearly visible in the Northern hemisphere. The observational tower on screen forms the crucial link between the fantastic architecture of the world below and the portent-filled starry sky above.

---

[3] Among the myriad texts on this topic, see especially Carl Christian Bry, *Verkappte Religionen* (Gotha/Stuttgart: Friedrich Andreas Perthes, 1925) and Oscar A. H. Schmitz, *Der Geist der Astrologie* (Munich: Georg Müller, 1922) for two contemporary accounts of this resurgence.

Most discussions of the set of this film concentrate on the style of the buildings and interiors that the film's architect, Hans Poelzig, designed.[4] However, I argue that the sky itself plays an equally important role in both the film's content and its technical production. Observation and interpretation of the stars drives the action of the film.

Wegener conceived the third and final Golem film as a prequel to his first Golem film, yet the general movements of the plot are similar to the first film—the Golem is awakened, falls in love with the Rabbi's daughter (who herself falls in love with a Junker from the nearby castle), wreaks havoc when she is courted by others, and is eventually deactivated by a small child outside the huge gate of the Jewish ghetto.[5] There are significant differences as well.

The film opens at an indeterminate moment set in the early modern Prague of Rudolph II. The Jewish community, living in a ghetto, receives a decree from the Emperor that they must evacuate the realm. The well-respected Rabbi Löw has already gleaned from his astrological interpretations that a threat looms over his community, and he has set in motion a plan to combat it, namely, by bringing the legendary Golem to life to save his people. He succeeds in creating the Golem, first by using ritual magic to invoke the demon Astaroth, who reveals to him the life-giving word, and then by placing a capsule with the life-giving word in the Golem's chest. As long as the capsule is in place the Golem will be animated with life. Once the creature has saved the Jewish people from

---

[4] See, among other examples, Claudia Dillman, 'Die Wirkung der Architektur ist eine magische. Hans Poelzig und der Film', in *Hans Poelzig. Bauten für den Film.* Exhibition catalogue Kinematograph 12 (Frankfurt am Main: Deutsches Filmmuseum, 1997), pp. 20–75; Theodor Heuss, *Hans Poelzig. Bauten und Entwürfe. Das Lebensbild eines deutschen Baumeisters* (Berlin: Wasmuth, 1939); Heike Hambrock, *Hans und Marlene Poelzig. Bauen im Geist des Barock. Architektur-phantasien, Theaterprojekte und moderner Festbau (1916–1926)* (Berlin: Aschenbeck & Holstein, 2005); Sabine Röder, 'Traumspiel mit Kulissen', in *Der dramatische Raum. Hans Poelzig. Malerei Theater Film*, ed. Sabine Röder and Gerhard Storck (Krefeld: Kat. Museum Haus Lange, Haus Esters, 1986), pp. 8–22.

[5] In 1917 Wegener filmed a second Golem film, *Der Golem und die Tänzerin* (*Golem and the Dancer*)—a comedy set in Wilhelmine Germany, wherein the Golem falls in love with a dancer. This film, like his first Golem film, has been lost. It branches off significantly enough from the synergies of the first and third films that I'll refrain from discussing it here.

eviction, it turns on the Rabbi and threatens to destroy the community from within.

The legend of the Golem enjoyed renewed popularity leading up to and during the first World War, and although it has been suggested that Wegener borrowed material from Gustav Meyrink's famous serialised novel *Der Golem* (1913–1914), a contemporary dissertation by Beate Rosenfeld has shown that Wegener's version of the legend contains strains of more than one kind of Golem legend circulating at the time.[6] Rosenfeld proposes that Wegener's mixture of Golem legends was designed to speak to two types of audience members: on the one hand, those who were familiar with the emerging trend in astrology and esotericism would have been stimulated by the references to Rudolphine Prague. On the other hand, simple elements of occult practice were present for the less educated viewers, who could appreciate the magic without knowing historical details about it.[7]

Gershom Scholem's scholarly account of the idea of the Golem traces the tradition of Golem creation myths throughout history.[8] Recently, Moshe Idel has critiqued Scholem's tendency to see the progressive evolution of one specific technique of creating the Golem, a technique that was common to different varieties of European texts. Idel describes many interpretations of the creation of Golem that do not fit into Scholem's conception.[9] Along with Rosenfeld's work, contemporary reviewers of Wegener's film echo Idel's analysis: On November 7, 1920, a reviewer noted in *Der Kinematograph*: 'When considering the plot, initial reminiscences, for example of Meyrink's *Golem*, can be completely eliminated. The figure of the Golem has surfaced several times in the last years, but is based on such little documentary evidence, that a traditional treatment is eliminated. The only thing they all have in common is the

---

[6] Beate Rosenfeld, *Die Golemsage und ihre Verwertung in der deutschen Literatur*, Sprache und Kultur der germanischen und romanischen Völker. B. Germanistische Reihe, vol. 5 (Breslau: H. Priebatsch, 1934).

[7] Rosenfeld, *Die Golemsage*, p. 149.

[8] Gershom Scholem, 'The Idea of the Golem', in Gershom Scholem, *On Kabbalah and its Symbolism*, trans. R. Manheim (New York: Schocken, 1996), pp. 158–204.

[9] Moshe Idel, *Golem: Jewish Magical and Mystical Traditions on the Artificial Anthropoid* (Albany, NY: SUNY Press, 1990), pp. xxii–xxiii.

Jewish milieu'.[10]

One of the diverse forms of the Golem myth that is important for my reading of Wegener's film is Idel's identification of the possibility of an astrological infusion into the Golem literature as early as the thirteenth century.[11] Scholem insisted that Hebrew sources of the creation of artificial men were not connected to astrology but to the combinatorial possibilities of the *Sefer Yetzirah* alone. However, Idel reveals a thirteenth century text, *Sefer ha-Hayyim*, composed by an anonymous author, that proposes a legend of the creation of an artificial man independent of the *Sefer Yetzirah*. This text uses the astrology of Abraham ibn Ezra, and 'the correspondence between higher and lower, the affinity between peculiar elements and supernal entities [...] rather than the magical power inherent in the combinations of Hebrew Letters' to awaken the Golem.[12] In addition, R. Yoḥanan Alemanno (1435/8–c. 1510) in northern Italy, understands Golem creation to be a Jewish counterpart to Ptolemaic astral magic, and he explicitly infuses it with an astrological interpretation wherein the Golem is created both by a specific combination of letters as well as the influx of astral forces into the clay matter.[13]

Around the seventeenth century, the Golem legend enters the realm of Jewish folklore, and innovations begin to appear. This cultural transfer brings us the first examples of the Golem becoming dangerous when it 'escape[s] the control of the human creator'.[14] Further, the mid-seventeenth century tale about R. Eliyahu contains the first instance of the word *'emet* being placed on the Golem's neck rather than attached to or carved into the forehead.[15] The famous Rabbi Löw legend also belongs to the Jewish folklore tradition

---

[10] 'Bei der Betrachtung der Handlung sind zunächst Reminiszenzen, wie z. B. an Meyrinks "Golem" vollständig auszuschalten. Die Gestalt des Golem ist ja in den letzten Jahren verschiedentlich aufgetaucht, fußt aber dokumentarisch auf so geringen Unterlagen, daß eine traditionelle Behandlung ausgeschlossen ist. Gemeinsam ist allen nur das jüdische Milieu'. Reprinted in *Film und Presse*, no. 18, 441.

[11] Idel, *Golem*, pp. 88–91.

[12] Ibid., 89.

[13] Ibid., pp. 169–75, esp. 171–72.

[14] Ibid., p. 261.

[15] Ibid., p. 209.

rather than the magical tradition.[16] By embedding his Golem within astrological discourse, Wegener invokes both the magical and the mystical facets of the legend.

Wegener's insistence on the suitability of the Golem for film carries another valence: muteness.[17] In early versions, the Golem was a mute creature with whom it was nearly impossible to communicate. Early cinema itself had to grapple with techniques to bridge the silence inherent in early film technology. Historically the creation of a Golem seemed to be more a confirmation of the power of the Hebrew language than the specific actions or instrumentality of the Golem doing anything for the Rabbi who created him. The Golem was created for the purpose of experiencing the process of creating it. Idel suggests that the Golem techniques reinforced the 'peculiar power of the Hebrew language' in competing polemics (pagan, Christian, and others) during the last eight hundred years.[18] Kocku von Stuckrad's recent work on the plurality of esotericism in Europe agrees with this hypothesis of identity formation.[19] The double sense of Wegener creating the Golem just to experience the power of creating it, as well as the film's inescapable involvement in questions of German-Jewish identity are illuminated by this facet of the historical Golem material.[20] One of the central problems of the transformation of the Jewish magical practice into cinema is the loss of the specific linguistic, that is, Hebrew, nature of the practice.

---

[16] Ibid., p. 252.

[17] Paul Wegener, 'Die künstlerlischen Möglichkeiten des Films', in Rudolph S. Joseph, *Paul Wegener. Der Regisseur und Schauspieler* (Munich: Photo- u. Filmmuseum im Münchner Stadtmuseum, 1967), p. 6.

[18] Idel, *Golem*, p. 262.

[19] See Kocku von Stuckrad, *Locations of Knowledge in Medieval and Early Modern Europe: Esoteric Discourse and Western Identities* (Leiden & Boston: Brill, 2010), esp. p. 34.

[20] For a reading of the film in its myriad Jewish contexts, see: Omar Bartov, *The 'Jew' in Cinema from* The Golem *to* Don't Touch my Holocaust (Bloomington, IN: Indiana University Press, 2005) on the film in the context of Jews depicted in film history; S. S. Prawer *Between Two Worlds: The Jewish Presence in German and Austrian Film, 1910–1933* (New York: Berghan, 2005), esp. pp. 33–41; and the most recent reading by Noah Isenberg, 'Of Monsters and Magicians: Paul Wegener's *The Golem how he came into the World* (1920)', in *Weimar Cinema: An Essential Guide to Classic Films of the Era*, ed. Noah Isenberg (New York: Columbia University Press, 2009), pp. 33–54.

Linguistic contemplation cedes ground to visual observation.

But Wegener, historically known as a somewhat dominant individual, was not content to leave his personal vision of the Golem to the free play of the literary imagination. He sought to create a film, and for that he needed someone who could help him give the right shape to his vision. He chose Carl Boese, to direct the film, and architect Hans Poelzig to design the set.

Hans Poelzig spent the early part of his architectural career traveling to Austria, the Rheinland, Franconia, and Bavaria. He was impressed by the late gothic architecture in Regensburg and the high baroque of Melk.[21] Poelzig viewed the theatre as an analogue to a modern church-like building, a new temple. In his speech about his design for the Salzburger Festival building, Poelzig emphasized the historical relationship between spectatorship and the night sky. He saw the open-air theatre set as an instrumental component to his architectural vision: 'sky and clouds hover over the audience; they are surrounded by mountains and trees. All of that must be *built* here. A sky must be built over the people in the room, a sky in an architectonic form, which continues out from the stage'.[22] Poelzig places emphasis on the spatial relationships inimical to the theatrical experience.

Poelzig executed this vision in his design for the Großes Schauspielhaus theatre in Berlin. On the ceiling of this theatre, Poelzig hung over 1,200 stalactites in dripping concentric rings. In a gorgeous inversion of Plato's cave, the grotto interior had light bulbs installed on the tips of the stalactites which, when the lights were turned off, could be arranged to look like constellations in the night sky. The claustrophobic interiority of the cave opens to the illusion of communion with the cosmos.[23] The stalactites hid an

[21] Heuss, *Hans Poelzig. Bauten und Entwürfe*, p. 13.

[22] Hans Poelzig, 'Festspielhaus in Salzburg', *Das Kunstblatt* (1921; repr., Nendeln/Liechtenstein: Kraus Reprint, 1978), p. 84: 'über den Zuschauern schweben der Himmel und die Wolken, um sie herum stehen die Berge und die Bäume. All das muß hier *gebaut* werden. Über den Menschen im Raum muß ein Himmel gebaut werden, ein Himmel in architektonischer Form, der sich über die Bühne fortsetzt'. Emphasis in original.

[23] 'Im Schauspielhaus konnten die an den 1200 Rabitzzapfen befestigten Glühbirnen zu Sternbildern geschaltet werden'. Wolfgang Pehnt, 'Wille zur Ausdruck. Zu Leben und Werk Hans Poelzigs', in *HANS POELZIG Architekt Lehrer Künstler. 1869–1936*,

organically embedded instrumental effect: they also worked as acoustic buffers to dampen echoes in the great building.[24]

Later, in 1927, Poelzig also designed the Deli-Kino in Breslau to look as though it had the night sky in its ceiling. He not only arranged the lights in concentric elliptical forms on the domed ceiling, but he also stylized them as ten pointed stars.[25] Unlike the wild stalactites of the Großes Schauspielhaus, which reference the drippy cave formations, the lights in the Deli-Kino assume a more literal and rationalized form, both by taking the shape of stars but also in their clearly mathematical arrangement on the ceiling. Although the dome of the ceiling is off-center, the ellipses of lights appear as rational points in the white of the ceiling plan, making overtures as a mathematical map of the night sky.

Poelzig's interest in stars wasn't limited to architecture alone. He became fascinated with astrology and the occult in Breslau, where he lived between 1903 and 1916.[26] Wolfgang Pehnt notes that Poelzig belonged to a circle of friends who dabbled in 'astrology, chiromancy, spiritism, hypnosis, autosuggestion, divination' and reported occult experiences.[27] In a letter from November 11 1918, Poelzig wrote that he considered himself to be a visionary medium.[28] Theodor Heuss points out that while Poelzig seemed to have a knack for the supernatural, it was actually Wegener who had command of historical knowledge of the occult.[29] When Wegener and director Carl Boese tapped Poelzig to create the sets for *Der Golem wie er in die Welt kam*, Poelzig finally got a chance to infuse his knowledge of esotericism and his interest in early modern architecture into the set of *Golem*.

---

ed. Wolfgang Pehnt and Matthias Schirren (Munich: DVA, 2007), pp. 10–51, here p. 30.

[24] Röder, 'Traumspiel mit Kulissen', p. 11.

[25] Hambrock, *Hans und Marlene Poelzig*, p. 137. Pictures of the plans for the Deli-Kino lights can be found on p. 163.

[26] Heuss, *Poelzig*, pp. 40–41. Heuss, writing in 1939, mentions astrology apart from 'okkulten Wissenschaften' showing that the trends witnessed in Oscar A. H. Schmitz's group continued into other spheres. Although affiliated, for Heuss, and therefore for Poelzig, astrology and the occult still distinct.

[27] Pehnt, 'Wille zum Ausdruck', pp. 37–38.

[28] Heuss, *Poelzig*, p. 37.

[29] Theodor Heuss, 'Begegnungen mit Paul Wegener', in Möller, *Paul Wegener. Der Regisseur und Schauspieler*, p. 2.

Immediately after *Golem*'s release, reviewers took note of the starry skies it depicted.[30] On 31 October 1920, a reviewer remarked in the journal *Der Vorwärts*: 'Indeed, the middle ages come alive before our eyes, with all of its fantastic beliefs and absorbing mysticism; astrology and kabbala captivate the people'.[31] Astrological knowledge drives the film from the start. The establishing shot cuts to the first title, which announces that 'High Rabbi Löw reads in the stars that a terrible misfortune threatens the Jewish community'.[32] The next shot presents the Rabbi at his telescope in a circle frame that fades at its edges. The telescope points to the top left of the screen, as Rabbi Löw looks through it. Then a jump cut shows a stark circular frame of the stars, which move from left to right. The seven bright stars again pass through the circular frame as it pans to the right. The sharp contrast between the faded frame of the Rabbi using his telescope and the stark circular frame surrounding stars creates the impression that the viewer sees what the Rabbi sees through his telescope. The panning of the stars also adds to this impression that the viewer is participating in the Rabbi's point of view. The camera will assume the perspective of Rabbi Löw reading again and again throughout the film as the viewer is led to read other textual documents along with him.

The opening sequence continues with the Rabbi adjusting the telescope, sitting to read a book, and looking up again at the sky with squinted eyes, before checking the book one last time and going down his windy staircase to tell his assistant and his daughter to warn the community about what he has just read in the stars.

The seven stars depicted in the film's sky look eerily similar to the constellation Cygnus, which is visible in northern Europe during the summer and autumn. The film was shot on an open-air set in Berlin Tempelhof during the

[30] Andrej in *Film Kurier*, Oct. 30, 1920.

[31] 'Ja, das Mittelalter wird lebendig vor unseren Augen, mit all seinem Wunderglauben und seiner mystischen Versunkenheit; Astrologie und Kabbala berücken die Menschen'; *Der Vorwärts*, Oct. 31, 1920, reprinted in *Film und Presse*, no. 17 (1920): p. 418.

[32] 'Der Hohe Rabbi Löw liest in den Sternen, daß der Judengemeinde ein schweres Unheil droht', Akt I, Title 1., Zensurkarte Prüf-Nr. 29156 (Berlin, 1931), 3, in the Schriftgutarchiv of the Stiftung Deutsche Kinemathek. Hereafter as 'Zensurkarte (1931)'.

summer of 1920 while Cygnus was visible in the night sky. Although the shots of stars at night are not of the actual sky, the film crew must have modeled the stars they filmed from some kind of pattern, and the pattern of Cygnus is clearly discernable.

A fade-in dissolve at minute 8:12 shows the view of the Jewish ghetto at night with the starry universe and these extra bright stars. Another dissolve brings the seven stars into view, but another dissolve quickly follows to a shot of the decree banishing the Jews from the empire. The dissolves work to associate the community to the stars to the decree, showing first the community, then the message, and finally a representation of the actual disaster that is threatening the community. The montage forces the audience to see the correspondences between these images. In addition, the decree itself uses terminology associated with timing based on astronomical observation instead of an arbitrary date on a calendar: 'noch ehe der Mond wechselt',[33] that is to say by the new moon. Putting the deadline in terms of the phase of the moon and not a randomly selected numerical calendar date shows that the kingdom was still operating by the same principles used by the Rabbi to predict the doom of his community—a lunar calendar.

Just as in Wegener's first film, scenes of the third film depict early modern astronomical and astrological texts. After the emperor sends Junker Florian to deliver the decree banning the Jews from the empire, a jump cut reveals a book with Rabbi Löw's hand on the right side. The page shows a line drawing of the Golem with a five-pointed star in the middle of his chest, surrounded by Hebrew characters. The hand then turns the page containing a medieval horoscope (fig. 10.1).[34] The hand of the Rabbi points to the words 'Dom VII', which indicates the seventh house of the medieval natal chart depicted here. The glyph for Libra appears next to it. The glyphs flanking the seventh house are the appropriate traditional rulers of those houses: starting from the bottom of the image going counter-clockwise, the fourth house by the glyph for Cancer; the fifth house by Leo; then the eighth by Scorpio; the ninth by Sagittarius; and tenth by Capricorn.

---

[33] Zensurkarte (1931), p. 4.

[34] For more on the shape of natal charts see recent discussions in Nicholas Campion and Liz Greene, eds., *Astrologies: Plurality and Diversity* (Ceredigion, Wales: Sophia Centre Press, 2011).

However, the houses are empty; no planetary glyphs adorn the space in the triangles of the houses, which is a critical component of drawing an astrological chart. Despite the fact that the chart is generic, it serves to show the viewer that there are rules and codes to the symbolic language the Rabbi needs to use in order to decipher the messages he is reading in the configuration of the planets in the sky.

Fig. 10.1: Rabbi Löw reads an astrological manuscript in Golem, wie er in die Welt kam *(1920)*

As the camera pans, one can read the words 'Theoricae Planetarium' above the horoscope. This title which is similar to the title of Campanus of Navarre's thirteenth-century textbook, *Theorica Planetarum*, which included instructions on how to build an equatorium—a scale model of the Ptolemaic system used to compute the positions of the planets.[35] One can think of this as a type of primitive planetarium, which calls Poelzig's star-filled theatre ceilings to mind. Perhaps Poelzig used historical astronomical texts as inspiration for his starry constructions? Campanus also devised a system of astrological houses (the division of the sky into various symbolic parts) that bears his name. But the film's textbook may refer to another text. In 1473 Georg von Peurbach's *Theoricae Novae Planetarium* became the first printed astronomical textbook and compendium of astronomical information to that date. Either Campanus or Peurbach could

---

[35] Margaret J. Osler, 'Book Note on *Campanus of Novara and medieval planetary theory. Theorica planetarum*', *Journal of the History of Philosophy* 14, no. 4 (October 1976): p. 498.

have been the inspiration for this fictional text.

The moments after this still reveal a sketch of an eclipse and mathematical notes scrawled next to the archetypal horoscope in the Rabbi's text. The numbers and the drawings are presented right next to the zodiac signs arranged in a prototypical astrological chart and suggest that the Rabbi was versed in both mathematical astronomy and astrology, as would be expected of an early modern 'mathematicus'. Given that the film is set in a fantasy Prague, it may be most likely that these are fantasy astronomical texts based on an amalgamation of historical astronomical texts.

The texts in this film highlight a significant shift in the use of media. Scenes of 'reading along' with the Rabbi reading medieval texts are accompanied by scenes of 'watching along' with him as he practices observational astronomy and astrology. Despite the advances in film technology, scenes of reading are still critical for the story of the third film to unfold. These textual moments are shrouded through the typography of the intertitles and subject matter; their intense convergence into the visual fold of the narrative shows the extent to which textuality is still paramount in film at this time.

The instructions in the books are acted out in the action of the film. During the sculpting of the Golem, Rabbi Löw invites his assistant, or famulus, into the secret that the viewers already know. He points to the images on the walls of the basement to indicate to the famulus where the star (effectively, a celestial variant on what should be Golem's heart) must go. These pointing gestures resonate with the following scene, and also the scene of supreme magic (waking the Golem to life by instructing the viewers to look away from the mechanics of the cinematic trick). Even though the scene of Golem's awakening so pointedly addresses the spectators, it is embedded in a network of scenes of educating the famulus. The role of the famulus then is to act as a foil to allow the magical arts to be conveyed as well to the viewers; the Rabbi educates the famulus in such a way as the audience also gets educated. He shows him the star and the strip of papyrus that must go inside it; then they look through two different books. We read the directions for waking the Golem in the first book, and then we read along with them in the book titled *Nekyomantie, die Kunst, Totes lebendig zu machen* (*Necromancy, the Art of Bringing the Dead to*

Life).[36] The book's subtitle explaining the meaning of *Nekyomantie* serves to bolster the claim of educating the viewing public. In this grimoire, we are shown the instructions for invoking Astaroth, which are acted out in the next scene.

Despite the film's seventeenth-century setting, or perhaps because of it, Wegener and his crew were challenged to make their magic believable. By 1920, film had been around for two decades, and the viewing public was no longer easily tricked by special effects. Four years after the debut of *Golem,* Béla Balázs discussed the relationship between magic and special effects in films in his essay 'Filmwunder':

> The dreams of a child or of a poet have their own believability that the fantastic inventions of a rationalist do not have. One believes to feel the false front, and is annoyed. Unfortunately it seems to have gone the same way for film. Otherwise it would be impossible to grasp, why in a time when in literature, the fantastical, the occult, the magical was the great fashion, the fantastical could hardly permeate film or, in the films of fairy tales, not at all. And yet, one should think, that no other art on the basis of its technical possibilities for representation of such magic would be as appropriate as film. Oddly exactly the known technical possibilities appear to awaken our rational skepticism.[37]

---

[36] Akt II, Title 11, Zensurkarte (1931), p. 5. The title *Nekyomantie* is represented as two lines, '— —' indicating perhaps that it was illustrated in the intertitle, which is how it appears in the 2004 DVD version of the film. *Nekyomantie* is a synonym for the more popular word *Nekromantie,* or in English, necromancy.

[37] Béla Balázs, 'Filmwunder', *Film Kurier* (Berlin), 12.12.1924. Reprinted in Balázs, *Schriften I,* p. 322–25, here p. 323. 'Die Träume eines Kindes oder eines Dichters haben eine eigene Glaubwürdigkeit, die die phantastischen Erfindungen eines Rationalisten nicht haben. Man meint, die Mache zu spüren und wird verstimmt. Leider scheint es dem Film ähnlich ergangen zu sein. Sonst wäre es ja gar nicht zu begreifen, warum in einer Zeit, wo in der Romanliteratur das Phantastische, Okkulte, Märchenhafte die große Mode war, der phantastische in Film kaum und der Märchenfilm gar nicht durchdringen konnte. Und doch sollte man meinen, daß keine andere Kunst auf Grund ihrer technischen Möglichkeiten zur Darstellung solcher Zaubereien so geeignet wäre wie eben der Film. Seltsamerweise scheinen aber gerade die bekannten technischen Möglichkeiten unsere rationelle Skepsis zu erwecken'.

Balázs argues here that to be successful, film must use a sense of realism to make the common world seem strange. Therefore a film like *Golem* would not necessarily have inspired wonder or horror in its audience, even though it deals with bringing inanimate matter to life. *Golem*'s ability to make audiences believe its magic is stunted because of its fantasy setting in a distant past, which the audience already knows is not real. In fact, a reviewer from the *Berliner Morgenpost* on October 31, 1920 noted: 'Astaroth's appearance seems like the great bluff of a spiritist séance'.[38] In films like these, Balázs asserts that 'the representational possibilities of ancient myths appear as the most modern achievement. Not what happens, but "how it is done" interests the people'.[39] Echoing Balázs, Carl Boese recalled that 'there was hardly a person who spoke to me who didn't ask: how did you guys actually do that?'[40] In the middle of the last century, he made sure to leave a manuscript in the Stiftung Deutsche Kinematek discussing just that.[41]

Boese's unpublished typescript about the making of *Der Golem* recounts the specific technological means by which he and his film crew staged various aspects of magical practice in the film. These effects were path-breaking in their own time, evidenced not only by the immediate critical reception of the film, but also by the sheer stylistic differences between this and Wegener's two previous Golem films. Boese reveals himself to be a jack-of-all-trades problem-solver in his approach to achieving various cinematic effects by way of simple, yet genius, application of techniques of meteorology, physics, and chemistry. Considering that the early modern period is often seen as a time of great scientific experimentation, I suggest that these effects were dependent on the film's early modern setting for the conditions of their own possibility. Later films of the period continue to contain echoes of this homage to early modern magico-scientific

[38] *Berliner Morgenpost* on October 31, 1920 (reprinted in *Film und Presse*, no. 17 [1920]), p. 415. 'wie der große Bluff einer spiritistischen Seance wirkt die Erscheinung Astaroths'.

[39] Balázs, 'Filmwunder', pp. 323–24: 'Die Darstellungsmöglichkeit uralter Mythen erscheint als modernste Errungenschaft. Nicht was geschieht, sondern 'wie wird's gemacht' interessiert die Leute'.

[40] Boese, *Erinnerungen*, p. 20. 'Es hat kaum jemand gegeben, der mich sprach und nicht gefragt hätte: wie habt Ihr das eigentlich gemacht?'

[41] Ibid.

practice, as I'll discuss with the conjuration of Mephisto in F. W. Murnau's 1926 *Faust* below.

In one instance, Boese recounts how they solved a complicated lighting issue using astronomical knowledge. In the storyboards for the film, Hans Poelzig had drawn a sketch of the Rabbi inside his laboratory with strong diagonal streams of sunlight flooding through the window. Boese wanted to make sure that the film captured this scene exactly as Poelzig had drawn it, but finding a light source capable of illuminating at that specific angle strongly enough for the camera to pick it up proved difficult. Boese recalls, that, even after talking to the main lighting technician, they were unable to find a source strong enough. The powerful and popular spotlights—aptly titled 'Jupiter lights'—were not strong enough.[42] After deliberating for some time, it finally occurred to Boese that 'the sun itself!' would do the trick just perfectly![43] However, because the sun appears to move across the sky throughout the day, it only created the angle Poelzig envisioned for approximately two hours a day, and Boese had to reposition the window to capture this light.[44] Thus, they had to pay attention to the sun along the ecliptic in order to realize Poelzig's diagonal beam of light. And if the day happened to be cloudy, they couldn't film at all. In short, the filmmakers were just as observational of the sun's path across the sky as an astronomer or astrologer would be. They had to get the timing just right in order to make the image have a specific character. Even a detail as small as this illuminates the synergy of filmmaking and solar observation.

The 'how it is done' sensibility aptly identified by Balázs is satisfied by Boese's account of the making of the film; the magical effects are negated when we learn about their technical creation. However, the technical effects do not account for the agency of astrological symbolism in the film's plot. As if a blind spot, with the exception of Beate Rosenfeld, scholarship has overlooked this critical presence in the film. This is precisely where the enchantment resides.

As discussed above, astrological language appears in the film's first intertitle, when the Rabbi reads the stars. This form of reading is mundane astrology, or astrology applied to nations and populations as a whole, and this 'threat' is

[42] Ads for these lights can be found in film journals of the time.
[43] Boese, *Erinnerungen*, p. 8. My emphasis.
[44] Ibid.

actually not read as a certainty, but a possibility. Human action of some kind must accompany the vision he saw; and the Rabbi asks his community to pray to avert this, suggesting that by the action of praying, whatever stands in the sky might be averted on earth. In addition to asking the community to act through prayer, the Rabbi acts secretly and deploys electional astrology—or the astrology of timing the beginning of an event to achieve a specific outcome—to bring the Golem to life at a fortuitous moment and counterbalance the forces he sees at work in the sky. Returning for a moment to Beate Rosenfeld's argument, the astrological emphasis allows the motivation for the central catastrophe to be neither psychological or kabbalistic, and thus further establishes Wegener's unique intervention into the Golem material.[45] Astrology becomes the central driver of the plot.

At the end of the first act, the Rabbi states, 'Venus enters the constellation of Libra. The time is nigh for the conjuration to succeed'.[46] Thus he has viewed the conditions necessary to elect a time to perform his conjuration of Astaroth to awaken his clay figure. In the second act, an intertitle attributed to the Rabbi, who has just finished sculpting the face of Golem to the likeness of Paul Wegener, looks out of the window at the sun and determines the time with more specificity, 'The hour has arrived. The alignment of the stars aids the magic'.[47] The German word for 'aid' used here, *begünstigen*, also means to promote, abet, favor, or benefit. The sense of the word then is generally to assist, encourage, or advantage to the benefit of someone or something. The filmmakers' choice of language indicates an interpretation that the move of Venus into the sign of Libra does not determine that the Rabbi's magic would be successful, but that it would 'contribute to' or 'encourage' the success. Human action, or inaction, ultimately establishes the determining factor.

The timing of the conjuring of Astaroth is essential to the success of bringing the Golem to life. The crucial instruction for wresting the word from Astaroth appears at the end of an intertitle posing as an ancient grimoire: the practitioner desiring the word will be successful, 'as long as he considers

---

[45] Rosenfeld, *Die Golemsage*, pp. 148–49. Emphasis mine.

[46] Akt I, Title 6, Zensurkarte (1931), p. 4. 'Die Venus tritt in das Sternenbild der Wage. Die Zeit ist da, wo die Beschwörung glücken muß'.

[47] Akt II, Titel 4, Zensurkarte (1931), p. 4. 'Die Stunde ist da. Der Lauf der Gestirne begünstigt den Zauber'.

the auspicious constellation of the planets'.[48] The German word, *günstig*, which I translate here as auspicious, also means beneficial, opportune, advantageous, and this word choice reinforces that the constellation does not determine the outcome, only that specific constellations are more beneficial than others. Man's powers of observation of the sky and of carrying out the 'great magic' 'großen Zauber'[49] are the critical elements in achieving a successful result. In this case, the demon that the Rabbi seeks to invoke, Astaroth, has been traditionally associated with Venus, the moon, and the goddesses Inanna, Ishtar, and Astarte.[50] Although the demon appears in male form in the film, Jake Stratton-Kent gives one possible explanation for the permutation of Astaroth's historically female form: 'the gender of Astaroth on a particular occasion may be determined by the position of Venus in relation to the Sun. As a male when Venus rises before the Sun and as a female when Venus rises after the Sun'.[51] More likely, though, the appearance of Astaroth as a male demon in *Golem* has to do with a tradition of translation errors that associated the male form with demonic magic and the female form with pagan religion.[52]

Astaroth's connection to Venus carries with it other powerful symbolic elements. For one, the pentagram symbol features prominently. Over the course of eight years the conjunction cycle of the Sun and Venus form a sort of five-petalled shape, akin to a pentagram, when traced around the zodiac. This helps to explain the dual presence of five and six pointed stars throughout Wegener's 1920 Golem film. In addition, by appealing to Astaroth and not to Lucifer, the Rabbi was able to escape the worn trope of invoking the devil. In this respect, the presence of Astaroth makes Wegener's allusion to black magic more authentic. One of Astaroth's more important demonic capabilities includes conferring the favor of the great and powerful, which applies directly to the Rabbi's plight.[53] Astaroth's demonic associations were evidenced by her inclusion in the

---

[48] Akt II, Titel 12, Zensurkarte (1931), p. 5. 'so er die günstige Konstellation der Planeten beachtet'.
[49] Ibid.
[50] Jake Stratton-Kent, 'Astaroth: Lady of the Crossroads', in his book *The True Grimoire* (Scarlet Imprint, 2009), pp. 185–200, here p. 185.
[51] Ibid.
[52] Ibid., p. 198.
[53] Ibid., p. 186.

demonology in Carl Kiesewetter's popular 1893 book *Faust in der Geschichte und Tradition* (*Faust in History and Tradition*).[54] However, Venus is not the only planet that gets specific mention in this film. In later scenes, the mention of Uranus in conjunction with Golem's rebellion and the subsequent chaos for the Jewish ghetto is astrologically accurate, and yet startlingly anachronistic.

Sir William Herschel discovered Uranus on March 13 1781 in Bath, England. Given that the plot of the 1920 version of *Golem* takes place in Rudolphine Prague, that is, just before the start of the Thirty Years War, ancient texts that discussed Uranus in a planetary context did not exist. Thus the following intertitle's prediction that the Golem will rebel when Uranus switches planetary houses is a significant anachronism: 'If you have brought an inanimate being to life through magic, beware of your creation. When Uranus enters the house of the planets, Astaroth will reclaim his instrument. Then the lifeless clay will turn against its master, intent on deceit and destruction'.[55]

Astrologers in the early 1920s were still in the process of discovering what the new planets Uranus and Neptune might mean. Oscar A. H. Schmitz suggested that 'observation compels us to bring the planet Uranus, discovered shortly before the French Revolution, in relationship with the sign

[54] Carl Kiesewetter, *Faust in der Geschichte und Tradition* (Leipzig, 1893; repr. New York: Georg Olms Verlag, 1978), p. 367: '*Astaroth* is a great and powerful duke: he appears in a very ugly angel form [...] he confers true answers with respect to past, present, and future and secret things. [...] The exorcist should be wary of coming too close to him because of the unbearable stink that issues forth from him'. ['*Astaroth* ist ein großer und mächtiger Herzog: er erscheint in sehr häßlicher Engelsgestalt [...] Er giebt wahre Antworten bezüglich vergangener, gegenwärtiger, zukünftiger und geheimer Dinge. [...] Der Exorcist möge sich wegen des von ihm ausgehauchten unerträglichen Gestanks hüten, ihm zu nahe zu kommen.']. Kiesewetter takes his material from a Latin source by Johann Wier, a student of Agrippa von Nettesheim, who published *Pseudomonarchia Daemonum* in 1660.

[55] Akt IV, Titel 7, Zensurkarte (1931), p. 6. 'Hast Du durch Zauberwort Totes zum Leben erweckt, sei auf der Hut vor Deinem Geschöpf. Tritt der Uranus ins Planetenhaus, fordert Astaroth sein Werkzeug zurück, dann spottet der tote Lehm seinem Meister, sinnet auf Trug und Zerstörung'. English translation from the TRANSIT Film edition of *Golem*.

Aquarius'.[56] And that, 'the entire world of cinema and those possessed by it should be ruled by the sign Aquarius'.[57] He uses the following adjectives to describe the meaning of Uranus, all of which carry significant resonances with Wegener's Golem: 'explosive, volcanic, energetic, awakening, headstrong, stubborn, unpredictable, high-minded, bisexual, asexual, enigmatic, heroic, creative, critical, sarcastic, messy, quirky, lightning-like, paradoxical, aphoristic'.[58] Later, in 1926, Munich astrologer Heinz Artur Strauß added nuance to the emerging definition of Uranus's meaning, associating it both with media and 'sudden "coincidences", the irruption of "unexpected" fates'.[59] The film suggests that Uranus shifting to a new planetary house correlates to a surprising change in the Golem's behavior. The intertitle in the film simply states 'Planetenhaus' (or, house of the planets), which doesn't technically exist, leaving an immense degree of uncertainty (again a Uranian trait!) about just what house, or area of life, Uranus needs to move into to activate a rebellion. However, what is important here is that the Rabbi uses ancient texts to read information concerning the creation and management of the Golem. No ancient text contained information in the early 1600s about Uranus and its interpretation vis-à-vis the Golem. Therefore we are dealing with a very modern Golem here. Once the circulating astrological context becomes clear, the presence of Uranus in the film suddenly establishes that Wegener's film was participating in the emerging discourses surrounding astrology taking place at the time.

In turn, the immense symbolic role played by the stars and planets in *Golem* serve to highlight the fascinating

---

[56] Schmitz, *Der Geist der Astrologie*, p. 257: 'Die Beobachtung zwingt dazu, den kurz vor der französischen Revolution entdeckten Planeten Uranus [...] in Beziehung mit dem Wasserman zu bringen'.

[57] Schmitz, *Der Geist der Astrologie*, p. 254: 'Die gesamte Welt der Kino und der von ihm Besessenen dürfte stark unter Wasserman stehen'.

[58] Schmitz, *Der Geist der Astrologie*, p. 259: 'explosiv, vulkanisch, energisch, erwachend, eigensinnig, verbohrt, unberechenbar, hoch-geistig, bisexuell, asexuell, rätselhaft, heroisch, schöpferisch, über-kritisch, sarkastisch, unordentlich, verschroben, blitzhaft, paradox, aphoristisch'.

[59] Heinz Artur Strauß, *Astrologie: Grundsätzliche Betrachtungen* (München: Kurt Wolff, 1927), p. 40: 'plötzlichen "Zufälle", die "unerwartet" hereinbrechenden Schicksale'.

debates about cinema raging in Weimar Berlin. These *Kino-debatte* concerned the negotiation of the role of text in film.[60] In *Golem* they compliment one another: Rabbi Löw shifts smoothly between looking at a book (reading a mediated text) to looking at the stars (reading a visual constellation, which is really a cinematographic representation of a celestial scene). Intertitles add a whole new level of textual dispersion in the film that mirror the historical appearance of other forms of cultural continuity. The language of the stars mediates textual legibility and visual legibility. Over the next few years, F. W. Murnau engaged the occult revival, but the differences in his approach highlight the many changes taking place in both film technology and the occult revival itself in Weimar Germany.

FAUST

On October 14, 1926, F. W. Murnau's *Faust* premiered at the UfA-Palast am Zoo in Berlin, where *Golem* also had premiered six years earlier. The Faust material has always functioned as a battle cry for disillusionment with religion and science, but in addition to the plethora of the publications on the occult emerging around the turn of the twentieth century, many monographs started to appear about Faust, and even about Goethe's relationship to the occult arts.[61] Despite the many cinematic treatments of the Faust material that came before Murnau's version, this film remains iconic in German silent cinema.

Murnau's film begins with the familiar Prologue in Heaven found in Goethe's *Faust*, wherein an archangel and Mephisto wager on Faust's soul. If Mephisto can destroy what is divine in Faust, namely his soul, then the Earth will be his. The film then departs from Goethe's version to depict Faust as an academic and alchemist in a medieval town that is struck with the plague. After failing to cure his people of the plague Faust summons Mephisto, who promises Faust eternal youth. Faust is seduced by this offer, goes on many

---

[60] Anton Kaes, *Kino-Debatte: Texte zum Verhältnis von Literatur und Film* (Munich: dtv, 1978).

[61] See, Carl Kiesewetter's 1893 *Faust in der Geschichte und Tradition: mit besonderer Berücksichtigung des occulten Phänomenalismus und des mittelalterlichen Zauberwesens*, or Max Seiling's 1919 *Goethe als Okkultist* (Berlin: Joh. Baum Verlag, 1919; repr., Leipzig: Bohmeier Verlag, 2008).

adventures, and the second half of the film rehearses the familiar Gretchen story from Goethe's *Faust*. Central to my discussion here are the ways Murnau stylises the opening scenes of the film culminating with the conjuration of Mephisto. Murnau's treatment of ritual magic differs in important ways from Wegener's.

In the initial scenes of the film, with his town fully stricken with plague, Faust encounters two specific deaths that walk the audience through his disillusionment with religious faith and scientific knowledge. First, after praying to God for a cure to the plague, he administers medicine to a woman who immediately dies. Faust dashes the vial to the ground. Then, in the next scene, Murnau sets up a contrast between religious folk chanting in the streets to repent, with revelers who couldn't care less that their world is coming to an end. The allegory appears quite simple: within the reference to the medieval experience of plague, with wanton pleasure flaunting itself in the face of mass casualties, one can easily envision the mass death of the Great War contrasting with the excesses of the Golden Twenties. After claiming that only sinners will die, one of the faithful men holding a cross on the street is stricken. Seeing this, Faust returns to his laboratory, less ornate than that of Rabbi Löw, and begins to throw his books into the fire. He pauses for a minute at his bible, and then casts it into the fire with the others. As he tries to throw another stack of books in, a burning text opens, seemingly by itself, and reveals to Faust the 'Threefold Key' for commanding the powers of the dark spirits. All he must do is go to a crossroads and call for Mephisto three times. With religion and science of no use to him anymore, Faust grasps for the occult. In this way, the opening scenes of the film ground the audience in a familiar magical tradition in order to surpass it. The magical conjuration scene invokes the early roots of the magic Murnau and his crew are able to create through technology. As in *Golem*, the very tricks that create the illusion of this old magic's success are the tricks that then go on to create even wilder illusions of fantastical reality. Staging this plague scene as he does, embedded within the Faust material and functioning as an allegory of the occult revival, Murnau tells the story of how in early twentieth century Germany, cinema is now the rightful heir of magic and all it can achieve. Yet, without astrology as a central thematic element, how does Murnau engage the sky to tell his story?

Many scenes of *Golem* were shot on an open air set in Berlin Templehof. The entirety of Faust was filmed inside at the UfA studios. This structural shift between open and closed space in the making of the films is witnessed in the content of the films themselves. The establishing shots of the main cities in the two films show us that the space of the sky is utilized in two different ways (fig. 10.2). *Golem* shows an open sky with an observational tower that acts as a hinge between the starry heavens and the earthly realm of the Jewish ghetto. *Faust*'s medieval city is shrouded in clouds; it's not the earth's rotation that makes the sun disappear, but the massive cloak of Mephisto as he brings plague to the city. *Golem*'s setting is more realistic, the stars dotting the night sky. *Faust*'s setting is more fantastical and mythic with an impossibly large Mephisto occupying the canvas of the sky above the town.

Fig. 10.2: Establishing shots of Golem *(left) and* Faust *(right).*

These shots resonate with the two early shots in the films that establish the threats coming from the sky. In *Golem,* Rabbi Löw reads a threat from the positions stars. In *Faust,* three of the four writers of the apocalypse are shown writing in a clouded space. *Golem* depicts a hermetic practice where the misfortune isn't given a kind of visibility to the audience. It is not clear to the viewer exactly which astrological data allows Rabbi Löw to come to his interpretation that the Jews will be threatened. In *Faust,* the trope of the riding horsemen is familiar through the popularity of the book of revelations. To an audience in a generally Christian nation, this threat is obvious. Thus two kinds of celestial hermeneutics are at work: a cryptic one and a mythical, religious one.

A further set of early shots reinforces the differences in the treatment of the sky. Even though Faust lacks overt astrological references, from the outset both main characters are shown being involved in practices concerning the stars. Rabbi Löw practices astrology and visual astronomy in his open-air observatory and *Faust* lectures on astronomy in a dark classroom in front of an illuminated orb that has rings around it—presumably an odd conflation of the sun and Saturn (fig. 10.3).

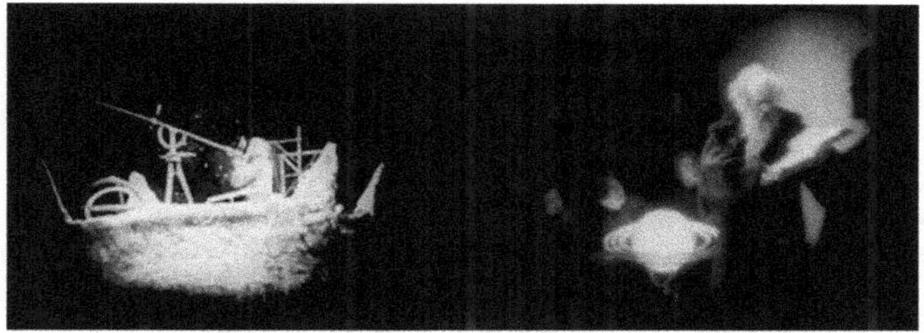

Fig. 10.3: Left, Rabbi Löw observes the stars and, right, Faust teaches using an astronomical model.

In addition to these establishing shots, the sky plays a critical role in the conjuration scenes of both films, and the differing ways they are depicted reveals a stark shift in both film culture and the status of the occult revival in Germany. Before we begin to read the scenes themselves, we can learn from the instructions given to Rabbi Löw and Faust.

Murnau's *Faust* appeared one year before the first sound film, so, like *Golem*, Murnau was dependent on intertitles to convey information to his audience. Unlike *Golem*, however, *Faust* conveys much less information in the texts shown in the film. The instructions for the creation of the Golem and the summoning of Astaroth span multiple pages of fictional medieval manuscript. Only one page of instructions shows Faust how to conjure Mephisto, stating simply, 'Go to a crossroads and call him three times'.[62] There are no further indications for what the ritual situation requires. Comparing the amount of instruction shown the audience alongside the depiction of celestial symbolism helps to illuminate why Lotte Eisner found Wegener's conjuration scene more effective than Murnau's: 'the scene of the appeal to the demon with its circles of flames is even more poignant than the corresponding scene in Murnau's *Faust*'.[63]

---

[62] Murnau, *Faust*. English quote from the Kino Lorber edition.

[63] Lotte Eisner, *The Haunted Screen: Expressionism in the German*

In preparation for his conjuration, Rabbi Löw uses astrology to figure out the precise moment his ritual should begin. When Venus is in Libra in the seventh house, his magic should succeed.[64] The film's intertitles declare when the time is right, and the ritual can proceed. Alternating between medium and long shots, the Rabbi is shown inside his indoor laboratory with his adept in the middle of a burning circle. He is wearing a hat with many stars and he holds a star in his hands that functions as a wand. He summons the spirit Astaroth and speaks directly to it, asking for the magical word that will allow the Golem to come to life. Once the word has been revealed, lighting strikes more than 13 times in less than 3 seconds of film, and the scene flashes, then goes dark. When the smoke lifts, the Rabbi and his adept are splayed out in the center of the circle. The conjuration was successful.

In Murnau's conjuration scene, Faust is shown underneath an open sky standing at the center of a crossroads. Instead of a star-wand, he holds up a book, but the book doesn't have anything legible written on it; it's an abstract picture, almost like a Wassily Kandinsky sketch. He appears as a tiny human in the midst of a massive sky during the scene of the most intense magic of the film. And yet this sky is actually inside a film studio. As Faust dutifully calls out Mephisto's name, the camera pans out nearly as far as it can go, a flash appears, and a small burst of light descends from the top of the screen to the bottom. Mephisto has arrived. As the demon looks at the camera, a gleam of light appears in his eyes, seeming to reference his celestial providence. In a pamphlet by Gerhart Hauptmann handed out to audience members at the premiere, Hans Kyser, one of the screenwriters, describes Mephisto as a star that falls to earth.[65] Rather than the esoteric space of a laboratory, Mephisto's conjuration takes place exoterically, out in the open, at the crossroads underneath the heavens.

Pausing for a moment to consider the contemporary experience of watching these films gives an added element to the role the sky in influencing their effect upon the spectator.

Cinema and the Influence of Max Reinhardt (1969; repr., Berkeley, CA: University of California Press, 1973), p. 56.

[64] For a contemporary German account of Venus, see, Schmitz, *Der Geist der Astrologie* (1922).

[65] Gerhart Hauptmann, *Worte zu Faust*: *eine deutsche Volksage* (Berlin: Universum-Film Aktiengesellschaft, 1926), p. 12.

Audiences in a movie theatre occupy an interior space, and even though Hans Poelzig attempted to make theatres that gave people the illusion of being under a starry sky, those were only two theatres out of the hundreds that existed at the time, so we have to revert to the sense of interiority of the cinematic experience. With the conjuration scene in the Rabbi's laboratory in *Golem,* the audience is treated to an intimate front row seat to witness a ritual forbidden to all but the initiated. They might as well be present in the room with him. In Murnau's *Faust,* the tension between the epic outdoor spaces on-screen and the interior conditions of the movie theatre allow for a certain remove from the illusion of participation. The frame is made critically present in *Faust,* both by the open sky, and by the extreme long shot during the arrival of Mephisto, where Faust is reduced to an impossibly small size. Yet, it is entirely possible that members of the audience could forget their surroundings in the dark, and be mentally transported to this outdoor scene. The magic is no longer one of an initiation into ritual magic, the spell has been cast by Murnau directly on the audience itself.

In general, *Golem* contains the overwhelming presence of accurate occult knowledge being shared through the public medium of film, yet depicted in an esoteric space. In this way, Wegener's film serves as an attempt to initiate people into one of the arts described. The audience is included as the adept in the darkness of the cinema, compounding the 'esoteric' nature of the magical scene. Yet, when Wegener divulges esoteric knowledge in context in a mass medium, the esoteric becomes exoteric. What was passed on for generations between master and apprentice is now revealed for all to see. Murnau shields the contents of esoteric arts from the vagaries of this most exoteric medium, without sacrificing its magical appeal.

Faust contains nearly no specific occult knowledge, with the magical ritual being depicted in a very public space. The books in Faust's lab and in the conjuration scene have generic text and drawings, some from the Bible, some from sketches in the magical text resembling script and hangings. These are not symbols that correspond to any occult practice. Even the commands of the ritual are quite generic. In Murnau's world, words and symbols have ceased to be effective tools by which magic can be worked. The audience is at once exposed to the idea of magical practice, but it is made safe, disarmed, shrouded in an esoteric illusion created by textual idiosyncracy instead of tradition; magical tradition cedes its power for the tricks of the cinema to take over. Thus the spectator, unlike *Golem*, is never really genuinely included as an adept.

There are multiple levels as to how one can read this shift in disclosure from 1920 to 1926. A film historical conclusion might say that by the time Murnau's film was made the audience was less dependent on 'reading' as earlier film audiences had been, even though they were handed a twenty-five page pamphlet at the film's premiere! However, read against the background of the occult revival, it seems that Murnau created cinema *as* magic, instead of trying to put magic in cinema. There are a few hypotheses for why this is so. Either by 1926 audiences actually knew enough about esoteric practices that they didn't really need explicit instruction, or in order to keep the esoteric elite, Murnau obscured the explicit details of its functioning so that he could have an exoteric display of magic while actually keeping the esoteric secrets of its working private. As Thomas Elsaesser has pointed out, Murnau 'was said to have consulted astrologers, occupied himself with Eastern philosophy and strongly believed in the occult', but he was also 'difficult and a recluse'.[66] So the elitist hypothesis is strong, but it, of course, can only ever be conjecture. What is known, is that by the time Murnau created his conjuration scene, interest in the occult was near its peak in Weimar Germany. Less than a year after the appearance of *Faust*, the journal *Süddeutsche Monatshefte* dedicated a special issue on astrology spanning over 100 pages and 21 contributors. The demand for such coverage shows that interest—whether for or against—was greater than ever before. This trend dovetails with the use of space in both scenes: by 1926 the magic circle—appearing indoors in Wegener's *Golem* film—appears out in the open, underneath a moonlit sky. At its most basic level, I read this change of scene as an acknowledgement of the ubiquity of interest in the occult in Weimar Germany.

While Wegener depicted sky *and* symbol in *Golem*, that is, realistic depictions of the sky alongside explicit symbolic knowledge gleaned from the sky, Murnau used the sky *as*

[66] Thomas Elsaesser, *Weimar Cinema and After: Germany's Historical Imaginary* (New York: Routledge, 2000), p. 226.

symbol in *Faust*, that is, fantastical spatial scales without any explicit symbolic knowledge of the sky. Thus in examining the role of the sky and celestial symbolism in these two films, and thereby also comparing them along the double axes of knowledge transfer and spatial organization, one can see how Wegener's conjuration scene seems to be a more successful display of ritual magic than Murnau's. However, with new technological means at his disposal, Murnau was able to push the limits of his craft and create a wholly filmic magical experience. And in doing so, Murnau provides an answer to the question of the occult revival occurring all around him: Words and written symbols are no longer of consequence. Magic, for Murnau, must be made for the masses through the medium of film.

RECEIVING THE 'MESSENGERS': THE ASTROLOGY OF JUNG'S LIBER NOVUS

Liz Greene

ABSTRACT: From the time of its publication in 2009, C. G. Jung's *Liber Novus*, known as *The Red Book*, has generated a plethora of scholarly papers from both the analytical and academic communities, as well as from astrologers interested in psychological approaches to their work. As one of the twentieth century's greatest and most controversial thinkers, Jung has exercised a pervasive influence on many academic disciplines, including literary criticism, theology, and the history of religions. It is clear that this very personal 'diary' of Jung's journey into his troubled inner world provided most if not all of the raw material for the ideas which he later developed into his 'analytical psychology'. That Jung was well versed in astrology and used it in his analytical practice has never been a secret, as he constantly refers to astrological themes throughout the *Collected Works*. However, although Jung had been studying astrology for at least two years before he began work on *Liber Novus*, there has been no exploration within either the academy or the psychological community of the astrological motifs in the book, or the ways in which this work reflects Jung's profound and unique understanding of astrological symbolism within the context of the human psyche. Astrologers focusing on *Liber Novus* have tended to examine the astrological configurations in Jung's horoscope during the period in which he was writing the work, rather than the astrological symbolism presented, overtly or covertly, in the text and its images. This paper will examine the astrological themes in *Liber Novus* and the ways in which this remarkable work demonstrates the central importance of astrology as one of the building-blocks of Jung's understanding of human beings and the world in which they live.

THE ADVENT OF *LIBER NOVUS*

In the autumn of 2009, a new work by C. G. Jung entitled *Liber Novus* was finally released for publication after many decades of resistance on the part of Jung's heirs.[1] *Liber Novus*,

[1] For the history of the publication of *Liber Novus*, see C. G. Jung, *The Red Book: Liber Novus*, ed. Sonu Shamdasani, trans. Mark Kyburz,

usually known as *The Red Book* because of its red leather binding, is a compilation of private diaries covering the period from 1913—the time of Jung's break with Freud—to 1930. The existence of the diary had been an open secret once Jung's autobiographical memoir, *Memories, Dreams, Reflections*, was published in German in 1961, since he described the genesis of *Liber Novus* in this work; and unpublished copies of the diary had been circulating within the Jungian community for some time. But it had never been accessible to a wider reading public, nor even to the majority of analytical psychologists outside Jung's own intimate circle of colleagues and friends. Now its publication has inspired a proliferation of articles, workshops, interviews, lectures, and commentaries, ranging from reviews in newspapers and analytic journals to the websites of modern pagan groups, one of which, with considerable justification, declares online that *Liber Novus* is 'the most important grimoire of our modern age'.[2]

Jung himself described *Liber Novus* as the pursuit of his 'inner images':

> The years when I was pursuing my inner images were the most important in my life—in them everything essential was decided. It all began then; the later details are only supplements and clarifications of the material that burst forth from the unconscious, and at first swamped me. It was the *prima materia* for a lifetime's work.[3]

The dominant narrative of *Liber Novus* is Jung's journey from spiritual alienation to the restoration of his soul, through the long and painful process of healing a fundamental rift within his own nature: the seemingly irreconcilable conflict between reason and vision, outer and inner worlds, subjectivity and objectivity, and between the scientist and the mystic, both of whom he experienced as authentic, demanding, and mutually exclusive dimensions of his own being.

John Peck, and Sonu Shamdasani (New York: W. W. Norton, 2009), pp. viii–xii. For Jung's own description of the genesis of *Liber Novus*, see C. G. Jung, *Memories, Dreams, and Reflections*, ed. Aniela Jaffé, trans. Richard and Clara Winston (London: Routledge & Kegan Paul, 1963), pp. 194–225.

[2] <http://wildhunt.org/blog/tag/liber-novus>.

[3] Jung, *Memories, Dreams, Reflections*, p. 225.

Jung has not always fared well in psychology departments within the academy. A number of scholars have noted the religious connotations of his ideas and raised doubts about the validity of his psychological models: is Jung's work psychology or religion?[4] Sometimes viewed as a mystic or even, in the view of the Freudian analyst D. W. Winnicott, a schizophrenic,[5] Jung's theoretical models are often considered questionable because they are insufficiently 'scientific': it is difficult to demonstrate through repetitive experiments the existence and location of the archetypes of the collective unconscious. Departments of psychology within the academy currently favour cognitive approaches to the human mind, which are more amenable to the methodologies of the natural sciences.[6] The view of Jung as 'unscientific' has also been promulgated in the sphere of esoteric studies, an area of scholarship where one might reasonably expect a more methodologically neutral approach. This view of Jung suggests that, because he drew on esoteric sources and occult traditions in formulating his psychological models, he is suspect as a psychologist and should instead be viewed as an esotericist with a 'religionist' perspective and even, perhaps, a messianic mission.[7] Jung

---

[4] For a discussion on this issue, see the insightful paper by G. William Barnard, 'Diving Into the Depths: Reflections on Psychology as a Religion', in Diane Elizabeth Jonte-Pace and William B. Parsons, *Religion and Psychology: Mapping the Terrain. Contemporary Dialogues, Future Prospects* (London: Routledge, 2001), pp. 297–318.

[5] See D. W. Winnicott, 'Review of C. G. Jung, *Memories, Dreams, Reflections*', *International Journal of Psycho-analysis* 45 (1964): pp. 450–55. The term 'mysticism' was first applied to Jung's esoteric interests by Freud, who warned him, in a letter dated 12 May 1911: 'You will be accused of mysticism'. William McGuire, ed., *The Freud/Jung Letters*, trans. Ralph Manheim and R. F. C. Hull (London: Hogarth Press/Routledge & Kegan Paul, 1974), p. 422.

[6] Two notable exceptions in Great Britain are the Centre for Psychoanalytic Studies at the University of Essex (<http://www.essex.ac.uk/centres/psycho/>) and the Department of Psychoanalytic Studies at University College, London (<www.ucl.ac.uk/psychoanalysis/courses/phd/phd.htm>), which offer both clinical and academic approaches to Freudian and Jungian models.

[7] See, for example, Richard Noll, *The Jung Cult* (Princeton: Princeton University Press, 1994); Richard Noll, *The Aryan Christ* (New York: Random House, 1997). For a somewhat more nuanced approach, see

has also been interpreted by some scholars as a philosopher of religion who covertly utilised psychological models to support metaphysical claims.[8] This implies that his psychological theories are dubious because he attempted to conceal his belief in the existence of God under the cloak of the psychological analysis of religious experience. Such scholarly perspectives rely on particular assumptions about the nature of psychology as an 'empiric' science entirely unrelated to 'religion' and best pursued within the broader frameworks of medicine and sociology—assumptions which plagued Jung himself but which are as 'religionist' as any other scholarly agenda with a set of *a priori* beliefs. Such assumptions are open to serious challenge, not only by therapeutic practitioners of various persuasions, but by those historians of religion who have recalled that the term 'psychology' is in fact derived from the Greek word for soul. As G. William Barnard suggests: 'It is not so obvious that simply because Jung's work was more overtly esoteric and religious than was previously thought, that it should therefore automatically be discounted by scholars'.[9] The tensions around Jung's position in the academy, however, reflect with precision his own profound conflict between scientifically demonstrable 'truths' and visionary religious experience.

Despite the ambiguous reception of Jung within the world of academic psychology, his influence has been pervasive in the various currents of transpersonal psychology.[10] In theological circles, he has long been favoured by psychologically inclined theologians, and *Liber Novus* is

---

Wouter J. Hanegraaff, *New Age Religion and Western Culture: Esotericism in the Mirror of Secular Thought* (Leiden: Brill, 1996), pp. 496–513.

[8] See, for example, Robert A. Segal, 'Jung as psychologist of religion and Jung as philosopher of religion', *Journal of Analytical Psychology* 55 (2010): pp. 361–84.

[9] Barnard, 'Diving Into the Depths', p. 305.

[10] See, for example, Ira Progoff, *At a Journal Workshop: Writing to Access the Power of the Unconscious and Evoke Creative Ability* (New York: Penguin-Tarcher, 1992); Ira Progoff, *The Symbolic and the Real: A New Psychological Approach to the Fuller Experience of Personal Existence* (New York: McGraw-Hill, 1973); Barbara Somers, *The Fires of Alchemy: A Transpersonal Viewpoint* (Bourne: Archive Publishing, 2004).

likely to enhance rather than detract from that influence.[11] In the field of literature, Jung's ideas glow like flourescent threads throughout the work of novelists such as James Joyce, Hermann Hesse, and Thomas Mann,[12] and scholars of twentieth-century literature, in addition to exploring these threads, frequently utilise Jungian models as a methodology to explore the major themes in literary texts.[13] For many decades Jung also exercised an enormous influence on historians of religion such as Mircea Eliade, Gilles Quispel, Henry Corbin, and Pierre Riffard, who have examined the repeating themes of myth and ritual across cultures and historical epochs and have embraced Jung's idea of archetypal patterns as reflections of the deepest dynamics of the human religious imagination. Quispel, for example, refers to these patterns as 'basic structures of religious apperception'; Riffard calls them 'anthropological structures', implying a human predisposition to generate religious ideas according to specific patterns of thought.[14] The Eranos Conferences, which began in 1933 and were held annually in Ascona, Switzerland until 1976, were inspired by Jung's

---

[11] See, for example, John P. Dourley, *The Intellectual Autobiography of a Jungian Theologian* (Lampeter: Edwin Mellen Press, 2006); Brendan Collins, 'Wisdom in Jung's Answer to Job', *Biblical Theology Bulletin* 21 (1991): pp. 97–101.

[12] For Joyce, see, for example, Hiromi Yoshida, *Joyce and Jung: The 'Four Stages of Eroticism' in* A Portrait of the Artist as a Young Man (New York: Peter Lang, 2007). Joyce's daughter Lucia went into analysis with Jung in 1934. For Hesse, see Miguel Serrano, *C. G. Jung and Hermann Hesse: A Record of Two Friendships* (Einsiedeln, CH: Daimon Verlag, 1998); Emanuel Maier, *The Psychology of C. G. Jung in the Works of Hermann Hesse* (unpublished PhD dissertation, New York University, 1953). Hesse was the analytic patient of Jung's assistant J. B. Lang (1883–1945). For Mann, see Paul Bishop, 'Thomas Mann and C. G. Jung', in *Jung in Contexts: A Reader*, ed. Paul Bishop (London: Routledge, 1999), pp. 154–88.

[13] See, for example, Terence Dawson, 'Jung, Literature, and Literary Criticism', in *The Cambridge Companion to Jung*, ed. Polly Young-Eisendrath and Terence Dawson (Cambridge: Cambridge University Press, 1997), pp. 255–80; Bettina L. Knapp, *A Jungian Approach to Literature* (Carbondale, IL: Southern Illinois University Press, 1984); Richard P. Sugg, ed., *Jungian Literary Criticism* (Evanston, IL: Northwestern University Press, 1992).

[14] See Gilles Quispel, *Gnosis als Weltreligion: Die Bedeutung der Gnosis in der Antike* (Zürich: Origo Verlag, 1951), p. 39; Pierre Riffard, *L'esoterisme* (Paris: Laffont, 1990), p. 135.

work and attracted the participation of Eliade and Corbin as well as Gershom Scholem.[15] According to Stephen Wasserstrom, these scholars were all variously influenced by Jung's theory of archetypes, and focused on the generic features of religions and the centrality of mystical experience. 'The approach to religion that they epitomized', observes Wasserstrom, 'infiltrated scholarship on religion throughout the world'.[16] But all academic paradigms have a finite life. The culture-specific approach to the history of religions now dominant within the academy has served as a necessary corrective to the dangers of 'metanarratives',[17] but it can sometimes go to extremes, viewing Jung's theory of archetypes as 'universalist' or 'essentialist' and therefore tending to reject it—although Jung himself emphasised the cultural adaptability and fluidity of particular archetypal themes.

Commenting on the modern tendency within esoteric circles to validate occult philosophies and practices through psychology, Olav Hammer refers to 'Jungianism' as a pervasive current.[18] Nicholas Campion suggests this term as applicable to astrologers such as Dane Rudhyar and Alexander Ruperti, who selected particular themes they found in Jung's work and used those themes to justify their own Theosophical doctrines.[19] Hammer, like many other scholars in the academic field now known as 'Western esotericism', places Jung's work in the category of 'psycho-religion' and perceives him as an esotericist 'who adapted esoteric motifs to the requirements of a psychologizing and scientistic epoch'.[20] Jung's influence on contemporary astrologies has certainly been potent, not least because he himself was an astrologer: the influence flowed both ways. In a letter to Freud dated 8 May 1911, Jung stated:

---

[15] See Stephen Wasserstrom, *Religion after Religion: Gershom Scholem, Mircea Eliade, and Henry Corbin at Eranos* (Princeton, NJ: Princeton University Press, 1999).

[16] Wasserstrom, *Religion after Religion*, p. 3.

[17] For 'metanarratives' and their methodological problems, see Charlotte Aull Davies, *Reflexive Ethnography: A Guide to Researching Selves and Others* (London: Routledge, 1999), pp. 4–5.

[18] Olav Hammer, *Claiming Knowledge: Strategies of Epistemology from Theosophy to the New Age* (Leiden: Brill, 2004), pp. 67–70.

[19] See Nicholas Campion, 'Is Astrology a Symbolic Language?', in this volume, pp. 9–46, esp. p. 22.

[20] Hammer, *Claiming Knowledge*, p. 68.

> At the moment I am looking into astrology, which seems indispensable for a proper understanding of mythology. There are strange and wondrous things in these lands of darkness.[21]

Freud's reply was not antagonistic, but he expressed anxiety at this latest display of eccentricity in his favourite disciple: 'I am aware that you are driven by innermost inclination to the study of the occult and I am sure you will return home richly laden...You will be accused of mysticism'.[22] This observation turned out to be entirely prophetic. In another letter to Freud, dated 12 June of the same year, Jung commented further on his astrological studies, revealing an increasing emphasis on their importance to psychology:

> My evenings are taken up very largely with astrology. I make horoscopic calculations in order to find a clue to the core of psychological truth...It appears that the signs of the zodiac are character pictures, in other words libido symbols which depict the typical qualities of the libido at a given moment.[23]

In his reply, Freud avoided any mention of Jung's deepening involvement with astrology. The painful and acrimonious break between them came two years later, at which time Jung began work on *Liber Novus*.

That Jung was a competent astrologer is no secret in Jungian circles, although even in the milieu of analytical psychology there is some discomfort about this apparently dubious proclivity.[24] However, a perusal of the general index for Jung's *Collected Works* clearly indicates how extensively and profoundly astrology permeated his work. It may have even formed the basis for his formulation of the psychological types, as well as providing a symbolic framework for his ideas about complexes and

---

[21] *Freud-Jung Letters*, 254J, p. 421.

[22] Ibid., 255F, p. 422.

[23] Ibid., 259J, p. 427.

[24] See, for example, Andrew Samuels, *Jung and the Post-Jungians* (London: Routledge, 1985), where Samuels, an analytical psychologist, disregards Jung's repeated discussions about astral fate in the *Collected Works*, and makes only one dismissive reference to astrology (p. 123).

'individuation'.[25] The parallels Jung drew between his understanding of the processes of psychological development and the mythic ascent of the soul, recounted in Platonic, Hermetic, Mithraic, Gnostic, and Jewish works from late antiquity and in alchemical and Kabbalistic texts from the medieval and early modern periods, are central to his concept of individuation. In a work called *Mysterium Coniunctionis* (CW14), Jung declared:

> The journey through the planetary houses boils down to becoming conscious of the good and the bad qualities in our character, and the apotheosis means no more than maximum consciousness, which amounts to maximal freedom of the will.[26]

Many other references in various volumes of the *Collected Works* illustrate Jung's conviction that the soul's mythic planetary journey mirrors in imaginal form the movement of the individual psyche toward integration.

Liber Novus is full of astrological motifs. But although Jung had been studying astrology for at least two years before he began work on it, there has been little exploration within either the academy or the psychological community of the astrological themes in the book, or the ways in which this work reflects Jung's profound and unique understanding of astrological symbolism within the context of the human psyche. Astrologers focusing on *Liber Novus* have tended to examine the astrological configurations in Jung's horoscope during the period in which he was writing the work, rather than the astrological symbolism presented, overtly or covertly, in the text and its images.[27] Most striking is the

---

[25] See, for example, the reference to a female patient's natal horoscope in C. G. Jung, *The Archetypes and the Collective Unconscious*, CW9i (London: Routledge & Kegan Paul, 1959), ¶606 n. 166, where Jung refers to her lack of planets in the element of air along with a Moon-Mercury square as 'reflecting' a problematic animus. He mischievously uses the astrological glyphs for the Moon-Mercury configuration rather than spelling out the planetary names, thereby baffling the uninitiated.

[26] C. G. Jung, *Mysterium Coniunctionis*, CW14 (London: Routledge & Kegan Paul, 1963), ¶309.

[27] See, for example, <http://heavenlytruth.typepad.com/heavenly-truth/2009/09/carl-jungs-red-book-the-astrology-behind-the-publication-of-jungs-most-personal-work.html>;

parade of figures in *Liber Novus* with which Jung engaged in an imaginal dialogue. These encounters comprise a kind of soul journey through the planetary spheres, which unfolds a theurgic ritual in the sense that Crystal Addey defines it: the use of symbols to awaken the soul's connection to divinity.[28] The figures Jung met and interacted with are easily identifiable as planetary figures, through allusions to traditional symbolic associations (such as Mars with his colour, red, and his metal, iron) and sometimes through explicit references in the text or in the imagery (such as the figure of Salome painted with the crescent Moon beside her head, or the solar giant Izdubar, who is transformed into the Sun itself).[29] Although it might be argued that these figures are alchemical rather than astrological and embody the alchemical metals rather than the planets, Jung categorically declared that he did not recognise the importance of alchemical texts until 1928, fifteen years after he had begun work on *Liber Novus*;[30] and alchemy itself is rooted in an astrological cosmology informed by the ancient idea of *sympatheia*—correspondences or 'traces' of the celestial realms in all things, most eloquently proclaimed by the *Tabula smaragdina* or 'Emerald Tablet' of Hermes Trismegistus, beloved of alchemists from the seventh century onward.[31]

> *Quod est inferius est sicut quod est superius; et quod est superius est sicunt quod est inferius, ad perpetranda miracula rei unius.*

<http://carljungsredbook.blogspot.com/2010/03/carl-jungs-astrology-transits-in-red.html>.

[28] Crystal Addey, 'The Lotus and the Sunflower: The Role of Symbol and Soul in Late Antique Theurgic Ritual', paper given at the ninth annual Sophia Centre conference 'Sky and Symbol', 4–5 June 2011.

[29] For more on these symbols in *Liber Novus*, see below.

[30] Jung, *Memories, Dreams, Reflections*, p. 230.

[31] The first known references to the *Tabula smaragdina* are as an appendix to the *Kitab Ustuqus al-Uss al-Thani* ('Second Book of the Elements of Foundation'), attributed to Jabir ibn Hayyan, and in the *Kitab Sirr al-Khaliqua wa San'at al-Tabi'a* ('Book of the Secret of Creation and the Art of Nature'), both dated between 650 and 830CE. See H. E. Stapleton, G. L. Lewis, and F. Sherwood Taylor, 'The sayings of Hermes quoted in the *Ma Al-Waraqi* of Ibn Umail', *Ambix* 3 (1949): pp. 69–90; M. Plessner, 'Hermes Trismegistus and Arab Science', *Studia Islamica* 2 (1954): pp. 45–59; Michela Pereira, 'Heavens on Earth: From the Tabula Smaragdina to the Alchemical Fifth Essence', *Early Science and Medicine* 5, no. 2 (2000): pp. 131–44.

> What is above is like what is below, and what is below is like what is above, so that the miracle of the One may be accomplished.[32]

In alchemical texts, the seven planetary metals are the earthly symbols of the planetary spirits,[33] a perspective of which Jung himself was fully aware and which he in turn understood psychologically:

> The planetary names refer not only to metals but, as every alchemist knew, to the (astrological) temperaments, that is, to psychic factors.[34]

I am using the term 'symbol' here in the sense that Jung defined it:

> Symbols are *tendencies* whose *goal* is as yet unknown. ...Symbols are the best possible formulation of an idea whose referent is not clearly known.[35]

I have placed the words 'tendencies' and 'goal' in italics because both imply that the symbol is a dynamic process, a narrative, rather than a static equation of one thing representing another. In his later works, Jung frequently reiterates this correspondence of interior psychic states with astrological symbols and alchemical processes, stating unequivocally:

> Astrologically...this process [alchemy] corresponds to an ascent through the planets from the dark, cold, distant Saturn

---

[32] For various translations of the Emerald Tablet, see <http://www.levity.com/alchemy/emerald.html>.

[33] There are many scholarly works on alchemy and its symbolism, in addition to Jung's own three volumes—C. G. Jung, *Psychology and Alchemy*, CW12 (London: Routledge & Kegan Paul, 1953); C. G. Jung, *Alchemical Studies*, CW13 (London: Routledge & Kegan Paul, 1967); and Jung, CW14. See, among others, Mircea Eliade, *The Forge and the Crucible*, trans. Stephen Corrin (Chicago: University of Chicago Press, 1962); Allen G. Debus, *The Chemical Philosophy* (Mineola, NY: Dover, 1977); William R. Newman and Anthony Grafton, eds., *Secrets of Nature: Astrology and Alchemy in Early Modern Europe* (Cambridge, MA: MIT Press, 2001); Stanton J. Linden, ed., *The Alchemy Reader: From Hermes Trismegistus to Isaac Newton* (Cambridge: Cambridge University Press, 2003).

[34] Jung, CW13, ¶355.

[35] Jung, CW14, ¶668 and n. 54.

> to the sun....The ascent through the planetary spheres therefore meant something like a shedding of the characterological qualities indicated by the horoscope....The journey through the planetary houses, like the crossing of the great halls in the Egyptian underworld, therefore signifies the overcoming of a psychic obstacle, or of an autonomous complex, suitably represented by a planetary god or demon. Anyone who has passed through all the spheres is free from compulsion; he has won the crown of victory and become like a god.[36]

As a work of 'active imagination' and a recounting of a dramatic and difficult personal journey, *Liber Novus* offers no discussions on astrological technique or theory, although Jung sometimes uses the astrological glyphs in his drawings.[37] However, the major themes of Jung's own horoscope may be perceived in his dialogues with the various imaginal figures he meets on his inner journey and, given his familiarity with astrological language, there can be no question that he was fully aware of it.

'ACTIVE IMAGINATION' AND THEURGY

A 'measure of mystery' surrounds Jung's practice of active imagination and the sources from which he derived the technique.[38] Leon Hoffman, a Freudian psychoanalyst reviewing *Liber Novus*, suggests that active imagination is not rooted in Freud's technique of free association, as is sometimes assumed, but is 'a direct descendant of the manifestations of mediums in trance states at the end of the nineteenth century and the beginning of the twentieth'.[39]

---

[36] Jung, CW14, ¶308.

[37] See, for example, the solar and planetary glyphs in the original version of the 'Systema Munditotius' (discussed below), reproduced in Jung, *Liber Novus*, p. 363, and the astrological glyph for the Sun in the caption for the painting of the Orphic creator-god Phanes in Jung, *Liber Novus*, p. 113.

[38] Dan Merkur, *Gnosis: An Esoteric Tradition of Mystical Visions and Unions* (Albany, NY: SUNY Press, 1993), p. 37.

[39] Leon Hoffman, 'Varieties of Psychoanalytic Experience: Review of *The Red Book*', *Journal of the American Psychoanalytic Association* 58 (2010): pp. 781–85, on p. 783. For Jung's early involvement in mediumistic phenomena, see C. G. Jung, 'On the Psychology and Pathology of So-called Occult Phenomena', in C. G. Jung, *Psychiatric Studies*, CW1 (London: Routledge & Kegan Paul, 1957), pp. 3–92.

Robert Kugelman calls *Liber Novus* 'visionary writing', following a tradition emerging from the German Romanticism of the early nineteenth century,[40] although Kugelman does not consider the much older alchemical and Kabbalistic roots of German Romanticism itself.[41] *Liber Novus* indicates that Jung began to develop active imagination no later than 1913, and he published his early thoughts on it in 1921,[42] but he did not use the term itself until a series of lectures he gave at the Tavistock Clinic in London in 1935. Jung insisted that psychic energy, or libido, 'cannot appear in consciousness except in the form of images'.[43] This is apparently a 'modern' understanding of such images: they are products of the deeper levels of the unconscious, emerging either spontaneously in dreams and visions or through deliberate invocation, and they give shape to otherwise incommunicable psychic realities. But the practice itself long predates the modern era and forms an important aspect of alchemical and Kabbalistic magical techniques, as well as being evident in late antique Gnostic, Jewish, Hermetic, and Neoplatonic literature such as the work of Iamblichus.[44] Jung was familiar with all these sources,

---

This paper was originally published in Leipzig in 1902 as *Zur Psychologie und Pathologie sugennanter occulter Phänomene*. See also Roderick Main, ed., *Jung on Synchronicity and the Paranormal* (Princeton: Princeton University Press, 1997), which includes not only the early psychiatric paper but letters and extracts from other works dealing with the theme of mediumistic trance phenomena. See also F. X. Charet, *Spiritualism and the Foundations of C. G. Jung's Psychology* (Albany, NY: SUNY Press, 1993).

[40] Robert Kugelman, 'Review of *The Red Book*', *Journal of the History of the Behavioral Sciences* 47, no. 1 (2011): pp. 101–4, on p. 101.

[41] See Ernst Benz, The *Mystical Sources of German Romantic Philosophy*, trans. Blair R. Reynolds and Eunice M. Paul (Eugene, OR: Pickwick Publications, 1983). See also Wouter J. Hanegraaff, 'Romanticism and the Esoteric Connection', in *Gnosis and Hermeticism: From Antiquity to Modern Times*, ed. Roelof van den Broek and Wouter J. Hanegraaff (Albany, NY: SUNY Press, 1998), pp. 237–68.

[42] C. G. Jung, *Psychological Types*, CW6 (London: Routledge & Kegan Paul, 1971), ¶711–22. Merkur states that Jung wrote an early article about active imagination in 1916, although he did not publish it until 1957; see Merkur, *Gnosis*, pp. 37–54.

[43] Jung, CW6, ¶722.

[44] For the practice in Kabbalistic literature, see Elliot R. Wolfson, *Through a Speculum That Shines: Vision and Imagination in Medieval Jewish Mysticism* (Princeton, NJ: Princeton University Press, 1997).

including Iamblichus' *De mysteriis*.[45] Although Jung's expositions on active imagination are psychological, they often, albeit covertly, attribute the qualities of divinity to the collective unconscious, reflecting a panentheistic understanding of symbols as complex webs of associations that transcend the perceived opposites and paradoxically conjoin, in an apparently simple image, object, word, number, glyph, or physical structure, the inner and the outer, the material and the psychic, the corporeal and the spiritual.

Active imagination rests on Jung's conviction that fantasy is 'a real psychic process'. An image arising from a dream, a waking reverie, a visionary state, a meditation, or a magical invocation triggers a strong emotional response and opens the gateway; the individual must then engage with the images, which 'present themselves spontaneously'.[46] Jung then advises: 'You must enter into the process with your personal reactions...as if the drama being enacted before your eyes were real...It is as real as you—as a psychic entity—are real'.[47] Active imagination, as expressed in *Liber Novus*, may be viewed not only as a therapeutic process but also as a theurgic technique which opens the lines of communication between seen and unseen worlds, and between the individual and a universal life-force ('libido') in constant flux and creative emergence, which Jung referred to

---

For the practice in Gnostic currents, see Merkur, *Gnosis*. For the practice in Iamblichus, see Gregory Shaw, *Theurgy and the Soul: The Neoplatonism of Iamblichus* (University Park, PA: Penn State Press, 1971). For the practice in Hermeticism, see Dan Merkur, 'Stages of Ascension in Hermetic Rebirth', *Esoterica* 1 (1999): pp. 79–96.

[45] See Jung, CW9i, ¶573. A number of editions of Iamblichus' *De mysteriis* are found in Jung's private library, including a translation in French (1895) and a rare Latin edition published in Venice (1497); see *C. G. Jung Bibliothek Katalog*, ETH, Küsnacht-Zürich, 1967.

[46] Jung, CW9i, ¶334.

[47] Jung, CW14, ¶753. For more of Jung's many discussions on 'active imagination', see C. G. Jung, *The Structure and Dynamics of the Psyche*, CW8 (London: Routledge & Kegan Paul, 1960), ¶166–75; Jung, CW9i, ¶621; Jung, CW14, ¶752–55. See also Joan Chodorow, ed., *Jung on Active Imagination* (Princeton, NJ: Princeton University Press, 1997); Marie-Louise von Franz, *Alchemical Active Imagination* (Irving, TX: Spring Publications, 1979; repr. New York, NY: Shambhala, 1997); Jeffrey Raff, *Jung and the Alchemical Imagination* (York Beach, ME: Nicolas-Hays, 2000); Merkur, *Gnosis*, pp. 37–54; Benjamin Sells, ed., *Working with Images: The Theoretical Base of Archetypal Psychology* (Woodstock, CT: Spring Publications, 2000).

as the 'collective unconscious'. In *Liber Novus* he describes this magical process as necessary in order to 'receive or invoke the messenger'.[48] The starter motor for initiating a transformative relationship with the images is emotional alignment, which facilitates psychic participation between the individual and that formless potency which is symbolised by and embodied in the image. This is akin to Iamblichus' idea of 'ritual receptivity'.[49] The psychological and philosophical implications are profound: the human being is able to meet the archetypal realm—or, as Jung calls it in *Liber Novus*, the Pleroma and the realm of the star-gods—not as a helpless victim of astral or psychic fate, but as a co-creator who can inaugurate the transformation of both the archetypal powers and the conscious personality.

'The Red One'

One of the more obvious planetary figures Jung describes in *Liber Novus* is called 'The Red One'. Jung did not create a pictorial image of this entity, but he provided enough descriptive material in the text to allow the reader to recognise which astral potency he encountered.[50] The epithet, 'Red One', is not unique to Jung: the Egyptians referred to the planet Mars as *Har décher*, 'The Red One',[51] and the horned warrior-god Cocidius—known from inscriptional and iconographic evidence around the area of Hadrian's wall—and the Gaulish Rudiobus, both equated by the Romans with their battle-god Mars, were likewise called 'The

---

[48] Jung, *Liber Novus*, p. 314.

[49] Iamblichus discusses the idea of επιτεδειοτες (*epitedeiotes*), 'fitness', 'aptitude', or 'receptivity', a particular blend of human engagement and divine agency; see Iamblichus, *De mysteriis*, 105.1; 125.5; 127.9, trans. Emma C. Clarke, John M. Dillon, and Jackson P. Hershbell (Leiden: Brill, 2004). On 'receptivity' as a critical factor in theurgy, see also Crystal Addey, 'Divine Possession and Divination in the Graeco-Roman World: The Evidence from Iamblichus' *On the Mysteries*', in Bettina E. Schmidt and Lucy Huskinson (eds.), *Spirit Possession and Trance: New Interdisciplinary Perspectives* (London: Continuum, 2010), 171-181, on pp. 179-181.

[50] Jung, *Liber Novus*, pp. 259–61.

[51] Markus Hotakainen, *Mars: From Myth and Mystery to Recent Discoveries* (New York, NY: Springer, 2008), p. 13. In Egypt, Mars was also called 'Horus-Desher, 'Red Horus'; see Margaret Bunson, *Encyclopedia of Ancient Egypt* (New York, NY: Facts on File Inc., 2002), p. 58.

Red One'.[52] There is also a Celtic deity called The Dagda, most powerful of the Tuatha de' Danann ('peoples of the goddess Dannu'), known also as Ruadh Rofessa: the 'Red One of Great Knowledge', with power over life and death.[53] In his later explorations into alchemy, Jung encountered, and quoted, many descriptions of this Martial spirit and its associations with the colour red and the metal iron, particularly in the works of his favourite alchemist, Theophrastus Bombastus von Hohenheim, known as Paracelsus.[54] From Paracelsus and various other alchemical authors, Jung later extrapolated his own understanding of Mars:

> Astrologically, Mars characterizes the instinctual and affective nature of man. The subjugation and transformation of this nature seems to be the theme of the alchemical opus.[55]

It is also likely that Jung drew on the description of Mars provided by the English astrologer Alan Leo in *How to Judge a Nativity*—a work included, with many other books by Leo, in Jung's private library—first published in 1912, a year before Jung began work on *Liber Novus*.

> All the animal propensities, sensations, passions, desires, and appetites come under the vibration of Mars...Mars is the ruler over the animal nature in man; and the task set for humanity is not only that of subjecting, ruling and controlling the animal nature, but also its transmutation into a higher force than that which ministers to the animal soul.[56]

---

[52] For these deities and others with the same epithet see also <http://faculty.indwes.edu/bcupp/solarsys/Names.htm>; <http:planetarynames.wr.usgs.gov>; <http://www.celtnet.org.uk/gods_c/cocidius.html>.

[53]<http://www.tairis.co.uk/index.php?option=com_content&view=article&id=125:the-dagda-part-1&catid=45:gods&Itemid=8>.

[54] See, for example, Jung's discussion of Paracelsus' 'Ares' principle (the Greek name for Mars), which Jung equated with 'the principle of individuation', and the references to Johannes Braceschus' definition of Mars as 'Daemogorgon': a 'hot and bilious man' whose 'complexion is choleric' and who 'is iron'; in Jung, CW13, ¶176–78.

[55] Jung, CW13, ¶177 n. 39.

[56]Alan Leo, *How to Judge a Nativity* (London: Modern Astrology, 1912; repr., Edinburgh: International Publishing Co., 1965), pp. 30–31.

It is perhaps not coincidental that The Red One In *Liber Novus* initially appears on horseback, as Mars, at the time of Jung's birth, was placed in the zodiacal sign of Sagittarius, the Centaur or mounted Archer.[57]

> I look out into the distance. I see a red point out there. It comes nearer on a winding road, disappearing for a while in forests and reappearing again: it is a horseman in a red coat, the red horseman...There stands the Red One, his long shape wholly shrouded in red, even his hair is red. I think: in the end he will turn out to be the devil.[58]

The Red One's garments 'shine like glowing iron', and his passion and ferocity, as well as his mocking cynicism, initially convince Jung that he is a demonic rather than a daimonic being. But in the course of their encounter this entity changes, and transforms Jung's clothing into living foliage—in effect, bringing him to life and uniting him with the life of nature and the instincts—and reveals his secret, announcing: 'I am joy!' Jung discusses the nature of the joy this figure offers, describing him as 'of a fiery...passionate nature' and calling him 'red-coloured, red-scented, warm, red joy'. The transformation can only occur through emotional engagement with the figure, demonstrating the importance Jung attributed to an authentic, sympathetic encounter rather than a mere intellectual analysis. What is particularly revealing about 'The Red One' is that the figure does not merely represent simple character traits (anger, aggression) or concrete events (violence, war), as one might expect from the more traditional astrological sources available to Jung at the time. The Red One behaves as an ontically autonomous entity with a distinct individuality, a *daimon* or a god, capable both of transformation and of initiating transformation in Jung himself. An initially mocking and cynical being who becomes 'warm red joy' is not a typical portrait of Mars as found in astrological texts. The astrological symbols, for Jung, are clearly not mere signatures of temperament or tools for the divination of future events, but embody dynamic potencies and patterns of an archetypal nature, unfolding within both seen and unseen worlds.

---

[57] For Jung's horoscope and birth data, see below.

[58] Jung, *Liber Novus*, p. 259.

The Giant Izdubar and the Solar Apotheosis

Another figure whom Jung described in *Liber Novus,* and whom he did embody in pictorial form, is the giant Izdubar.

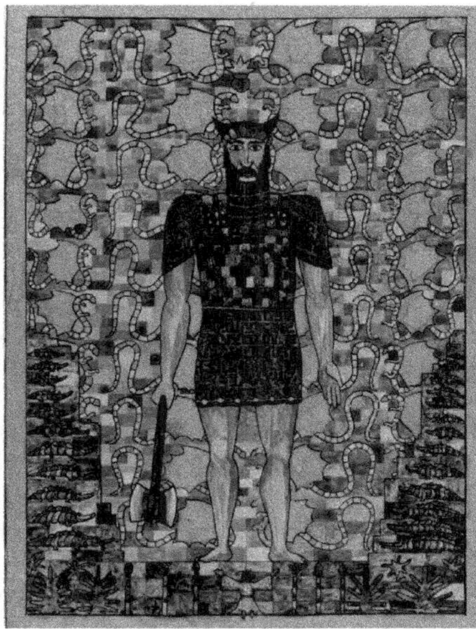

Fig. 11.1: The giant Izdubar[59]

Izdubar is a profoundly personal figure in the parade of inner images in *Liber Novus,* because Jung's dialogue with the giant dramatises the severe internal conflict that helped to precipitate his psychological collapse. The encounter with Izdubar, which extends for many pages, describes what Jung initially felt to be an irreconcilable antipathy between intuitive and rational faculties, and the apparently mutually exclusive world-views these faculties reflect. Jung's conflict was no mere intellectual debate; it was intensely felt and experienced, somatically as well as psychologically, and its importance to him must be viewed against the backdrop of his upbringing and the culture in which he lived and worked. It expresses a drastic and painful dilemma within a man who was both an artist communing with imaginal realms and a rational thinker determined to be accepted

[59] Image in Jung, *Liber Novus,* p. 36.

within the scientific community of his time and the claustrophobic social and religious structures of the late-nineteenth-century Swiss German milieu in which he was brought up. Jung called the pressure of this cultural background 'the spirit of this time' when he stated in *Liber Novus*:

> The spirit of the depths from time immemorial and for all the future possesses a greater power than the spirit of this time, who changes with the generations. The spirit of the depths has subjugated all pride and arrogance to the power of judgement. He took away my belief in science, he robbed me of the joy of explaining and ordering things, and he let devotion to the ideals of the time die out in me. He forced me down to the last and simplest things.[60]

The giant Izdubar bears bull's horns and wields a double-headed axe. The name Izdubar is based on a mistranslation in 1872 of the cuneiform ideograph for Gilgamesh.[61] By 1906 the mistake had been recognised in scholarly literature,[62] and Jung used the corrected form in 1912 in *Wandlungen und Symbole der Libido* (*Transformations and Symbols of the Libido*),[63] but for *Liber Novus* he chose the older, apparently 'wrong' name for his own imaginal figure. Sonu Shamdasani suggests that Jung's use of the earlier form 'was a way of

---

[60] Jung, *Liber Novus*, p. 229.

[61] See George Smith, 'The Chaldean Account of the Deluge', *Transactions of the Society of Biblical Archaeology* 1–2 (1872): pp. 213–34. The error was later perpetuated in Leonidas le Cenci Hamilton's translation of the epic, *Ishtar and Izdubar: The Epic of Babylon* (London: W. H. Allen, 1884), and in Alfred Jeremias, *Izdubar-nimrod: Eine altbabylonische Heldensage: Nach den Keilschriftfragmenten Dargestellt* (Leipzig: B. G. Teubner, 1891). The discovery of a lexicographic tablet by T. G. Pinches in 1890 eventually allowed the literal translation of the ideograph—*iz* (or *gish*), *du*, and *bar*—to be rendered phonetically as 'Gilgamesh'. See Morris Jastrow, *The Religion of Babylonia and Assyria* (Boston, MA: Athenaeum Press, 1898), p. 468.

[62] See Peter Jensen, *Das Gilgamesh-Epos in der Weltliteratur* (Strasbourg: Karl Trübner, 1906), p. 2.

[63] C. G. Jung, *Wandlungen und Symbole der Libido*, in *Jahrbuch für psychoanalytische und psychopathologische Forschungen*, III–IV (Leipzig, 1911–12), published in English as *Psychology of the Unconscious* (London: Kegan Paul, Trench, Trubner & Co., 1917) and then revised and published as *Symbols of Transformation*, CW5 (London: Routledge & Kegan Paul, 1967).

indicating that his figure, while related to the Babylonian epic, was a free elaboration'.[64] The giant certainly exhibits many of the attributes of the Sumerian hero in his persistent quest for immortality.[65] In *Symbols of Transformation* (CW5), Jung describes Gilgamesh as a 'sun-hero',[66] and in *Liber Novus*, Izdubar is explicitly linked with the symbolism of the Sun. He seeks the Western land 'where the Sun goes to be reborn', but Jung reminds him that he is mortal and consequently can never reach the Sun, which is, after all, merely a star circling in empty space, orbited by the Earth and the other planets: there is no 'Western land'.[67] In Jung's horoscope, which I will look at later, the Sun is placed in Leo precisely at the Western horizon, as he was born at sunset; the giant is none other than the questing solar spirit of Jung himself. When Izdubar invokes the truth of 'the astrologers and priests', Jung compares this so-called truth unfavourably with 'real' scientific truth. He asks Izdubar: 'Can you really not accept that you are a fantasy?' Izdubar's shock at recognising that he must accept his mortality nearly destroys him, and results in his being lamed. He accuses Jung of offering him poison, whereupon Jung replies:

> What you call poison is science. In our country we are nurtured on it from youth, and that may be one reason why we haven't properly flourished and remain so dwarfish....We had no choice. We had to swallow the poison of science....This poison is so insurmountably strong that everyone, even the strongest, and even the eternal Gods, perish because of it.[68]

The conflict between the truth of science and the truth of the imagination forms the dominant motif of this section of *Liber Novus*. Yet it is through compassion for this newly lamed figure, once so strong and proud, that it becomes apparent to

---

[64] Sonu Shamdasani, *C. G. Jung: A Biography in Books* (New York: W. W. Norton, 2012), p. 93.

[65] *The Epic of Gilgamesh* probably emerged in Sumeria in the third millennium BCE, but the surviving cuneiform tablets come from the seventh century BCE. See Andrew R. George, ed. and trans., *The Babylonian Gilgamesh Epic: Critical Edition and Cuneiform Texts* (Oxford: Oxford University Press, 2003).

[66] Jung, CW5, ¶251 n. 1.

[67] Jung, *Liber Novus*, p. 278.

[68] Ibid., pp. 278–79.

Jung that he is responsible for healing the giant. He begins to acknowledge that the 'wisdom of the astrologers...is that which comes to you from inner things'.[69] Jung carries the giant to a 'house of healing', where a series of magical incantations is performed.[70] Izdubar, who has shrunk and is now encapsulated within an egg, emerges transformed into the Sun itself, radiating light. Astonished, Jung declares in amazement: 'Oh Izdubar! Divine one! How wonderful! You are healed!' The giant replies: 'Healed? Was I ever sick? Who speaks of sickness? I was sun, completely sun. I am the sun.'[71]

Fig. 11.2: The solar apotheosis of Izdubar[72]

One of the most remarkable features about this narrative is that, from the perspective of Jung's understanding of

[69] Ibid., p. 278.
[70] For the text of the incantations, see Jung, *Liber Novus*, pp. 284–85.
[71] Ibid., p. 286.
[72] Image in Jung, *Liber Novus*, p. 64.

astrological symbols, the Sun is not only 'the centre of each individual character', as Alan Leo put it,[73] but also portrays a lifelong process embodying a fundamental archetypal conflict and an urgent quest for resolution of the conflict. A parallel to this perception of solar symbol as dialectical process can also be found in James Hillman's *The Soul's Code*, although no explicit astrological equations are given there.[74] *Liber Novus* reveals Jung's understanding of his own natal Sun in Leo, setting on the western horizon at his birth, embodied in a heroic questing figure whose nature, although contradictory to the 'spirit of this time', had to be consciously recognised and honoured before he could emerge as what Jung finally perceived him to be: the divine essence of the human soul. This presentation of the dynamic and developmental nature of astrological symbols, which, to use the words of William Butler Yeats, 'bring our soul to crisis',[75] is absent from the astrological texts to which Jung refers in the various volumes of the *Collected Works* as well as those listed in his private library—with the qualified exception of Alan Leo[76]—and seems to reflect not only his understanding

---

[73] Alan Leo, *How to Judge a Nativity* (London: Modern Astrology, 1903), p. 29.

[74] James Hillman, *The Soul's Code: In Search of Character and Calling* (London: Bantam, 1996). For Hillman's familiarity with astrological symbols, see James Hillman, 'The Azure Vault: The Caelum as Experience', in *Sky and Psyche: The Relationship Between Cosmos and Consciousness*, ed. Nicholas Campion and Patrick Curry (Edinburgh: Floris Books, 2005), pp. 37–54.

[75] William Butler Yeats, letter to Florence Farr, cited in Kathleen Raine, *Yeats, the Tarot, and the Golden Dawn* (Dublin: Dolmen Press, 1972), p. 44, and William Butler Yeats, *The Autobiography of William Butler Yeats* (New York: Macmillan, 1953), p. 272.

[76] In the *Collected Works*, Jung cites classical works such as Ptolemy's *Tetrabiblos*. In his private library, a more comprehensive list of astrological sources is revealed. Among these are classical, medieval, and early modern authors such as Julius Firmicus Maternus and Johannes Kepler as well as the modern British astrologers 'Raphael', Charles Carter, 'Sephariel', Alfred John Pearce, and Alan Leo, the American astrologers Dane Rudhyar and Francis J. Mott, and the German astrologers Adolph Drechsler, Heinz Arthur Strauss, and Alfons Rosenberg. None of these authors discusses astrological symbolism in the manner in which Jung presents it, although Leo viewed the horoscope as a pathway toward spiritual evolution rather than a simple map of character. Rudhyar dipped into Jung's works and extracted certain ideas which he adapted to support his Theosophical framework.

of the alchemical opus, but also the engagement with the gods described by Neoplatonic authors such as Iamblichus, the transformational processes of late antique Greco-Egyptian magical texts such as the so-called 'Mithras Liturgy', and the divinisation rituals of Hermetic treatises such as the *Poimandres*.[77] That Jung was inspired by such 'esoteric' sources does not contradict the possibility that his psychological models might be partly or even entirely valid and useful; these early religious sources exhibited profound insights into the deepest conflicts and aspirations of the human being long before the invention of the word 'psychology'. Their efficacy was certainly unquestioned by Jung, who believed, perhaps justifiably, that his efforts to engage with the imaginal realm had healed him.

The Anima and the Moon

Jung's theories about the anima—the feminine 'soul' in men—are now sometimes viewed as seriously limited by the particular cultural values of his time.[78] Switzerland was, after all, the last European country to offer women the vote, only conceding defeat in 1971, and the stern suppression of all those qualities which Jung understood as 'feminine'—feeling, imagination, receptivity to the invisible world—contributed to his formulation of a sharp psychological

[77] For the 'Mithras Liturgy', see Hans Dieter Betz, *The 'Mithras Liturgy': Text, Translation and Commentary* (Tübingen: Mohr Siebeck, 2003). For the *Poimandres*, see Brian P. Copenhaver, ed. and trans., *Hermetica: The Greek Corpus Hermeticum and the Latin Asclepius* (Cambridge: Cambridge University Press, 1992), pp. 1–7. Jung relied heavily on the English translations of Gnostic and Hermetic texts published by G. R. S. Mead (e.g. G. R. S. Mead, *Thrice-Greatest Hermes*, London: Theosophical Publishing House, 1906) and had Mead's translation of 'A Mithras Liturgy' (G. R. S. Mead, *The Mysteries of Mithra*, Vol. V of G. R. S. Mead, *Echoes From the Gnosis*, London: Theosophical Publishing House, 1907) as well as the original German translation by Albert Dieterich, *Ein Mithrasliturgie* (Leipzig: Tübner, 1910). 'Mithras Liturgy' is an erroneous name given to the magical papyrus by Dieterich; the second-century text is syncretic and refers to itself as a 'ritual of divinisation'.

[78] For Jung's essays on the anima, see C. G. Jung, *Two Essays on Analytical Psychology*, CW7 (London: Routledge & Kegan Paul, 1972), ¶296–340; Jung, CW13, ¶57–63; Jung, CW9i, ¶111–47; C. G. Jung, *Aion: Researches into the Phenomenology of the Self*, CW9ii (London: Routledge & Kegan Paul, 1959), ¶20–42.

dichotomy between the sexes which may be somewhat less applicable to early twenty-first-century Western society. In Jung's time and country the stereotypical roles against which the feminist movement of the late twentieth century fought so vigorously were still firmly in place, and Jung was acutely aware of the imprisonment of his own imaginal faculties by the scientific straitjacket of his time.

Fig. 11.3: Salome[79]

There is more than one feminine figure in *Liber Novus*, but they are linked by their association with night, darkness, and the Moon, and they all bear the name of Salome. She first appears as the beautiful daughter of Elijah, the Old Testament prophet whom Jung later described as 'a living archetype' symbolising the collective unconscious.[80] Although this early manifestation of Salome is blind and apparently bears the name of a bloodthirsty temptress who once asked for the head of John the Baptist on a platter, she

[79] Image in Jung, *Liber Novus*, p. 155.

[80] See C. G. Jung, *The Symbolic Life*, CW18 (London: Routledge & Kegan Paul, 1977), ¶1518–31.

may be more closely related to the Gnostic Salome, not to be confused with the daughter of Herodias, who in the second-century *Gospel of Thomas* and other Gnostic texts appears as the confidante and disciple of Jesus.[81] Jung's Salome personifies *sophia*, the wisdom of the unconscious, as Elijah informs Jung: 'My wisdom and my daughter are one'.[82] The same figure appears again, nameless but with 'a beautiful and unworldly soul', as the daughter of an old scholar living in solitude in a castle in a forest; in this guise she is imprisoned by her father in 'unbearable captivity', and Jung feels responsible for her release and redemption. At the end of their dialogue, 'dim moonlight' penetrates the room and, as she disappears into the darkness, she informs Jung: 'I bring you greetings from Salome'.[83] Toward the end of *Liber Novus*, after Jung's transformative meeting with the magician Philemon, Salome, once again Elijah's daughter, reappears as 'a great irridescent serpent' whom Jung addresses as his sister and soul;[84] he has at last realised that 'Salome is where I am'.[85] Her voice is now identifiable as the voice of the soul which, at the very beginning of *Liber Novus*, instructed him in the visionary path he would have to follow to fulfil the requirements of the 'Spirit of the Depths'.[86] Like The Red One, Salome herself undergoes transformation as she transforms Jung and unlocks the potency of the archetypal realms through his emotional engagement with her.

---

[81] *Gospel of Thomas*, 61, trans. Thomas O. Lambdin, in James M. Robinson, ed., *The Nag Hammadi Library in English* (Leiden: Brill, 1988), pp. 117–30. Jung would not have known this text at the time he worked on *Liber Novus*, as it was only discovered in 1945. However, Salome also appears as a disciple in the *Secret Gospel of Mark*, the *Greek Gospel of the Egyptians*, the *Protevangelium of James* (XIX–XX), and the *Pistis Sophia* (54, 58, 132, 144), a second-century Gnostic text with which Jung was familiar through the English translation of G. R. S. Mead (London: J. M. Watkins, 1921). Jung also had access to the *Greek Gospel of the Egyptians* and the *Protevangelium of James* through M. R. James, *The Apocryphal New Testament* (Oxford: Clarendon Press, 1924). For the *Secret Gospel of Mark*, see Morton Smith, *The Secret Gospel: The Discovery and Interpretation of the Secret Gospel According to Mark* (London: Victor Gollancz, 1974).

[82] Jung, *Liber Novus*, p. 246.

[83] Ibid., pp. 262–63.

[84] Ibid., pp. 317.

[85] Ibid., p. 327.

[86] Ibid., pp. 233–37.

Elaborating on the image of Salome in *Liber Novus* in an essay written nearly forty years later, Jung, in the Mercurial spirit which has annoyed critics such as Wouter Hanegraaff and Richard Noll,[87] attributed the image of Salome to the dream of a patient rather than conceding that it was his own, and declared: 'Then she [the anima] appears in a church, taking the place of the altar...[Dreams] restore the anima to the Christian church, not as an icon but as the altar itself.'[88] The inscription Jung inscribed along the border of the image states: 'The wisdom of God is a mystery, even the hidden wisdom, which God ordained before the world unto our glory....For the Spirit searcheth all things, yea, the deep things of God.' Along the right side of the image, Jung wrote: 'The Spirit and the Bride say, Come.' This quote from Revelation 22:17 may be linked with Jung's understanding of the Kabbalistic *Shekhinah* as the Bride or feminine face of the godhead, the manifestation of God's glory in the world.[89] In this sense Jung's Salome is, to quote the elegant phrase Elliot Wolfson used in relation to the *Shekhinah*, 'a mirror through which the hidden God is disclosed'.[90] Jung was already familiar with the Kabbalah when he worked on *Liber Novus* and, although he avoids referring to it in his early publications, following the trend of the time in his efforts to

---

87 See Hanegraaff, *New Age Religion*, p. 507 and n. 429; Noll, *The Jung Cult*, pp. 181–84.

88 C. G. Jung, 'The Psychological Aspects of the Kore', in C. G. Jung and C. Kerényi, *Essays on a Science of Mythology: The Myth of the Divine Child and the Mysteries of Eleusis* (Princeton, NJ: Princeton University Press, 1963), pp. 175–76.

89 For the *Shekhinah* as the Bride, the manifestation of God's glory in the world (e.g. Nature), the personification of God's wisdom, and the Moon, see Gershom Scholem, 'The Feminine Element in Divinity', in Gershom Scholem, *On the Mystical Shape of the Godhead: Basic Concepts in the Kabbalah*, trans. Joachim Neugroschel (New York, NY: Schocken Books, 1991), pp. 140–96; Arthur Green, 'Shekhinah, the Virgin Mary, and the Song of Songs: Reflections on a Kabbalistic Symbol in Its Historical Context', *AJS Review* 26, no. 1 (2002): pp. 1–52. The *Shekhinah* is often associated with the Moon: see Wolfson, *Through a Speculum*, pp. 267n, 359; Moshe Idel, *Kabbalah and Eros* (New Haven, CT: Yale University Press, 2005), pp. 69, 91, 261n; Gershom Scholem, *On the Kabbalah and Its Symbolism*, trans. Ralph Mannheim (New York, NY: Schocken Books, 1965), pp. 107–8, 151–53.

90 Elliot Wolfson, 'Theosis, Vision, and the Astral Body in Medieval German Pietism and the Spanish Kabbalah', in this volume, p. 123.

dissociate himself from the Jews (especially Freud),[91] it is evident from the knowledge he exhibited in private seminars given in English in Zürich between 1928 and 1930 and finally published in 1984,[92] and from a letter written to him by G. R. S. Mead, dated 19 November 1919, which discusses the Kabbalistic proclivities of the Austrian novelist Gustav Meyrink,[93] that Jung had encountered the Kabbalah no later than 1919 and possibly earlier, through Mead's work.[94]

The altar may be understood as the place of sacrifice and of transformation, the liminal zone where the ineffable becomes visible and human and divine meet and recognise each other through the mediation of symbols. It is on the altar that the wine is transformed into blood, the wafer into flesh, and the smoke of the burnt offering transformed into food for the gods. Salome, in Jung's active imagination, is a highly ambivalent figure: she is both a dark and potentially destructive force and, as the noetic daughter of Elijah and a personification of the wisdom of the unconscious, an image of both the individual soul and the world-soul, the *anima mundi*. There is an interesting family resemblance between the image of Salome, with her association with 'hidden wisdom' and the 'deep things of God', and the image of the High Priestess in A. E. Waite's Tarot deck, published in 1910.[95] Both figures are personifications of secret wisdom and

---

[91] For Jung's pre-war foray into antisemitism, see Aryeh Maidenbaum and Stephen A. Martin, eds., *Lingering Shadows: Jungians, Freudians, and Anti-Semitism* (Boston: Shambhala, 1991); Aryeh Maidenbaum, ed., *Jung and the Shadow of Anti-Semitism: Collected Essays* (Berwick, ME: Nicolas-Hays, 2002); Léon Poliakov, *The Aryan Myth: A History of Racist and Nationalist Ideas in Europe*, trans. Edmund Howard (New York: Basic Books, 1971), pp. 286–90.

[92] William C. McGuire, ed., *Dream Analysis: Notes of the Seminar Given in 1928–1930 by C. G. Jung* (London: Routledge & Kegan Paul, 1984). See especially pp. 504–6.

[93] Letter to C. G. Jung from G. R. S. Mead, Archives, ETH-Bibliothek Zürich, Hs 1056.29826.

[94] For Jung's early acquaintance with the Kabbalah, see Sanford L. Drob, 'Towards a Kabbalistic Psychology: C. G. Jung and the Jewish Foundations of Alchemy', *Journal of Jungian Theory and Practice* 5, no. 2 (2003), pp. 77–100.

[95] A. E. Waite, *The Pictorial Key to the Tarot: Being Fragments of a Secret Tradition under the Veil of Divination* (London: William Rider & Son, 1910). The cards themselves were published both separately and as illustrations in the book. For a detailed history of this Tarot deck,

are shown with lunar crescents; both appear between two columns beneath an arch.

Fig. 11.4: Waite's 'High Priestess'[96]

Although Waite's description of his High Priestess differs from Jung's in a number of pictorial details (for example, the crescent Moon is beneath her feet instead of overhead), the similarities in their interpretations are striking. According to Waite,

> The scroll in her hands is inscribed with the word Tora, signifying the Greater Law....It is partly covered to show that some things are implied and some spoken....She is the spiritual Bride and Mother, the daughter of the stars and the Higher Garden of Eden. She is the moon nourished by the milk of the Supernal Mother.[97]

---

see K. Frank Jensen, *The Story of the Waite-Smith Tarot* (Melbourne: Association of Tarot Studies, 2006).

[96] Waite, *Pictorial Key*, p. 77.

[97] Ibid., pp. 76–79.

Waite, who was a member of the Hermetic Order of the Golden Dawn, was a prolific author whose scholarly volumes on esoteric traditions such as alchemy and Kabbalah are referenced in a number of Jung's *Collected Works* and were included in Jung's private library.[98] Probably through the work of Waite and other occultists of the late nineteenth and early twentieth centuries,[99] Jung became acquainted with, and respectful of, the imagery of the Tarot:

> If one wants to form a picture of the symbolic process, the series of pictures found in alchemy are good examples...It also seems as if the set of pictures in the Tarot cards were distantly descended from the archetypes of transformation.[100]

It seems that Jung even progressed to offering 'keyword' interpretations of the Major Arcana of the Tarot, which he evidently borrowed from Papus' *Tarot of the Bohemians*.[101] The interpretation he gives for the High Priestess is:

> Sitting Priestess. She wears a veil. On her knees is a book. This book is open. She stands in connection with the moon. Occult wisdom. Passive, eternal woman.

---

[98] See, for example, Jung's citing of various of Waite's works in C. G. Jung, *Psychology and Alchemy*, CW12 (London: Routledge & Kegan Paul, 1953), ¶490; Jung, CW14, ¶18, 27, 312; C. G. Jung, *The Practice of Psychotherapy* (London: Routledge & Kegan Paul 1954), CW16, ¶417, 500.

[99] Jung's private library contained several works by the British Theosophists H. P. Blavatsky, G. R. S. Mead, Anna Bonus Kingsford, and C. W. Leadbeater, as well as a number of books on occultism and magic by British authors such as Israel Regardie, Algernon Charles Blackwood, and Arthur Machen, all members of the Hermetic Order of the Golden Dawn, and several volumes (1920–49) of the British occult journal, *The Occult Review*.

[100] Jung, CW9i, ¶81.

[101] See <http://marygreer.wordpress.com/2008/04/18/carl-jung-on-the-major-arcana/>. Greer states that Jung's definitions of the cards consist of brief notes Hanni Binder took of Jung's descriptions in German when he spoke to her about the Tarot. According to Greer, 'The deck he used was based on the Grimaud Tarot de Marseilles, which he felt most closely contained properties he recognized from his reading of alchemical texts'. She does not mention the similarities between the image of Salome and Waite's High Priestess, as *Liber Novus* was not published at the time her material was placed on the web.

Jung was more deeply involved than this apparently simple level of interpretation. In the *Visions* seminars, given between 1930 and 1934—just after completion of *Liber Novus*—he states:

> They [the Tarot cards] are psychological images, symbols with which one plays, as the unconscious seems to play with its contents. They combine in certain ways, and the different combinations correspond to the playful development of events in the history of mankind....For example, the symbol of the sun, or the symbol of the man hung up by the feet, or the tower struck by lightning, or the wheel of fortune, and so on. Those are sort of archetypal ideas, of a differentiated nature, which mingle with the ordinary constituents of the flow of the unconscious, and therefore it is applicable for an intuitive method that has the purpose of understanding the flow of life, possibly even predicting future events, at all events lending itself to the reading of the conditions of the present moment.[102]

Salome, the High Priestess who weaves her serpentine way through the various chapters of *Liber Novus*, eventually reveals herself as the wise avatar of the 'Spirit of the Depths', reflecting Jung's unique and profoundly psychological understanding of the astrological Moon.

PHILEMON THE MAGICIAN

There are other recognisable planetary images in *Liber Novus*, such as the Saturnian figure of The Anchorite, who is familiar with Jewish esoteric literature, quotes the first-century Jewish Platonist philosopher Philo, and lives in solitude in the barren desert where he 'longs for the sun'.[103] Unfortunately I do not have time to discuss all these figures here. I will focus instead on the most important figure whom Jung encountered, and to whom he gave the name Philemon after the old man described in Ovid's *Metamorphoses*. According to Ovid's tale, Philemon exhibits exceptional kindness and generosity to Zeus and Hermes, who are disguised as poor travellers, without realising they are gods; when the deities reveal themselves and promise that his

---

[102] C. G. Jung, *Visions: Notes of the Seminar Given in 1930–1934 by C. G. Jung*, ed. Claire Douglas, Vol. 2 (Princeton, NJ: Princeton University Press, 1997), p. 923.

[103] Jung, *Liber Novus*, pp. 267–70.

dearest wish will be granted, Philemon tells them that he and his wife Baucis wish nothing more than to spend the remainder of their days serving at Zeus' shrine, and to die at the same moment so that neither is left bereft.[104] They become the sole survivors of the great flood Zeus sends in anger at the cruelty and stupidity of humans, and become servants at Zeus' temple as they have requested. At the moment of their simultaneous death, they are transformed into intertwined oak and linden trees. The old couple also appears in Goethe's *Faust*, Part II, in which Faust, aided by Mephistopheles, attempts to evict them from their sacred shrine, and murders them by burning their house to the ground. Jung felt in some way responsible for this archetypal event through his German ancestry, and believed he had inherited the necessity of expiating Faust's crime. In a letter to Paul Schmitt written in January 1942, he declared:

> All of a sudden with terror it became clear to me that I have taken over Faust as my heritage, and moreover as the advocate and avenger of Philemon and Baucis, who, unlike Faust the superman, are hosts of the gods in a ruthless and godforsaken age.[105]

In the autobiographical *Memories, Dreams, Reflections*, Jung reiterates this realisation:

> When Faust, in his hubris and self-inflation, caused the murder of Philemon and Baucis, I felt guilty, quite as if I myself in the past had helped commit the murder of these two old people...I regarded it as my responsibility to atone for this crime, or to prevent its repetition.[106]

The stark dichotomy between the Faustian cold reason of modernity and the instinctive goodness of the mythic old couple, whom Jung described in a letter to Alice Raphael as 'close to the earth and aware of the Gods',[107] echoes Jung's

---

[104] Ovid, *Metamorphoses*, Book VIII, at Internet Classics Archive, <http://classics.mit.edu//Ovid/metam.html.

[105] *C. G. Jung Letters*, 2 volumes, ed. Gerhard Adler, trans. R. F. C. Hull (London: Routledge & Kegan Paul, 1973-76), Vol. 1, pp. 309–10.

[106] Jung, *Memories, Dreams, Reflections*, p. 260.

[107] Beinecke Library, Yale University, Stiftung der Werke von C. G. Jung, cited in Sonu Shamdasani, 'Who is Jung's Philemon? Unpublished Letter to Alice', *Jung History* 2, no. 2 (2011), at

remark in *Liber Novus* about the 'spirit of the depths' and the 'spirit of this time'. In *Liber Novus,* Philemon emerges as a powerful figure who eventually becomes Jung's *maggid* or inner guide and teacher: a magician who resembles the angel Raziel in late antique Jewish lore, who reveals the magical secrets of the divine realms.[108] Philemon's wings suggest that he is indeed an *angelos,* a heavenly 'messenger'.

Fig. 11.5: Philemon[109]

<https://www.philemonfoundation.org/resources/jung_history/volume_2_issue_2>.

[108] For the angel Raziel, whom Jung mentions when citing a passage from the *Zohar* (Jung, CW14, ¶572), see Rachel Elior, 'The Concept of God in Hekhalot Mysticism', in *Binah: Studies in Jewish History, Thought, and Culture,* Vol. 2, ed. Joseph Dan (NY: Praeger, 1989), pp. 97–120, esp. pp. 101 and 112. For the *Sefer ha-Raziel* ('Book of Raziel'), an early modern grimoire of magic with which Jung was also familiar (Jung, CW14, ¶572n), see Joseph Dan, 'Book of Raziel', in *Encyclopaedia Judaica* 13:1591–93. The *Sefer ha-Raziel* was first published in Amsterdam in 1701 but contains much older material, including portions of late antique Jewish literature concerning the conjuration of angels, and portions of works from the twelfth-century German Jewish Pietists, particularly Eleazar of Worms. The work also includes sections of Kabbalistic literature. The *Sefer ha-Raziel* was circulating in MS long before its publication in 1701, as it is mentioned by Renaissance Christian Kabbalists; see François Secret, 'Sur quelques traductions du Sefer Raziel', *REJ* 128 (1969), pp. 223–45. For an English translation of the *Sefer ha-Raziel,* see Steve Savedow, *Sepher Rezial Hemelach: The Book of the Angel Rezial* (York Beach, ME: WeiserBooks, 2001). Another English translation, Sloane MS 3826, can be found in the British Museum, entitled *Liber Salomonis: Cephar Raziel;* see Savedow's Appendix of *Sefer ha-Raziel* manuscripts in Savedow, trans., *Sepher Rezial Hemelach,* pp. 280–86.

[109] Image in Jung, *Liber Novus,* p. 154.

Philemon is a more complex form of the figure of Elijah:[110] a personification of the archetype of both the Saturnian *senex*, the 'wise old man', and the ever-youthful and self-renewing transformative agent Mercurius. Jung implies this identification in the stone carving he created at Bollingen Tower in 1950. The tower itself was dedicated to 'Philemonis Sacrum'.[111] The carving presents Philemon in the guise of the strange dwarf-god Telesphorus, who is portrayed beneath the astrological glyph of Saturn, with the glyph of Mercury on his body.[112] Jung invariably uses the Greek alphabet when he addresses Philemon in the text. He even used the Greek form of the name in his letter to Alice Raphael, as though he felt it necessary to honour this 'messenger' through using the alphabet belonging to the mythic realm out of which Philemon originally emerged. In *Liber Novus*, Philemon's magic rod and books are secreted in a cupboard along with the 'sixth and seventh books of Moses'. This real but pseudepigraphic collection of magical spells, probably originating in the sixteenth or seventeenth century and based on the influential thirteenth-century Hebrew magical text, *Sefer Raziel*, was published in 1849 by Johann Schiebel, who claimed that the spells came from ancient Talmudic sources.[113] Jung had a copy in his private library.

Philemon is also directly associated in the text of *Liber Novus* with Hermes Trismegistus, the semi-divine mythic teacher of magic, astrology, and alchemy and the pseudepigraphic author of the *Corpus Hermeticum*: the 'wisdom of Hermes Trismegistus' sits in Philemon's cupboard along with the *Sixth and Seventh Books of Moses*.[114] Jung later stated that Philemon 'represented superior insight' and was 'a spirit of nature' akin to the alchemical Mercurius.[115] Although Philemon is old, he is both *senex* and *puer*, Saturn and Mercury, paradoxically contained in each other as they are in alchemical texts: the Old King

---

[110] Jung, in *Memories, Dreams, Reflections*, p. 207, stated that Philemon 'developed out of the Elijah figure'.

[111] See Jung, *Memories, Dreams, Reflections*, p. 263, n. 5.

[112] For a further discussion of the equation of Philemon with Hermes, see Lance S. Owens, 'The Hermeneutics of Vision: C. G. Jung and *Liber Novus*', *The Gnostic* 3 (2010): pp. 23–46.

[113] See Joseph Peterson, ed., *The Sixth and Seventh Books of Moses* (Lake Worth, FL: Ibis Press, 2008).

[114] Jung, *Liber Novus*, p. 312.

[115] Jung, *Memories, Dreams, Reflections*, pp. 208–10 and n. 5.

transformed becomes the youthful Mercurius, who ages and becomes the old *senex* in need of transformation. As Jung points out: 'Mercurius is closely related not only to Venus but more especially to Saturn. As Mercurius he is *juvenis*, as Saturn, *senex*'.[116] In Philemon, as in Salome, Jung displays his tendency to view planetary symbols as complex figures with stories and goals, who secretly contain, generate, and transform into each other through invisible bonds and paradoxical associations, rather than as discrete characterological or prognosticative factors definable by keywords.

Philemon stands as a synthesis of opposites, a *coniunctio oppositorum* of age and youth, decay and death and rebirth into new life. He is also an image of the Sun as a central integrating power. In *Liber Novus*, recognising that Philemon and Izdubar, the primitive bull-slaying giant who seeks immortality, are one and the same solar *daimon*, Jung says to the magician: 'You lie in the Sun'.[117] Philemon is the chief symbol of the successful integration of opposites with which Jung was struggling throughout the period of the writing of *Liber Novus*. Philemon informs Jung: 'Magic happens to be precisely everything that eludes comprehension'.[118] The conflict between rationality and intuitive wisdom is presented in the dialogue once again, but by now, in 1927, twelve years after his tortuous encounter with Izdubar, Jung, having learned the nature of magic from Philemon, is firmly on the magician's side, and declares: 'We need magic to be able to receive or invoke the messenger and the communication of the incomprehensible'.[119]

THE 'SYSTEMA MUNDITOTIUS' AND JUNG'S NATAL HOROSCOPE

An image entitled 'Mandala of a Modern Man' appears as the frontispiece of Jung's *The Archetypes and the Collective Unconscious* (CW9i). Jung described this image as his 'first mandala', and stated in a commentary published in 1955 in a special edition of the Swiss journal *Du*, dedicated to the Eranos Conferences, that he had painted it in 1916, a year

[116] Jung, CW13, ¶250. See also Jung, CW13, ¶269; Jung, CW14, ¶298.
[117] Jung, *Liber Novus*, p. 315.
[118] Ibid., p. 313.
[119] Ibid., p. 314.

after his encounter with Izdubar and three years after beginning work on *Liber Novus*.[120] The title inscribed on the drawing is 'Systema Munditotius', or 'System of All Worlds'.

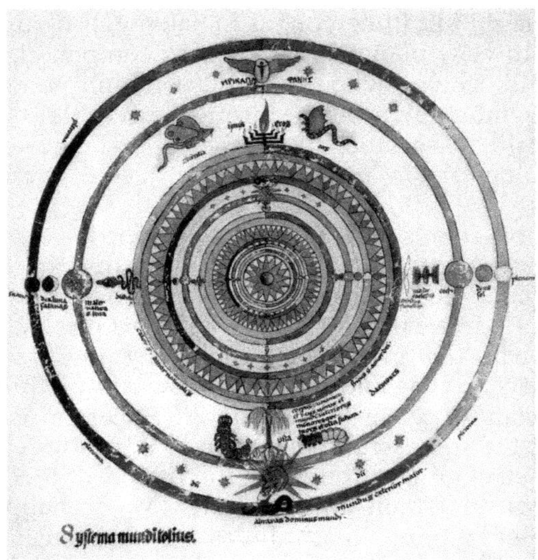

Fig. 11.6: Jung's 'Systema Munditotius', c. 1955

The image published in *Du* and in *The Archetypes and the Collective Unconscious* is in fact a polished version of the drawing produced in 1916. The original 'Systema', which was part of an earlier diary known as *The Black Book*, was not included in Jung's final version of *Liber Novus*.[121] However, it is reproduced in an Appendix in Sonu Shamdasani's published version, and it is more interesting from the perspective of Jung's astrology because it includes a good deal of astrological information which is absent from the reworked drawing.

[120] See Jung, *Liber Novus*, p. 364.
[121] *The Black Book* V, p. 169, reproduced in Jung, *Liber Novus*, p. 363.

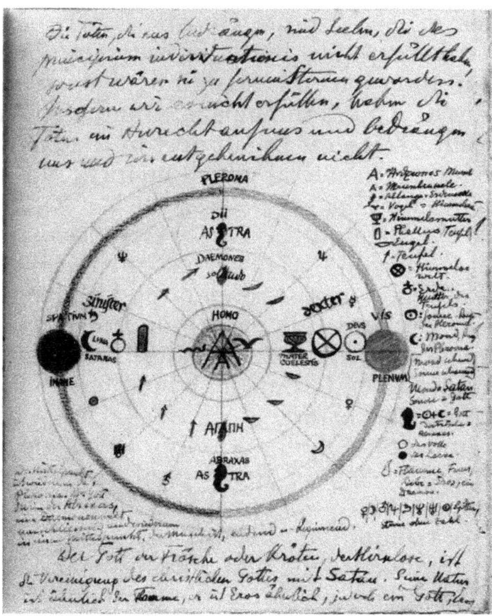

Fig. 11.7: The original 'Systema Munditotius', c. 1916

Shamdasani offers no comment on the astrological material included in the original 'Systema' image. Barry Jeromson, in a paper discussing the 'Systema', points out that it has been largely ignored by writers on Jung's symbolism.[122] Jeromson discusses the drawing comprehensively in the context of Jung's *Septem sermones ad mortuos* ('Seven Sermons to the Dead') and the Gnostic currents which influenced this work, which was written in 1916—simultaneous with the original 'Systema' drawing—and privately published as a small pamphlet which Jung circulated among his students and colleagues.[123] Jeromson suggests that the basis for the

---

[122] Barry Jeromson, 'Systema Munditotius and Seven Sermons: Symbolic Collaborators in Jung's Confrontation with the Unconscious', *Jung History* 1, no. 2 (2010), available at <https://www.philemonfoundation.org/resources/jung_history/volume_1_issue_2>.

[123] C. G. Jung, *Septem Sermones ad Mortuos: Written by Basilides in Alexandria, the City Where the East Touches the West*, trans. Stephan A. Hoeller, in Hoeller, *The Gnostic Jung*, pp. 44–58, on p. 58. The *Septem sermones* is also reprinted in Segal, ed., *The Gnostic Jung*, pp. 181–93, and at <http://www.gnosis.org/library/7Sermons.htm>. For the

'Systema' lies in late antique Gnostic concepts such as the dualism of a corruptible body and an immortal spirit—a perspective on Jung which is shared by a number of scholars who view him as a modern Gnostic,[124] and also a perspective which Jung himself seems to support in the commentary on the new version of 'Systema' which he included in a letter to Walter Corti in 1955.[125]

There are certainly clear references to Gnostic cosmological themes in the 'Systema', such as the Pleroma or abode of the unknown transcendent god above the realm of the planetary archons (*dii astra*) and *daimones*, and the demiurge Abraxas, lord of the manifest world, at the base of the drawing. However, Jeromson, who refers to the 'Systema' as 'Jung's psychocosmology', does not discuss the astrological underpinnings, not only of Gnostic ideas about the descent and ascent of the soul through the planetary spheres, but also of the original drawing of the 'Systema' itself, with its voluminous lists of planetary glyphs. Given that it the 'Systema' was the first of many mandalas that Jung produced over many decades, the drawing implicates an astrological cosmology as one of the primary inspirations for Jung's passionate commitment to the cyclical, circular, and centrifugal 'shapes' of psychic processes. Although it has sixteen 'houses' instead of the usual twelve divisions of the zodiacal wheel,[126] the 'Systema' is in fact drawn as an astrological rather than a geographical map. The east point or Ascendant is on the left, and the west point or Descendant is on the right; the south point of the culmination of the Sun, or *Medium Coeli*, is at the top, and the north point of the Sun's nadir, or *Immum Coeli*, is at the bottom. The planets are

---

history of the publication of the work, see Hoeller, *The Gnostic Jung*, pp. xxiii–xxiv, 8–9, 219–20.

124 See Gilles Quispel, 'C. G. Jung und die Gnosis', *Eranos-Jahrbuch* 37 (1968): pp. 277–98; Robert Segal, ed., *The Gnostic Jung* (Princeton: Princeton University Press, 1992), especially the references on pp. 19–20, n. 57; Hoeller, *The Gnostic Jung*.

125 See Jung, *Liber Novus*, p. 364. The letter does not appear in *C. G. Jung Letters*, but is reproduced in Shamdasani's Appendix to *Liber Novus*.

126 Jung does not seem to have placed great importance on the traditional twelve astrological 'houses', but considered the planets and their aspects, along with the zodiacal signs at the four angles of the horoscope, as the dominant themes in a natal chart. See Jung, CW9ii, ¶212.

placed around the mandala in specific locations, related to Jung's understanding of whether they were 'demonic' or 'angelic'. The Sun in the 'Systema', as I have mentioned, echoes its position in Jung's natal chart.

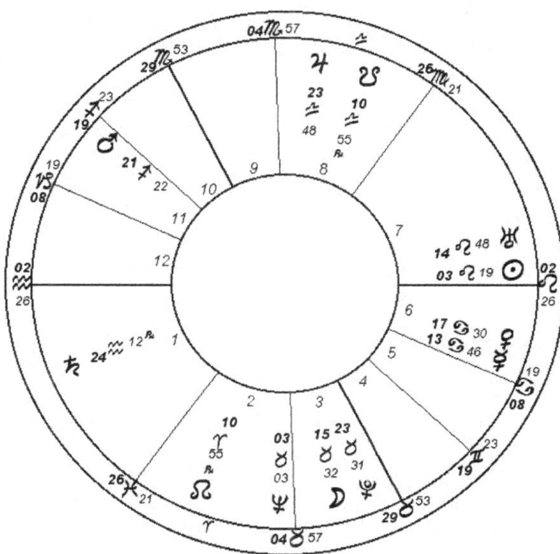

Fig. 11.8: Horoscope of C. G. Jung
26 July 1875, 7.27 pm SZOT, Kesswil, CH[127]

Not only is the 'Systema' a cosmological 'system of all worlds'; it is also the system of Jung's inner world presented as an idealised version of his own horoscope, with the Sun at the Descendant, just as it is in his natal chart. In the original 'Systema', this solar potency described as 'Deus Sol' is represented by the Sun's astrological glyph. This, like the planetary glyphs, has vanished from the later version, replaced by a simple gold disc. But although the complex astrological information of the original drawing was omitted from the polished drawing, it provides a vital key to the thinking behind the mandala.

That there are seven 'sermons' to 'the dead'—not eight, six, fourteen, or twenty-three—might, like the seven stars at

---

127 Computer-generated horoscope calculated by Io Edition programme, Time Cycles Research, http:www.timecycles.com.

the top and bottom of the 'Systema', immediately serve as a red flag to any scholar familiar with the importance of the planetary heptad in the symbolic thinking of antiquity. The focus of the 'Systema' on the Sun as the 'divine spark' shared by both humans and the deity links it directly to Jung's dialogue with Izdubar in *Liber Novus*. The idea of the 'spark' also echoes Kabbalistic concepts, especially those found in the texts of the Lurianic Kabbalah of the late sixteenth century,[128] which were published in a Latin translation by Christian Knorr von Rosenroth at the end of the seventeenth century in a work called *Kabbala denudata*—a text with which Jung was familiar and from which he frequently quotes in later volumes of the *Collected Works*.[129] In Sermon VII in the *Septem sermones*, Jung describes the divine world of the Pleroma as the macrocosmic form of the 'innermost infinity' within the human being, declaring:

---

[128] For the Kabbalah of Isaac Luria and its various offshoots, see Lawrence Fine, *Physician of the Soul, Healer of the Cosmos: Isaac Luria and His Kabbalistic Fellowship* (Stanford, CA: Stanford University Press, 2003); Louis Jacobs, 'Uplifting the Sparks in Later Jewish Mysticism', in *Jewish Spirituality, Vol. 2: From the Sixteenth Century Revival to the Present*, ed. Arthur Green (New York: Crossroad, 1987), pp. 99–126; Gershom Scholem, *Kabbalah* (New York: Keter Publishing House, 1974), pp. 128–44, 420–28, 443–48; Moshe Idel, *Hasidism: Between Ecstasy and Magic* (Albany, NY: State University of New York Press, 1995), pp. 33–43; R. J. Zvi Werblowsky, 'Mystical and Magical Contemplation: The Kabbalists in Sixteenth-Century Safed', *History of Religions* 1, no. 1 (1961): pp. 9–36; Shaul Magid, 'From Theosophy to Midrash: Lurianic Exegesis and the Garden of Eden', *Association for Jewish Studies Review* 22, no. 1 (1997): pp. 37–75; Don Karr, 'Notes on the Study of Later Kabbalah in English: The Safed Period and Lurianic Kabbalah', in Don Karr, *Collected Articles on the Kabbalah*, Vol. 2 (Ithaca, NY: KoM, No. 6, 1985), pp. 23–31.

[129] Christian Knorr von Rosenroth, *Kabbala denudata, seu, Doctrina Hebraeorum transcendentalis et metaphysica atque theologica: opus antiquissimae philosophiae barbaricae...in quo, ante ipsam translationem libri...cui nomen Sohar tam veteris quam recentis, ejusque tikkunim...praemittitur apparatus [pars 1–4]*, 3 vols. (Sulzbach/Frankfurt: Abraham Lichtenthal, 1677–1684). For Jung's many citations of the *Kabbala denudata*, particularly in his alchemical volumes, see, for example, Jung, CW12, ¶313; Jung, CW13, ¶411; Jung, CW14, ¶592–93; Jung, CW9i, ¶557n, ¶576n, ¶596n. A first edition of the *Kabbala denudata* was included in Jung's private library.

> In immeasurable distance there glimmers a solitary star on the highest point of heaven. This is the only God of this lonely one. This is his world, his Pleroma, his divinity....This star is man's God and goal. It is his guiding divinity; in it man finds repose. To it goes the long journey of the soul after death; in it shine all things which otherwise might keep man from the greater world with the brilliance of a great light. To this One, man ought to pray....Such a prayer builds a bridge over death.[130]

Earlier in the *Septem sermones,* in Sermon II, the identity of this 'solitary star' is made clear: it is the Sun-god Helios.[131] In the notes accompanying the original drawing of the 'Systema', Jung clarifies his understanding of the planets as symbols of divine potencies:

> The moon is the God's eye of emptiness, just as the sun is the God's eye of fullness. The moon that you see is the symbol, just as the sun that you see. Sun and moon, that is, their symbols, are Gods. There are still other Gods; their symbols are the planets.[132]

Jung's emphasis on the divine spark within the individual seems to have provided support for his later idea of individuation, which on a spiritual level represented, for him, oneness or integration with the inner or transpersonal Self.[133] The chief symbol for this inner spark, for Jung, is the Sun:

> The dynamic of the gods is psychic energy. This is our immortality, the link through which man feels inextinguishably one with the continuity of all life....The psychic life-force, the libido, symbolizes itself in the sun.[134]

It is worth emphasising that, of all the astrological authors appearing in Jung's personal library, Alan Leo's works take

---

[130] Jung, *Septem sermones,* in Hoeller, *The Gnostic Jung,* p. 58.

[131] Ibid., p. 50. Helios is also the Orphic god of light, Phanes, who appears as the central star in the group of seven at the top of the final version of the 'Systema', enclosed, like Izdubar, in the cosmic egg.

[132] Jung, *Liber Novus,* p. 371.

[133] For Jung's descriptions of individuation, see, *inter alia,* Jung, CW7, ¶266–406; Jung, CW9i, ¶489–524 and 525–626.

[134] Jung, CW5, ¶296–97.

pride of place. Although single works by various German astrologers can be found in the catalogue, and the usual suspects, such as Ptolemy's *Tetrabiblos*, are referenced in the *Collected Works*, virtually every volume Leo wrote was included in Jung's library in its original edition.[135] One of Leo's great innovations, as Nicholas Campion has pointed out in his *History of Western Astrology*, was an emphasis on the Sun-sign not only as the core of individual character but also, in accord with Theosophical doctrines, as the symbol of the 'Central Spiritual Sun'.[136] Leo understood the astrological Sun as 'the vehicle through which the Solar Logos is manifesting'.[137] His mentor, H. P. Blavatsky, in turn acquired her Theosophical ideas in large part from the solar-centred pagan monotheism of late antiquity, and drew on such sources as Thomas Taylor's translation of the Emperor Julian's *Oration to the Sovereign Sun* to promulgate the idea, as Blavatsky put it, of the 'invisible' Sun as 'the origin and end of the incorruptible and eternal spirit'.[138] As Blavatsky herself adapted many Gnostic and Neoplatonic themes for her Theosophical doctrines, it is hardly surprising that later Theosophical astrologers such as Rudhyar found much in Jung with which they felt affinity.

---

135 *C. G. Jung Bibliothek Katalog*, p. 45. The following works by Leo are included: Alan Leo, *Astrology: What It Is and How to Study It* (London: Modern Astrology, 1904); Alan Leo, *Astrology for All* (London: Modern Astrology, 1904); Leo, *How to Judge a Nativity;* Alan Leo, *The Key to Your Own Nativity* (London: Modern Astrology, 1910); Alan Leo, *The Progressed Horoscope* (London: Modern Astrology, 1905); Alan Leo, *Saturn: The Reaper* (London: Modern Astrology, 1916).

136 Nicholas Campion, *A History of Western Astrology, Vol. 2: The Medieval and Modern Worlds* (London: Continuum, 2009), p. 232.

137 Leo, *How to Judge a Nativity*, p. 29.

138 H. P. Blavatsky, *Isis Unveiled: A Master-Key to the Mysteries of Ancient and Modern Science and Theology,* 2 vols. (London: Theosophical Publishing Co., 1877), I:302. For Blavatsky's acknowledgement of Taylor, see Blavatsky, *Isis Unveiled*, I:284, 288, and II:108–9; H. P. Blavatsky, *The Secret Doctrine: The Synthesis of Science, Religion, and Philosophy,* 2 vols. (London: Theosophical Publishing Co., 1888), I:425, 453, and II:599. For Taylor's translation of Julian, see *The Emperor Julian's Oration to the Sovereign Sun,* in Thomas Taylor, *Collected Writings on the Gods and the World* (London: Edward Jeffrey, 1793; repr., Frome: Prometheus Trust, 1994), pp. 51–76.

Jung concludes his discussion of the cosmology of the 'Systema' with an unashamedly polytheistic perception of the world, which he attempts to reconcile with his interiorised and spiritualised but nevertheless fully committed Christianity:

> You need to recognize the multiplicity of the Gods. You cannot unite all into one being. As little as you are one with the multiplicity of men, just so little is the *one* God one with the multiplicity of the Gods. This one God is the kind, the loving, the leading, the healing. To him all your love and worship is due. To him you should pray, you are one with him, he is near you, nearer than your soul.[139]

Although some of his astrological interpretations were undoubtedly influenced by Theosophical astrologers such as Alan Leo, Jung was not a card-carrying Theosophist any more than he was a Gnostic, a Kabbalist, an Orphic, or a Neoplatonist; and his understanding of astrological symbols as presented in *Liber Novus* is fundamentally different and uniquely his own. While Leo emphasised the evolutionary journey of the soul reflected in the evolutionary potential of the birth chart, and encouraged the transcendence of the 'lower' aspects of the planets and zodiacal signs, Jung faced with greater honesty the apparently irreconcilable polarity of body and soul described in Orphic and Gnostic literature, and struggled with the integration of the opposites, believing that this was the only path to a direct experience of the hidden unity of life. There is no 'confrontation with the shadow' in Leo's work; Jung was concerned with integration and wholeness rather than transcendence and perfection.

The god of *Liber Novus* is unquestionably the Sun, but it is not the exclusively male and exclusively benefic Sun described in traditional astrological texts. Within this symbol, for Jung, are contained all the other planetary deities, or what he refers to as the 'multiplicity of gods'. The Red One, Izdubar, Salome, and even Philemon, are all, in the end, absorbed into the central symbolism of this *coniunctio oppositorum* represented as *deus sol*. These figures rose up as planetary images that Jung believed to be the living embodiments of archetypal patterns which exist both within and outside the human being, and to him they were teleological processes rather than characterological traits. He

---

[139] Jung, *Liber Novus*, p. 371.

does not seem to have viewed events and inner psychic states as separate things, but saw them instead as secretly, acausally conjoined in an archetypal spectrum which encompasses instinct at one end and image at the other, bound by a single core inaccessible to the intellect but expressed in the paradox of the symbol. Jung's figures are symbols in the most profound sense: messengers who communicate macrocosmic patterns and, at the same time, visual and tangible embodiments of that of which the patterns are the emanation. As Jung declared at the end of his dialogue with Philemon, 'We need magic in order to receive or invoke the messenger'. We might understand his use of the term 'magic' to describe the process of allowing the *mundus imaginalis* its reality as the liminal place, the altar on which all transformation occurs and through which all opposites meet and are reconciled through their secret unity.

Conclusion

The publication of *Liber Novus* has invoked a new and fertile stream of studies and explorations into Jung's psychological models and the sources from which he derived and developed his theories. Some of these studies seem to be intent on a deliberate scholarly form of 'Jung-bashing': if Jung's ideas are applied as a methodology for exploring the history of religions, the scholar is deemed to be 'religionist' and therefore suspect, and if Jung's models are viewed as an effective and useful method of psychological work, the analyst is deemed to be practising an 'esoteric' and therefore unscientific form of psychotherapy. Some discussions on Jung, such as the recent biography by Richard Noll,[140] are aggressively, if covertly, hostile, and focus on Jung as a seriously flawed and untrustworthy individual who could therefore only produce a flawed and untrustworthy system of psychology. Jung's flaws, especially his ferocious competitiveness and his complex and sometimes destructive attitudes toward women, were certainly flagrantly on display in many spheres of his life. But this is rather like declaring that Wagner's music is rubbish because he was such a nasty man, or that Michelangelo's sculptures are unworthy of consideration because he was so irritable and unpleasant.

---

[140] Noll, *The Jung Cult*.

My own paper has focused on Jung as one of a long line of thinkers stretching back to the cosmological and astrological authors and visionaries of antiquity. He promulgated, in twentieth-century language in a twentieth-century Western cultural milieu, a particular complex of ideas which appears to possess sufficient vitality and agency to ensure its survival in many cultural contexts over many centuries, and which encompasses astrology as one of its cornerstones. Given his difficult, autocratic, and tricky personality, I would not have liked to be married to Carl Gustav Jung. But whatever his personal failings, and whatever ontological perspective the scholar adopts toward his psychological models and their esoteric underpinnings, Jung was one of the greatest and most influential thinkers of the twentieth century, and also one of the most important modern exponents—evident most vividly in *Liber Novus*—of an extremely long-lived idea that the imagination is no mere spinner of wish-fulfilling fantasies, but is the human being's chief, and perhaps only, organ for 'receiving the messenger'.

CONTRIBUTORS

NICHOLAS CAMPION is Director of the Sophia Centre, University of Wales, Trinity St David, Senior Lecturer in the Department of Archaeology, History and Anthropology, and course director of the MA in Cultural Astronomy and Astrology. His books include *A History of Western Astrology* (London: Continuum, 2009) and *Astrology and Cosmology in the World's Religions* (New York: New York University Press, 2011).

PETER FORSHAW studied Sanskrit, Tibetan and Indian Philosophy at the School of Oriental and African Studies, University of London (1982–86), and took an MA in Renaissance Studies at Birkbeck, University of London, where he subsequently researched his doctorate in Early Modern Intellectual History, on the interplay of alchemy, magic and cabala in the *Amphitheatrum sapientiae aeternae* of Heinrich Khunrath of Leipzig (1560–1605). He was awarded a British Academy Postdoctoral Fellowship for research into the History of Ritual Magic in the Middle Ages and Renaissance, followed by fellowships at the universities of Strathclyde and Cambridge. In 2009 he was appointed Senior Lecturer/Assistant Professor for History of Western Esotericism in the Early Modern Period at the Center for History of Hermetic Philosophy and Related Currents, University of Amsterdam. He is Editor in Chief of *Aries: Journal for the Study of Western Esotericism*.

LIZ GREENE received a PhD in History from the University of Bristol (2010), an MA in Cultural Astronomy and Astrology from Bath Spa University (2007), and an MA and PhD in Psychology from Los Angeles University (1971). She is a qualified analytical psychologist (Association of Jungian Analysts, London, 1983) and a member of the International Association of Analytical Psychology. Her research interests include the Kabbalah, the British occult revival of the late 19th century, and Orphic, Gnostic, early Jewish, and Hermetic astrologies from late antiquity. She is an Honorary Research Fellow at the University of Bristol and a part-time tutor for the MA in Cultural Astronomy and Astrology at the University of Wales, Trinity St David, UK.

DARRELYN GUNZBURG is a teaching assistant for the Department of History of Art, University of Bristol, and will be co-teaching for the Department in 2014. She is also a tutor for the MA in Cultural Astronomy and Astrology at The University of Wales Trinity Saint David, UK. Before reading for her doctorate in the History of Art at the University of Bristol (anticipated completion December 2013) Darrelyn studied at the Open University (BA Hons in Art History) and came to the academic world as a professional astrologer and as a produced and published playwright with a Diploma of Directing from NIDA at the University of New South Wales, Sydney,

Australia. Her academic publications include: 'Looking Back: The transgression of social codes', *St Andrews Journal of Art History and Museum Studies* (2010), 'How Do Astrologers read Charts?' in *Astrologies: Plurality and Diversity* (Sophia Centre Press, 2011), as well as a forthcoming edited volume, *The Imagined Sky: Cultural Perspectives* (Equinox, 2014).

ANDREA D. LOBEL is a PhD candidate in Religion at Concordia University in Montreal, Canada, specializing in the history of astronomy, astrology, cosmology, and cosmogony in Second Temple and Late Antique Judaism. Her research also spans the history of religions, the history of science, ethnoastronomy, ancient Jewish magic and mysticism, religious identity formation and maintenance, otherness, and the anthropology of religion. She has designed and taught courses at both McGill University and Concordia, including courses on western religious traditions and the celestial myths and texts of the ancient Near East.

CHRISTEL MATTHEEUWS received her PhD from Aberdeen in 2008. Her main fields of interests are spirituality, astrology, intangible knowledge, development, and ecology (with a strong attraction for islands and mountainous regions), which she gradually explored not only in her academic work, but also during travels (to Sri Lanka, Thailand, Taiwan, Kenya, Guinea Conakry, and Madagascar) and visits to development projects, through making international contacts, and doing practical trainings in both development and biological agriculture. Her present direction is greatly influenced by Tim Ingold's work, Goethean Science, and her experiences in Madagascar.

ANTHONY THORLEY is a retired psychiatrist who graduated with an MA in Cultural Astronomy and Astrology in 2006. He has recently edited a book, *Legendary London and the Spirit of Place* and is slowly completing another book, *Sacred City, Secret City* on Masonic influences in the development of eighteenth-century Bath. He is currently pursuing an MPhil/PhD on Landscape Zodiacs at University of Wales, Trinity St David where he is a part-time tutor on the MA CAA teaching Sacred Geography. With his wife Celia Gunn he runs pilgrimages and with astrologer John Wadsworth he co-facilitates the workshop programme *The Alchemical Journey.* See www.earthskywalk.com and www.thealchemicaljourney.co.uk

GARY WELLS is an associate professor and chair of the Department of Art History at Ithaca College, Ithaca. New York. His areas of research and writing include the intersections of art and science in the nineteenth century, the visual culture of science, and the origins of Modernism in the early twentieth century. He has presented and published on artists such as Marcel Duchamp, Jules Breton and Paul Cézanne. He also teaches and writes about the material history of art and issues of technology in art history, and has developed technology projects on modern art in Paris and late–nineteenth century painting.

Ola Wikander (born 1981) is a doctor of Old Testament Exegesis at Lund University, Sweden. His main research interests are Ugaritic language, literature and religion, Hebrew philology, comparative Semitics and the relationship between the Indo-European cultures of the ancient world and that of the Hebrew Bible. He is also interested in ancient Gnosticism and Hermetism. He has worked with many ancient languages and has published a number of books and articles, including annotated translations from Ugaritic and Akkadian. His dissertation, *Drought, Death and the Sun in Ugarit and Ancient Israel: A Philological and Comparative Study,* concerns the relationship between the destructive sun and the netherworld in Ugaritic and Old Testament thought.

Elliot R. Wolfson is Abraham Lieberman Professor of Hebrew and Judaic Studies at New York University. He received his PhD in 1986 (Jewish mysticism and Philosophy) and his MA in 1983 (Jewish Mysticism and Philosophy) from Brandeis University, as well as a BA and MA in 1979 (Philosophy) from Queens College. To date he is the author of a hundred and twenty-nine published papers and many books, which include *Giving Beyond the Gift: Apophasis and Overcoming Theomania* (New York: Fordham University Press, 2013); *A Dream Interpreted Within a Dream: Oneiropoiesis and the Prism of Imagination* (New York: Zone Books, 2011) (Winner of the American Academy of Religion's Award for Excellence in the Study of Religion in the Category of Constructive and Reflective Studies, 2012); and *Through a Speculum That Shines: Vision and Imagination in Medieval Jewish Mysticism* (Princeton University Press, 1994).

Jennifer Zahrt received a PhD in German Literature and Film from the University of California, Berkeley (2012), with a dissertation on 'The Astrological Imaginary in Early Twentieth–Century German Culture'. She received her MA (also in German) from UC Berkeley in 2005, after graduating *summa cum laude* from New York University's Gallatin School of Individualized Study in 2004. Her research interests include the history of astrology in continental Europe, with an emphasis on Germany, as well as astrological discourses and the epistemological and interpretational issues they expose. She has served as the Deputy Editor of the journal *Culture and Cosmos* since 2010, and she is currently translating German astrological texts into English.

The Sophia Centre

The Sophia Centre was set up with funding from the Sophia Trust and is located within the University of Wales, Trinity St. David's School of Archaeology, History and Anthropology. It has a wide-ranging remit to investigate the role of cosmological, astrological and astronomical beliefs, models and ideas in human culture, including the theory and practice of myth, magic, divination, religion, spirituality, politics and the arts. The Centre teaches the MA in Cultural Astronomy and Astrology via distance learning online, and also supervises MPhil and PhD research. There is no need to live in the UK to study at the Sophia Centre.

Much of the Centre's work is historical, but it is equally concerned with contemporary culture and lived experience. If you are interested in the way the sky is used to create meaning and significance, then the Sophia Centre may be the best place for you to study. By joining the Sophia Centre, you enter a community of like-minded scholars whose aim is to explore humanity's relationship with the cosmos.

For further information about the Sophia Centre see the website at www.tsd.ac.uk/en/sophia or contact Nick Campion, the Course Director, at the details below.

The Sophia Centre
Department of Archaeology and Anthropology
University of Wales, Trinity Saint David
Ceredigion
Wales SA48 7ED
United Kingdom
Email: n.campion@tsd.ac.uk

BIBLIOGRAPHY

Abulafia, Abraham. *Ḥayyei ha-Olam ha-Ba*. Jerusalem, 1999.

———. *Imrei Shefer*. Jerusalem, 1999.

———. *Mafteaḥ ha-Ḥokhmot*. Jerusalem, 2001.

———. *Mafteaḥ ha-Sefirot*. Jerusalem, 2001.

———. *Mafteaḥ ha-Tokhaḥot*. Jerusalem, 2001.

———. *Sefer ha-Ḥesheq*. Jerusalem, 2002.

Adams, Douglas. 'The Hitchhiker's Guide to the Galaxy'. In *The Hitchhiker's Trilogy*. New York: Harmony Books, 1979.

Addey, John M. 'Astrology as Divination'. *Astrology* 56, no. 2 (1982).

Agrippa, Henry Cornelius. *Three Books of Occult Philosophy*. Facsimile of the 1651 translation. London: Chthonius Books, 1986.

Alexander, Jonathan. 'Labeur and Paresse: Ideological Representations of Medieval Peasant Labor'. *The Art Bulletin* 72, no. 3 (1990): pp. 436–52.

Alexander-Frizer, Tamar. *The Pious Sinner: Ethics and Aesthetics in the Medieval Hasidic Narrative*. Tübingen: J. C. B. Mohr, 1991.

———. 'Dream Narratives in "Sefer Hasidim"'. *Trumah* 12 (2002): pp. 65–78 [Hebrew].

Ames-Lewis, Francis. *Tuscan Marble Carving 1250–1350: Sculpture and Civic Pride*. Aldershot: Ashgate, 1997.

Andrej. Review of *Der Golem* in *Film Kurier*, Oct. 30, 1920.

Anon. *Tractatus alter De Lapide Philosophico*. In Martin Ruland, *Lapidis Philosophici Vera Conficiendi Ratio*. Frankfurt, 1606.

Anon. Review of *Der Golem* in *Der Vorwärts*, Oct. 31, 1920. Reprinted in *Film und Presse*, no. 17 (1920): p. 418.

Anon. Review of *Der Golem* in *Berliner Morgenpost* on October 31, 1920. Reprinted in *Film und Presse*, no. 17 (1920): p. 415.

Anon. Review of *Der Golem* in Der *Kinematograph*: November 7, 1920. Reprinted in *Film und Presse*, no. 18 (1920): p. 441.

Anon. 'Vom Sternen zum Stern', *Die Filmwoche* 19 (1925): p. 449.

Anon. 'Astrology As a Symbolic Language', (2005): http://www.democraticunderground.com/discuss/duboard.php?az=view_all&address=245x5288

Anon., [Katharine Emma Maltwood]. *A Guide to Glastonbury's Temple of the Stars*. London: John M. Watkins, 1935.

Artefius. *Clavis maioris sapientiae*. Strasburg, 1699.

Artephius. *Clavis Sapientiae*. Leipzig, 1736.

Ashmole, Elias, ed. *Theatrum Chemicum Britannicum*. London, 1652.

Astour, Michael C. 'New Evidence on the Last Days of Ugarit'. *American Journal of Archaeology* 69, no. 3 (1965): pp. 253–58.

Azzolini, Monica. 'Reading Health in the Stars: Politics and Medical Astrology in Renaissance Milan'. In K. von Stuckrad, G. Oestmann and H. Darrel Rutkin, eds., *Horoscopes and Public Spheres: Essays on the History of Astrology*, pp. 183–206. Berlin: Walter de Gruyter, 2005.

Balázs, Béla. 'Filmwunder'. *Film Kurier* (Berlin), 12.12.1924. Reprinted in Balázs, *Schriften I*, pp. 322–25.

———. *Schriften I*. Berlin: Henschelverlag Kunst und Gesellschaft, 1982.

Ballester, Luis Garcia, Roger French, Jon Arrizabalaga, and Andrew Cunningham, eds. *Practical Medicine from Salerno to the Black Death*. Cambridge: Cambridge University Press, 1994.

Bar-Ilan, Meir. 'Astrology In Ancient Judaism'. In J. Neusner, A. Avery-Peck and W.S. Green, eds., *The Encyclopaedia of Judaism*, vol. 1, pp. 129–41. Leiden: Brill, 2005.

———. 'Astronomy In Ancient Judaism'. In J. Neusner, A. Avery-Peck and W.S. Green, eds., *The Encyclopaedia of Judaism*, vol. 1, pp. 143–51. Leiden: Brill, 2005.

Barclay, Olivia. *Horary Astrology Rediscovered*. West Chester, PA: Whitford Press, 1990.

Barlet, Annibal. *La Theotechnie Ergocosmique*. Paris, 1653.

Barnard, G. William. 'Diving Into the Depths: Reflections on Psychology as a Religion'. In Diane Elizabeth Jonte-Pace and William B. Parsons, *Religion and Psychology: Mapping the Terrain. Contemporary Dialogues, Future Prospects*, pp. 297–318. London: Routledge, 2001.

Barnes, Jonathan, ed. *The Complete Works of Aristotle*, Bollingen Series LXXXI.2. Princeton, NJ: Princeton University Press, 1984.

Bartha, Paul. *By Parallel Reasoning: The Construction and Evaluation of Analogical Arguments*. Oxford: Oxford University Press, 2010.

Barton, Tamsyn. *Ancient Astrology*. London: Routledge, 1994.

Bartov, Omar. *The 'Jew' in Cinema from* The Golem *to* Don't Touch my Holocaust. Bloomington, IN: Indiana University Press, 2005.

Becher, Johann Joachim. *Chymischer Glücks Hafen*. Halle, 1726.

Beck, Roger. *The Religion of the Mithras Cult in the Roman Empire: Mysteries of the Unconquered Sun*. Oxford: Oxford University Press, 2006.

Ben-Dov, Jonathan. *Head of All Years: Astronomy and Calendars at Qumran in their Ancient Context*. Leiden: Brill, 2008.

Benin, Steven D. 'The Chronicle of Ahimaaz and its Place in Byzantine Literature'. *Jerusalem Studies in Jewish Thought* 4 (1984/85): pp. 237–50.

Ben-Sasson, Menahem, Robert Bonfil, and Joseph R. Hacker, eds. *Culture and Society in Medieval Jewry: Studies Dedicated to the Memory of Haim Hillel Ben-Sasson*. Jerusalem, 1989. [Hebrew].

Benz, Ernst. *The Mystical Sources of German Romantic Philosophy*. Translated by Blair R. Reynolds and Eunice M. Paul. Eugene, OR: Pickwick Publications, 1983.

Berthelot, Marcellin and C.E. Ruelle. *Collection des anciens alchimistes grecs*. Vol. 1 . Paris, 1887.

Betz, Hans Dieter. *The 'Mithras Liturgy': Text, Translation and Commentary*. Tübingen: Mohr Siebeck, 2003.

Bishop, Paul. 'Thomas Mann and C. G. Jung'. In Paul Bishop, ed., *Jung in Contexts: A Reader*, pp. 154–88. London: Routledge, 1999.

Blavatsky, H. P. *Isis Unveiled: A Master-Key to the Mysteries of Ancient and Modern Science and Theology*. 2 volumes. London: Theosophical Publishing Co., 1877. Reprinted by Pasadena, CA: Theosophical University Press, 2 Vols., 1976.

———. 'Stars and Numbers'. *The Theosophist* (June 1881) at http://www.blavatsky.net/...StarsAndNumbers.html [accessed 25 May 2010].

———. *The Secret Doctrine: The Synthesis of Science, Religion, and Philosophy*. 2 volumes. London: Theosophical Publishing Co., 1888. Reprinted by Pasadena, CA: Theosophical University Press, 1974.

Bocchi, Francesca. 'Regulation of the Urban Environment by the Italian Communes from the Twelfth to the Fourteenth Century'. *Bulletin of the John Rylands Library* 72 (1990).

Bockemühl, Jochen. 'Transformations in the Foliage Leaves of the Higher Plants'. In D. Seamon and A. Zajonc, eds., *Goethe's Way of Science: A Phenomenology of Nature*, pp. 115–28. New York: State University of New York Press, 1998.

Boese, Carl. *Erinnerungen an die Enstehung und an die Aufnahmen eines der berühmtesten Stummfilme.* Stiftung Deutsche Kinemathek, Berlin. Schriftgutarchiv. Golem file: 4359. Folder 2. 21 pages.

Boime, Albert. 'Van Gogh's *Starry Night*: A History of Matter and a Matter of History'. *Arts Magazine* 59, no. 4 (December 1984): pp. 86–103.

Bonfil, Robert. 'Between Eretz Israel and Babylonia'. *Shalem* 5 (1987): pp. 1–30 [Hebrew].

———. 'Myth, Rhetoric, History? A Study in the *Chronicle of Ahima'az*'. In Menahem Ben-Sasson, Robert Bonfil, and Joseph R. Hacker, eds., *Culture and Society in Medieval Jewry: Studies Dedicated to the Memory of Haim Hillel Ben-Sasson*, pp. 99–136. Jerusalem, 1989. [Hebrew].

Bortoft, Henri. *The Wholeness of Nature: Goethe's Way of Science.* Hudson, NY: Lindisfarne Press, 1996.

Bottéro, Jean. *Mesopotamia: Writing, Reasoning and the Gods*. Chicago: University of Chicago Press, 1992.

Boyle, Robert. *Certain Physiological Essays*. London, 1669.

Bracesco, Giovanni. *De alchemia dialogi duo*. Lyon, 1548.

Brady, Bernadette. 'The Horoscope as an Imago Mundi: Rethinking the Nature of the Astrologer's Map'. In Nicholas Campion and Liz Greene, eds., *Astrologies: Plurality and Diversity*, pp. 47–61. Ceredigion, Wales: Sophia Centre Press, 2011.

Brook, Isis. 'Goethean Science as a Way to Read Landscape'. *Landscape Research* 3, no. 1 (1998): pp. 51–69.

Brown, D. *Mesopotamian Planetary Astronomy-Astrology*. Leiden: Brill, 2000.

Brown, R. A. *Katharine Emma Maltwood, Artist: 1878–1961*. Victoria: Sono Nis Press, 1981.

Brunschwig, Hieronymus. *Medicinarius. Das buch der Gesuntheit.*

Liber de arte distillandi Simplicia et Composita. Strasburg, 1505.
———. *Liber de arte Distillandi de Compositis*. Strasbourg, 1512.
Bry, Carl Christian. *Verkappte Religionen*. Gotha/Stuttgart: Friedrich Andreas Perthes, 1925.
Buck, August, ed. *Die okkulten Wissenschaften in der Renaissance*. Wiesbaden: Harrassowitz, 1992.
Burnett, Charles. 'The Astrologer's Assay of the Alchemist: Early References to Alchemy in Arabic and Latin Texts'. *Ambix* 39, no. 3 (1992): pp. 103–9.
C. G. Jung Letters. 2 volumes. Edited by Gerhard Adler. Translated by R. F. C. Hull. London: Routledge & Kegan Paul, 1973–76.
C. G. Jung Bibliothek Katalog. ETH. Küsnacht-Zürich, 1967.
Caine, M. *The Glastonbury Zodiac: Key to the Mysteries of Britain*. Kingston: Mary Caine, 1978.
Campion, Nicholas. 'Mythical Moments in the Rectification of History'. in Noel Tyl, ed., *Astrology Looks at History*, pp. 24-63. St. Paul, MN: Lllewllyn, 1995.
———. 'Babylonian Astrology: Its Origin and Legacy in Europe'. In Helaine Selin, ed., *Astronomy Across Cultures: The History of Non-Western Astronomy*, pp. 509–53. Dordrecht: Kluwer Academic Publishers, 2000.
———. *Dawn of Astrology: A Cultural History of Western Astrology: The Ancient and Classical Worlds*. London: Continuum International Publishers, 2008.
———. *A History of Western Astrology*, Vol. 2. The Medieval and Modern Worlds. London: Continuum, 2009.
———. 'Astrology's Place in Historical Periodisation: Modern, Premodern or Postmodern?' In Campion and Greene, eds., *Astrologies: Plurality and Diversity*, pp. 217–54.
———. *Astrology and Cosmology in the World's Religions*. New York: New York University Press, 2012.
———. *Astrology and Popular Religion in the Modern West: Prophecy, Cosmology and the New Age Movement*. Abingdon: Ashgate, 2012.
——— and Patrick Curry, eds., *Sky and Psyche: The Relationship Between Cosmos and Consciousness*. Edinburgh: Floris Books, 2005.
——— and Liz Greene, eds., *Astrologies: Plurality and Diversity*. Ceredigion, Wales: Sophia Centre Press, 2011.
Carlson, John B. 'Venus-Regulated Warfare and Ritual Sacrifice in Mesoamerica'. In Clive Ruggles and Nicholas Saunders, eds., *Astronomies and Cultures* pp. 202–52. Niwot CA: University of Colorado Press, 1993.
Cassirer, Ernst, *The Philosophy of Symbolic Forms*, Vol. 3., The Phenomenology of Knowledge. 1955. Reprinted by New Haven, CT: Yale University Press, 1971.
———. *The Myth of the State*. New Haven, CT: Yale University Press, 1967.
Charet, F. X. *Spiritualism and the Foundations of C. G. Jung's Psychology*. Albany, NY: SUNY Press, 1993.
Charlesworth, James H. 'Jewish Astrology in the Talmud,

Pseudepigrapha, the Dead Sea Scrolls, and Early Palestinian Synagogues'. *The Harvard Theological Review* 70, nos. 3-4 (1977): pp. 183–200.

———, ed., *The Old Testament Pseudepigrapha*. Vol. 1. New York: Doubleday, 1983.

———, ed., *The Old Testament Pseudepigrapha*. Vol. 2. New York: Doubleday, 1985.

———. 'Jewish Interest in Astrology during the Hellenistic and Roman Periods'. *Aufstieg und Niedergang des römischen Welts* 2, no. 20.2 (1987): pp. 926–50.

Chazan, Robert. 'The Early Development of Hasidut Ashkenaz'. *Jewish Quarterly Review* 75 (1985): pp. 199–211.

Chodorow, Joan, ed., *Jung on Active Imagination*. Princeton, NJ: Princeton University Press, 1997.

Christie, J. R. R. 'The Paracelsian Body'. In Ole Peter Grell, ed., *Paracelsus: The Man and his Reputation, his Ideas and their Transformation*, pp. 269–91. Leiden: Brill, 1998.

Clark, Charles. 'The Zodiac Man in Medieval Medical Astrology'. *Journal of the Rocky Mountain Medieval and Renaissance Association* 3 (1982): pp. 13–38.

Clericuzio, Antonio. 'Norton'. In Claus Priesner and Karin Figala eds., *Alchemie: Lexikon einer hermetischen Wissenschaft*, pp. 258–59. Munich: Verlag C. H. Beck, 1998.

Clulee, Nicholas H. 'Astonomia inferior: Legacies of Johannes Trithemius and John Dee'. In William R. Newman and Anthony Grafton, eds., *Secrets of Nature: Astrology and Alchemy in Early Modern Europe*, 173–233. Cambridge, MA: MIT Press, 2001.

———. 'The Monas hieroglyphica and the Alchemical Thread of John Dee's Career'. *Ambix* 52, no. 2 (2005): pp. 197–215.

Coldstream, Nicola. *Medieval Architecture*, Oxford History of Art. Oxford: Oxford University Press, 2002.

Collingwood, R.G., *The Idea of History*. Oxford: Clarendon Press, 1946.

Collins, Brendan. 'Wisdom in Jung's Answer to Job'. *Biblical Theology Bulletin* 21 (1991): pp. 97–101.

Collins, John J. *Between Athens and Jerusalem: Jewish Identity in the Hellenistic Diaspora*. New York: The Crossroad Publishing Company, 1983.

Colquhoun, Margaret. 'An Exploration in the Use of Goethean Science as a Methodology for Landscape Assessment: the Pishwanton Project'. *Agriculture, Ecosystems, and Environment* 63 (1997): pp. 145–57.

Consilium Coniugii, seu De Massa Solis & Lunae. In Jean-Jacques Manget, ed., *Bibliotheca chemica curiosa*, 2 volumes. Vol. 2, pp. 235–66. Geneva, 1701–1702.

Constantine of Pisa. *The Book of the Secrets of Alchemy*. Introduction, Critical Edition, Translation, and Commentary by Barbara Obrist. Leiden: Brill, 1990.

Corbin, Henry. *Mundus Imaginalis, or the Imaginary and the Imaginal.* 1964. http://www.hermetic.com/bey/mundus_imaginalis.htm

———. 'The Visionary Dream in Islamic Spirituality'. In G. E. von Grunebaum and Roger Caillois, eds., *The Dreams and Human Societies*, pp. 381–408. Berkeley: University of California Press, 1966.

Cornelius, Geoffrey, *The Moment of Astrology: Origins in Divination.* London: Arkana 1994.

———. *The Moment of Astrology: Origins in Divination.* Second edition. Bournemouth: The Wessex Astrologer, 2003.

———. 'Verity and the Question of Primary and Secondary Scholarship in Astrology'. In Nicholas Campion, Patrick Curry and Michael York, eds., *Astrology and the Academy*, pp. 103–13. Bristol: Cinnabar Books, 2004.

Crollius, Oswaldus. *Philosophy Reformed & Improved in Four Profound Tractates*. London, 1657.

Crosland, Maurice P. *Historical Studies in the Language of Chemistry.* London: Heinemann, 1962.

Cryer, Frederick H. *Divination in Ancient Israel and its Near Eastern Environment*. Sheffield, England: Sheffield Academic Press, 1994.

———. *Prophecy and Power: Astrology in Early Modern England.* Oxford: Polity Press, 1989.

Curry, Patrick. 'The Historiography of Astrology: A Diagnosis and a Prescription'. In K. von Stuckrad, G. Oestmann and H. Darrel Rutkin, eds., *Horoscopes and Public Spheres: Essays on the History of Astrology*, pp. 261–74. Berlin: Walter de Gruyter, 2005.

D'Espagnet, Jean. *Arcanum Hermeticae Philosophia Opus, in Enchiridion Physicae Restitutae*. Paris, 1623.

———. *L'Ouvrage secret de la Philosophie d'Hermez*, in *La Philosophie Naturelle restablie en sa Pureté*. Paris, 1651.

da Nono, Giovanni. 'Visio Egidij Regis Patavie'. *Bollettino del Museo Civico di Padova* N.S. X–XI [XXVII–XXVIII] and XIII–XVII (1934–39): pp. 1–20.

Dale, Sharon, Alison Williams Lewin, and Duane J. Osheim, eds. *Chronicling History: Chroniclers and Historians in Medieval and Renaissance Italy*. University Park, PA: The Pennsylvania State University Press, 2007.

Dan, Joseph. *The Esoteric Theology of Ashkenazi Hasidism*. Jerusalem: Bialik Institute, 1968. [Hebrew].

———. 'On the Teaching Concerning the Dream in Ḥasidei Ashkenaz'. *Sinai* 68 (1971): pp. 288–293 [Hebrew].

———. 'The Emergence of Mystical Prayer'. In Joseph Dan and Frank Talmage, eds., *Studies in Jewish Mysticism: Proceedings of Regional Conferences Held at the University of California, Los Angeles and McGill University in April, 1978*, pp. 85–120. Cambridge, MA: Association for Jewish Studies, 1982.

———. '*Pesaq ha-Yirah veha-Emunah* and the Intention of Prayer in Ashkenazi Hasidic Esotericism'. *Frankfurter Judaistische Beiträge* 19 (1991–1992): pp. 185–215.

———. 'Prayer as Text and Prayer as Mystical Experience'. In Ruth Link-Salinger, ed., *Torah and Wisdom-Studies in Jewish Philosophy, Kabbalah, and Halacha: Essays in Honor of Arthur Hyman*, pp. 33–47. New York: Shengold Publishers, 1992.

———. 'Ashkenazi Hasidim, 1941–1991: Was There Really a Hasidic Movement in Medieval Germany?' In Peter Schäfer and Joseph Dan, eds., *Gershom Scholem's Major Trends in Jewish Mysticism 50 Years After: Proceedings of the Sixth International Conference on the History of Jewish Mysticism*, pp. 87–101. Tübingen: J. C. B. Mohr, 1993.

———. *Jewish Mysticism in the Middle Ages*. Northvale: Jason Aronson, 1998.

———. 'Book of Raziel'. In *Encyclopaedia Judaica* (2007), 13:1591–93.

———, and Frank Talmage, eds. *Studies in Jewish Mysticism: Proceedings of Regional Conferences Held at the University of California, Los Angeles and McGill University in April, 1978*. Cambridge, MA: Association for Jewish Studies, 1982.

Daniel, Dane T. 'Invisible Wombs: Rethinking Paracelsus's Concept of Body and Matter'. *Ambix* 53, no. 2 (2006): pp. 129–42.

Davies, Charlotte Aull. *Reflexive Ethnography: A Guide to Researching Selves and Others*. London: Routledge, 1999.

Davison, R. C. *The Technique of Prediction*. 1955. Reprinted by Romford, Essex: L.N.Fowler & Co. Ltd., 1983.

Dawson, Terence. 'Jung, Literature, and Literary Criticism'. In Polly Young-Eisendrath and Terence Dawson, eds., *The Cambridge Companion to Jung*, pp. 255–80. Cambridge: Cambridge University Press, 1997.

de Bonneau, Jean [I. D. B.]. *Abrege de l'Astronomie Inferieure, Expliquant le Systeme des Planetes; les douze signes du Zodiac & autres Constellations du Ciel Hermetique. Avec un Essay de l'Astronomie Naturelle, contre les Systemes de Ptolomée, Copernic, & Ticho Brahé*. Paris, 1644.

de Jong, T. and van Soldt, W.H. 'Redating an Early Solar Eclipse Record (KTU 1.78): Implications for the Ugaritic Calendar and for the Secular Accelerations of the Earth and Moon'. *Jahrbericht Ex Oriente Lux* 30 (1987–1988): pp. 65–77.

de la Riva, Bonvesin. *De Magnalibus Mediolani = Le Meraviglie Di Milano*. Translated by Giuseppe Pontiggia, Nuova Corona. Milano: Bompiani, 1974.

de Montanor, Guido. *Scala Philosophorum*. In Jean-Jacques Manget, ed., *Bibliotheca chemica curiosa*. 2 volumes. Vol. 2, pp. 134–47. Geneva, 1701–1702.

de Planis Campy, David. *Traite des playes faites par des mousquetades*. Paris, 1623.

———. *Bouquet Composé des plus belles fleurs Chimiques*. Paris, 1629.

———. *L'Hydre Morbifique exterminée par l'Hercule Chimique*. Paris, 1629.

———. *L'Ouverture de L'Escolle de Philosophie Transmutatoire Métallique*. Paris, 1633.

de Saint-Didier, Limojon. *Le Triomphe hermétique, ou La Pierre Philosophale victorieuse*. Paris, 1699.

de Vadis, Egidius. *Dialogus inter Naturam & Filium Philosophiae*. In *Theatrum Chemicum*, Vol. 2 (1602), pp. 95–123.

———. *Dialogus inter Naturam & Filium Philosophiae*. In Jean-Jacques Manget, *Bibliotheca chemica curiosa*. 2 vols. Vol. 2, pp. 326–35. Geneva, 1702.

Dean, Geoffrey and Mather, Arthur. *Recent Advances in Natal Astrology: A Critical Review 1900–1976*. Subiaco, Australia: Analogic, 1977.

——— and Ivan W. Kelly. 'Is Astrology Relevant to Consciousness and Psi?'. *Journal of Consciousness Studies* 10, no. 6-7, (June–July 2003): http://www.imprint.co.uk/pdf/Dean.pdf

Dean, Trevor. *The Towns of Italy in the Later Middle Ages*. Translated by Trevor Dean, Manchester Medieval Sources Series. Manchester: Manchester University Press 2000.

Debus, Allen G. *The Chemical Philosophy*. Mineola, NY: Dover, 1977.

Dever, William G. *Who* Were *the Early* Israelites *and Where Did They Come From?* Grand Rapids: Wm. B. Eerdmans Publishing, 2003.

DeVun, Leah. *Prophecy, Alchemy, and the End of Time: John of Rupescisssa in the Late Middle Ages*. New York: Columbia University Press, 2009.

Dickson, Donald R. *The Tessera of Antilia: Utopian Brotherhoods & Secret Societies in the Early Seventeenth Century*. Leiden: Brill, 1998.

Dieterich, Albert. *Ein Mithrasliturgie*. Leipzig: Tübner, 1910.

Dietrich, M., Loretz O., and Sanmartín, J., eds. *The Cuneiform Alphabetic Texts from Ugarit, Ras Ibn Hani and Other Places (KTU: Second, Enlarged Edition)*, Abhandlungen zur Literatur Alt-Syrien-Palästinas und Mesopotamiens 8. Münster: Ugarit-Verlag, 1995.

Dillman, Claudia. 'Die Wirkung der Architektur ist eine magische. Hans Poelzig und der Film'. In Reichmann, *Hans Poelzig. Bauten für den Film*, pp. 20–75.

Dorn, Gerard. *Congeries Paracelsicae chemiae de transmutationibus metallorum*. In *Theatrum Chemicum*, Vol. 1. 2nd edition, 1659.

Dourley, John P. *The Intellectual Autobiography of a Jungian Theologian*. Lampeter: Edwin Mellen Press, 2006.

Drob, Sanford L. 'Towards a Kabbalistic Psychology: C. G. Jung and the Jewish Foundations of Alchemy'. *Journal of Jungian Theory and Practice* 5, no. 2 (2003): pp. 77–100.

Eamon, William. *Science and the Secrets of Nature: Books of Secrets in Medieval and Early Modern Culture*. Princeton, NJ: Princeton University Press, 1994.

Edgar Laird, 'Christine de Pizan and Controversy Concerning Star Study in the Court of Charles V'. *Culture and Cosmos* 1, no. 2 (Autumn/Winter 1997): pp. 35–48.

Eisner, Lotte. *The Haunted Screen: Expressionism in the German Cinema and the Influence of Max Reinhardt*. 1969. Reprinted by Berkeley, University of California Press, 1973.

Eleazar of Worms. *Ḥokhmat ha-Nefesh*. Benei Beraq, 1987.
Eliade, Mircea. *The Sacred and the Profane: The Nature of Religion*. Translated by Willard R. Trask. Orlando: Harcourt Inc., 1957.
———. *The Forge and the Crucible*. Translated by Stephen Corrin. Chicago: University of Chicago Press, 1962.
Elior, Rachel. 'The Concept of God in Hekhalot Mysticism'. In Joseph Dan, ed., *Binah: Studies in Jewish History, Thought, and Culture*, Vol. 2, pp. 97–120. NY: Praeger, 1989.
———, and Peter Schäfer, eds. *Creation and Re-Creation in Jewish Thought: Festschrift in Honor of Joseph Dan on the Occasion of his Seventieth Birthday*. Tübingen: Mohr Siebeck, 2005.
Elkins, James. *Six Stories from the End of Representation: Images in Painting, Photography, Astronomy, Microscopy, Particle Physics, and Quantum Mechanics, 1980–2000*. Stanford: Stanford University Press, 2008.
Elsaesser, Thomas. *Weimar Cinema and After: Germany's Historical Imaginary*. New York: Routledge, 2000.
Emerton, J.A. 'A Difficult Part of Mot's Message to Baal in the Ugaritic Texts'. *Australian Journal of Biblical Archaeology* 2 (1972): pp. 50–71.
Epstein, S. R. 'The Rise and Decline of Italian City States'. Working Paper No.51/99 (1999).
Epstein, Steven A. *Genoa and the Genoese, 958–1528*. Chapel Hill, NC: University of North Carolina Press, 1996.
Erlanger, Gad. *Signs of the Times: the Zodiac in Jewish Tradition*. New York: Feldheim, 2001.
Ertan, Denis. *Dane Rudhyar: His Music, Thought and Art*. Rochester: University of Rochester Press, 2009.
Evans, S. *A High History of the Holy Graal*. London: J. M. Dent, 1898.
Eysenck, H. B., and Nias, D. K. B. *Astrology: Science or Superstition*. Second edition. Harmondsworth: Penguin, 1984.
Fabre, Pierre-Jean. *Opus Pan-Chymici seu Anatomiae totius universi*. Frankfurt, 1651.
Faivre, Antoine. *The Eternal Hermes: From Greek God to Alchemical Magus*. Grand Rapids, MI: Phanes Press, 1995.
Ficino, Marsilio. *Three Books on Life: A Critical Edition and Translation with Introduction and Notes* by Carol V. Kaske and John R. Clark. Binghamton, NY: Medieval and Renaissance Texts and Studies and the Renaissance Society of America, 1989.
Figuier, Louis. *L'Alchimie et les Alchimistes: Essai historique et critique sur la Philosophie hermétique*. Paris, 1854.
Figulus, Benedictus. *Thesaurinella Olympica Aurea Tripartita*. Frankfurt, 1682.
Fine, Lawrence. *Physician of the Soul, Healer of the Cosmos: Isaac Luria and His Kabbalistic Fellowship*. Stanford, CA: Stanford University Press, 2003.
Flamel, Nicolas. *Le Livre des Figures Hieroglifiques de Nicolas Flamel Escrivain*. In *Philosophie Naturelle de Trois Anciens Philosophes Renommez Artephius, Flamel, & Synesius, Traitant de l'Art occulte, &*

de la Transmutation metallique, pp. 45–88. Paris, 1682.

Forman, Simon. Oxford, Bodleian Library, Ms. Ashmole 389. *The Astrologicalle Judgmentes of Phisick and other Questions writen by Simon Forman Dr of Astronomy and Phisick 1606—Lambeth the 16 of September in which is comprised his experience for 20 yeares before And many yeares after.*

Forrest, Steven and Jeffrey Wolf Green. *Measuring the Night: Evolutionary Astrology and the Keys to the Soul.* Boulder CO: Daemon Press, 2000.

Forshaw, Peter J. 'The Early Alchemical Reception of John Dee's Monas Hieroglyphica', *Ambix* 52, no. 3 (2005): pp. 247–69.

———. 'Alchemical Exegesis: Fractious Distillations of the Essence of Hermes'. In Lawrence M. Principe, ed., *Chymists and Chymistry: Studies in the History of Alchemy and Early Modern Chemistry,* pp. 25–38. Sagamore Beach: Science History Publications, 2007.

Frank, Dennis. 'Dane Rudhyar's reformulation of astrological theory'. http://cura.free.fr/xx/17frank4.html, accessed 11 November 2012.

French, Chris. 'Astrologers and other inhabitants of parallel universes'. *The Guardian,* 7 February 2012.

French, Peter. *John Dee: The World of an Elizabethan Magus.* 1972; reprinted by London: Routledge, 2002.

French, Roger. 'Astrology in Medical Practice'. In Luis Garcia Ballester, Roger French, Jon Arrizabalaga, and Andrew Cunningham, eds., *Practical Medicine from Salerno to the Black Death,* pp. 30–59. Cambridge: Cambridge University Press, 1994.

Freundel, Barry. *Contemporary Orthodox Judaism's Response to Modernity.* Jersey City: Ktav Publishing House, 2004.

Frey, Pierre A. and Lise Grenier. *Viollet-le-Duc et la montagne.* Grenoble, Editions Glénat: 1993.

Fulco, William J., S.J. *The Canaanite God Rešep.* American Oriental Series 8. New Haven, CT: American Oriental Society, 1976.

Fuller, Benjamin. *A History of Philosophy.* New York: H. Holt & Co., 1938.

Gabriele, Mino. *Alchimia e Iconologia.* Udine: Forum, 1997.

Gadbury, John. *Collectio Geniturarum: Or, A Collection of Nativities.* London, 1662.

Gadd, C.J. *Ideas of Divine Rule in the Ancient East.* Schweich Lectures of the British Academy. 1945. Reprinted by Munich: Kraus Reprint, 1980.

Gansten, Martin. 'Reshaping Karma: An Indic Metaphysical Paradigm in Traditional and Modern Astrology'. In Nicholas Campion, ed., *Cosmologies: Proceedings of the Seventh Annual Sophia Centre Conference 2009,* pp. 52–68. Ceredigion, Wales: Sophia Centre Press, 2011.

Gardner, Gregg. 'Astrology in the Talmud: An Analysis of Bavli Shabbat 156'. In Eduard Iricinschi and Holger M. Zellentin, eds., *Heresy and Identity in Late Antiquity,* pp. 314–38. Tübingen: Mohr

Siebeck, 2008.

Geber. *De Alchimia libri tres*. Strasburg, 1531.

Geertz, Clifford. 'Religion as a Cultural System'. In Michael P. Banton, ed., *Anthropological Approaches to the Study of Religion*. London: Frederick A. Praeger Press, 1966.

Glazerson, Matityahu. *Above the Zodiac: Astrology in Jewish Thought*. Northvale, NJ: Aronson, 1997.

Goldstein-Jacobsen, Ivy. *Here and There in Astrology*. no publisher, no place, 1961.

Goodenough, Erwin R. *Jewish Symbols in the Greco-Roman Period*. Vols. 1–12. Bollingen Series XXXVII. New York: Pantheon Books, 1953–1965.

Goodrick-Clarke, Nicholas. *Paracelsus: Essential Readings*. Berkeley: North Atlantic Books, 1999.

Goodwin, Clare. 'The Language of Astrology'. 2000. http://www.abgoodwin.com/therapy-and-readings/language-of-astrology.html, accessed 4 November 2012.

Grafton, Anthony, and Nancy Siraisi. 'Between the Election and My Hopes: Girolamo Cardano and Medical Astrology'. In William R. Newman and Anthony Grafton, eds., *Secrets of Nature: Astrology and Alchemy in Early Modern Europe*, pp. 69–131. Cambridge, MA: MIT Press, 2001.

Grasseus, Johannes [Johann Grasshoff]. *Physica Naturalis Rotunda Visionis Chemicae Cabalisticae*. In *Theatrum Chemicum*, Vol. 6, pp. 343–81. 1661.

Green, Arthur. 'Shekhinah, the Virgin Mary, and the Song of Songs: Reflections on a Kabbalistic Symbol in Its Historical Context'. *AJS Review* 26, no. 1 (2002): pp. 1–52.

Greene, Liz. 'Signs, Signatures, and Symbols: the Languages of Heaven'. In Campion and Greene, eds., *Astrologies: Plurality and Diversity*, pp. 17–45. Lampeter: Sophia Centre Press, 2010.

Greenfield, J.C. and M. Sokoloff, M. 'Astrological and Related Omen Texts in Jewish Palestinian Aramaic'. *JNES* 48 (1989): pp. 201–14.

Grell, Ole Peter, ed., *Paracelsus: The Man and his Reputation, his Ideas and their Transformation*. Leiden: Brill, 1998.

Greverus, Jodocus. *Secretum Nobilissimum et Verissimum*. In *Theatrum Chemicum*, Vol. 3, pp. 735–59. 1613.

Grosseteste, Robert. *Die Philosophischen Werke des Robert Grosseteste, Bischofs von Lincoln*. Edited by Ludwig Baur. Münster: Aschendorffsche Verlagsbuchhandlung, 1912.

Grossman, Abraham. 'The Migration of the Kalonymos Family from Italy to Germany'. *Zion* 40 (1975): pp. 154–85 [Hebrew].

Grözinger, Karl Erich and Joseph Dan, eds. *Mysticism, Magic and Kabbalah in Ashkenazi Judaism: International Symposium Held in Frankfurt a. M. 1991*. Berlin: Walter de Gruyter, 1995.

Gundissalinus, Dominicus. *De Divisione Philosophiae*. Edited by Ludwig Baur. Münster: Druck und Verlag der Aschendorffschen Buchhandlung, 1903.

Gunning, Tom. 'Der frühe Film und das Okkulte'. In *Okkultismus und Avantgarde: von Munch bis Mondrian 1900–1915*, pp. 558–61. Frankfurt: edition tertium, 1995.

Hachlili, Rachel. *Ancient Jewish Art and Archaeology in the Land of Israel*. Leiden: Brill, 1988.

———. 'The Zodiac in Ancient Synagogal Art: A Review'. *Jewish Studies Quarterly* 9 (2002): pp. 219–58.

———. *Ancient Mosaic Pavements: Themes, Issues, and Trends*. Leiden: Brill, 2008.

Halbertal, Moshe. *Concealment and Revelation: Esotericism in Jewish Thought and its Philosophical Implications*. Translated by Jackie Feldman. Princeton, NJ: Princeton University Press, 2007.

Halevi, Judah. *Sefer ha-Kuzari*. Translated by Yehuda Even Shmuel. Tel-Aviv: Dvir, 1972.

Hall, Nor. *The Moon and the Virgin*. London, The Women's Press, 1980.

Hamarneh, Sami K. 'Arabic-Islamic Alchemy—Three Intertwined Stages'. *Ambix* 29, no. 2 (1982): pp. 74–87.

Hambrock, Heike. *Hans und Marlene Poelzig: Bauen im Geist des Barock. Architekturphantasien, Theaterprojekte und moderner Festbau (1916–1926)*. Berlin: Aschenbeck & Holstein, 2005.

Hames, Harvey J. *Like Angels on Jacob's Ladder: Abraham Abulafia, the Franciscans, and Joachimism*. Albany, NY: State University of New York Press, 2007.

Hamilton, Leonidas le Cenci, trans., *Ishtar and Izdubar: The Epic of Babylon*. London: W. H. Allen, 1884.

Hamlin, Cyrus and John Michael Krois, eds. *Symbolic Forms and Cultural Studies: Ernst Cassirer's Theory of Culture*. New Haven, CT: Yale University Press, 2004.

Hammer, Olav. *Claiming Knowledge: Strategies of Epistemology from Theosophy to the New Age*. Leiden: Brill, 2004.

Hand, Robert. 'The Proper Relationship of Astrology and Science'. *Astrological Journal* 31, no. 6 (Nov./Dec. 1989): pp. 307–16.

Hanegraaff, Wouter J. *New Age Religion and Western Culture: Esotericism in the Mirror of Secular Thought*. Leiden: Brill, 1996.

———. 'Romanticism and the Esoteric Connection'. In Roelof van den Broek and Wouter J. Hanegraaff, eds., *Gnosis and Hermeticism: From Antiquity to Modern Times*, pp. 237–68. Albany, NY: SUNY Press, 1998.

Hannaway, Owen. *The Chemists & the Word: The Didactic Origins of Chemistry*. Baltimore: The Johns Hopkins University Press, 1975.

Harding, Michael. *Hymns to the Ancient Gods*. London: Penguin, 1992.

Harris, Monford. 'Dreams in Sefer Hasidim'. *Proceedings of the American Academy of Jewish Research* 31 (1963): pp. 51–80.

———. *Studies in Jewish Dream Interpretation*. Northvale: Jason Aronson, 1994.

Harrison, Charles. *An Introduction to Art*. New Haven: Yale University Press, 2009.

———, et al., ed. *Art in Theory 1815–1900*. Oxford: Wiley-Blackwell,

1998.

Hasolle, James [Elias Ashmole]. *Fasciculus Chemicus: Or Chymical Collections. Expressing The Ingress, Progress, and Egress, of the Secret Hermetick Science, out of the choisest and most Famous Authors.* London, 1650.

Hauptmann, Gerhart. *Worte zu Faust: eine deutsche Volksage.* Berlin: Universum-Film Aktiengesellschaft, 1926.

Hayes, John. 'Introduction to the Symbolic Language of Astrology'. http://www.johnhayes.biz/Articles/symbolicLanguageAstrology.htm, accessed 4 November 2012.

Henry, John. 'Boyle and Cosmical Qualities'. In Michael Hunter ed., *Robert Boyle Reconsidered,* pp. 119–38. Cambridge: Cambridge University Press, 1994.

Hermes Trismegiste. *La Table d'Émeraude et sa Tradition alchimique.* Edited by Didier Kahn. Paris: Les Belles Lettres, 1995.

Heselton, P. 'A Provisional List of Terrestrial Zodiacs in Britain'. *Terrestrial Zodiacs Journal* 1 (1989): pp. 30–31.

Hesse, Mary. *Models and Analogies in Science.* Notre Dame: University of Notre Dame Press, 1966

Heugten, Sjaar van, Joachim Pissarro, and Chris Stolwijk. *Van Gogh and the Colors of the Night.* New York: The Museum of Modern Art, New York, 2008.

Heuss, Theodor. *Hans Poelzig. Bauten und Entwürfe. Das Lebensbild eines deutschen Baumeisters.* Berlin: Wasmuth, 1939.

———, 'Begegnungen mit Paul Wegener'. In Rudolph S. Joseph, *Paul Wegener. Der Regisseur und Schauspieler,* p. 2. Munich: Photo- u. Filmmuseum im Münchner Stadtmuseum, 1967.

Heydon, John. *Elhavareuna or the English Physitians Tutor in the Astrobolismes of Mettals.* London, 1665.

Hickey, Isabel. *Astrology, a Cosmic Science.* 1970. Reprinted by Sebastopol, CA, 1992.

Hildebrand, Wolffgang. *Ein new außerlesen Planeten-Buch.* Erfurt, 1613.

Hillman, James. *The Soul's Code: In Search of Character and Calling.* London: Bantam, 1996.

———. 'The Azure Vault: The Caelum as Experience'. In Nicholas Campion and Patrick Curry, eds., *Sky and Psyche: The Relationship Between Cosmos and Consciousness,* pp. 37–54. Edinburgh: Floris Books, 2005.

Hoeller, Stephan A. *The Gnostic Jung and the Seven Sermons to the Dead.* Wheaton, IL: Theosophical Publishing House, 1982.

Hoffman, Leon. 'Varieties of Psychoanalytic Experience: Review of *The Red Book*'. *Journal of the American Psychoanalytic Association* 58 (2010): pp. 781–85.

Hoffmann, Nigel. 'The Unity of Science and Art: Goethean Phenomenology as a New Ecological Discipline'. In D. Seamon and A. Zajonc, eds., *Goethe's Way of Science, : A Phenomenology of Nature,* pp. 129–76. New York: State University of New York Press, 1998.

Holdredge, Craig. 'Doing Goethean Science'. *Janus Head* 8, no. 1 (2005): pp. 27–52.

Hone, Margaret. *The Modern Textbook of Astrology*. 1951. Fourth edition by London: L.N. Fowler, 1973.

Howe, Ellic. *Astrology: A Recent History Including the Untold Story of Its Role in World War II*. New York: Walker and Company, 1967.

Hunter, Michael, ed., *Robert Boyle Reconsidered*. Cambridge: Cambridge University Press, 1994.

Husserl, Edmund. *Ideas: General Introduction to Pure Phenomenology*. 1913. English translation 1931; Reprinted by London: Collier-MacMillan, 1972.

Hyde, J. K. 'Italian Social Chronicles in the Middle Ages'. In *Literacy and Its Uses: Studies on Late Medieval Italy*, edited by D. Waley, pp. 33–57. Manchester: Manchester University Press, 1993.

———.'Medieval Descriptions of Cities.' In *Literacy and Its Uses: Studies on Late Medieval Italy*, edited by D. Waley, pp. 1–32. Manchester: Manchester University Press, 1993.

———. *Padua in the Age of Dante*. Manchester: Manchester University Press, 1966.

Iamblichus. *On the Mysteries of the Egyptians, Chaldeans, and Assyrians*. Translated by Thomas Taylor. 1821. Reprinted by Frome: Prometheus Trust 1999.

Ibn al-'Arabī, Muḥyīddīn. *The Bezels of Wisdom*. Translated by and introduction by Ralph W. J. Austin. New York: Paulist Press, 1980.

Idel, Moshe. *Studies in Ecstatic Kabbalah*. Albany, NY: State University of New York Press, 1988.

———. *The Mystical Experience in Abraham Abulafia*. Albany, NY: State University of New York Press, 1988.

———. *Language, Torah, and Hermeneutics in Abraham Abulafia*. Albany, NY: State University of New York Press, 1989.

———. *Golem: Jewish Magical and Mystical Traditions on the Artificial Anthropoid*. Albany, NY: SUNY Press, 1990.

———. *Hasidism: Between Ecstasy and Magic*. Albany, NY: State University of New York Press, 1995.

———. *Messianic Mystics*. New Haven, CT: Yale University Press, 1998.

———. *Le Porte della Giustizia, Sa'are Sedeq*. Milano: Adelphi Edizioni, 2001.

———. *Absorbing Perfections: Kabbalah and Interpretation*. New Haven, CT: Yale University Press, 2002.

———. *Les Kabbalistes de la nuit*. Paris: Editions Allia, 2003.

———. *Kabbalah and Eros*. New Haven, CT: Yale University Press, 2005.

———. 'On Še'elat Ḥalom in Ḥasidei Aškenaz: Sources and Influences'. *Materia Giudaica* 10 (2005): pp. 99–109.

Ingold, Tim. *The Perception of the Environment: Essays on Livelihood, Dwelling and Skill*. London: Routledge, 2000.

———. 'Ethnography is not Anthropology'. *Proceedings of the British*

Academy 154 (2008): pp. 69–92.

Iricinschi, Eduard and Holger M. Zellentin, eds., *Heresy and Identity in Late Antiquity*. Tübingen: Mohr Siebeck, 2008.

Isenberg, Noah. 'Of Monsters and Magicians: Paul Wegener's *The Golem how he came into the World* (1920)'. In Noah Isenberg, ed., *Weimar Cinema: An Essential Guide to Classic Films of the Era*, pp. 33–54. New York: Columbia University Press, 2009.

Ivakhiv, A. *Claiming Sacred Ground: Pilgrims and Politics at Glastonbury and Sedona*. Bloomington: Indiana University Press, 2001.

Jacobs, Louis. 'Uplifting the Sparks in Later Jewish Mysticism'. In Arthur Green, ed., *Jewish Spirituality, Vol. 2: From the Sixteenth Century Revival to the Present*, pp. 99–126. New York: Crossroad, 1987.

James, M. R., ed. and trans., *The Apocryphal New Testament*. Oxford: Clarendon Press, 1924.

Janacek, Bruce. 'A Virtuoso's History: Antiquarianism and the Transmission of Knowledge in the Alchemical Studies of Elias Ashmole'. *Journal of the History of Ideas* 69, no. 3 (2008): pp. 395–417.

Jastrow, Morris. *The Religion of Babylonia and Assyria*. Boston, MA: Athenaeum Press, 1898.

Jeffers, Ann. *Magic and Divination in Ancient Palestine and Syria*. Leiden: Brill, 1996.

Jennings, H. *The Rosicrucians: Their Rites and Mysteries*. 1870. Reprinted by London: George Routledge, 1887.

Jensen, K. Frank. *The Story of the Waite-Smith Tarot*. Melbourne: Association of Tarot Studies, 2006.

Jeremias, Alfred. *Izdubar-nimrod: Eine altbabylonische Heldensage: Nach den Keilschriftfragmenten Dargestellt*. Leipzig: B. G. Teubner, 1891.

Jeromson, Barry. 'Systema Munditotius and Seven Sermons: Symbolic Collaborators in Jung's Confrontation with the Unconscious'. *Jung History* 1, no. 2 (2010), at <https://www.philemonfoundation.org/resources/jung_history/volume_1_issue_2>

Johnston, Sarah Iles. *Ancient Greek Divination*. Malden, MA, & Oxford, UK: Wiley-Blackwell, 2008.

Joly, Bernard. *La Rationalité de l'Alchimie au XVIIe siècle*. Paris: Vrin, 1992.

Jonte-Pace, Diane Elizabeth and William B. Parsons. *Religion and Psychology: Mapping the Terrain. Contemporary Dialogues, Future Prospects*. London: Routledge, 2001.

Joseph, Rudolph S. *Paul Wegener. Der Regisseur und Schauspieler*. Munich: Photo- u. Filmmuseum im Münchner Stadtmuseum, 1967.

Josten, C. H. 'Elias Ashmole, F.R.S. (1617–1692)'. *Notes and Records of the Royal Society of London* 15 (July 1960): pp. 221–30.

———. 'A Translation of John Dee's "Monas Hieroglyphica"

(Antwerp, 1564), With an Introduction and Annotations'. *Ambix* 12, no. 2–3 (1964): pp. 83–222.

Judah the Pious. *Sefer Ḥasidim*. Edited by Jehuda Wistinetzki and Jacob Freimann. Frankfurt am Main: Wahrmann Verlag, 1924.

Julius Theodor and Chanoch Albeck, ed. *Genesis Rabbah*. Jerusalem: Wahrmann Books, 1965.

Jung, C. G. *Wandlungen und Symbole der Libido*. In *Jahrbuch für psychoanalytische und psychopathologische Forschungen*, III–IV. Leipzig, 1911–12.

———. *Psychology of the Unconscious*. Translated by Beatrice M. Hinkle. London: Kegan Paul, Trench, Trubner & Co., 1917.

———. *Psychology and Alchemy*. CW12. Translated by R. F. C. Hull. London: Routledge & Kegan Paul, 1953.

———. 'The Importance of Dreams'. In C.G. Jung, ed., *Man and his Symbols*. New York: Dell Publishing, 1954.

———. *The Practice of Psychotherapy*. CW16. Translated by R. F. C. Hull. London: Routledge & Kegan Paul, 1954.

———. *The Archetypes and the Collective Unconscious*. CW9i. Translated by R. F. C. Hull. London: Routledge & Kegan Paul, 1959.

———. *Aion: Researches into the Phenomenology of the Self*. CW9ii. Translated by R. F. C. Hull. London: Routledge & Kegan Paul, 1959.

———. *Psychiatric Studies*. CW1. Translated by R. F. C. Hull. London: Routledge & Kegan Paul, 1957.

———. *The Structure and Dynamics of the Psyche*. CW8. Translated by R. F. C. Hull. London: Routledge & Kegan Paul, 1960.

———. *Memories, Dreams, and Reflections*. Edited by Aniela Jaffé. Translated by Richard and Clara Winston. London: Routledge & Kegan Paul, 1963.

———. *Mysterium Coniunctionis*. CW14. Translated by R. F. C. Hull. London: Routledge & Kegan Paul, 1963.

———. 'The Psychological Aspects of the Kore'. In C. G. Jung and C. Kerényi, *Essays on a Science of Mythology: The Myth of the Divine Child and the Mysteries of Eleusis*. Princeton, NJ: Princeton University Press, 1963.

———. *Civilisation in Transition*. CW10. Translated by R. F. C. Hull. London: Routledge and Kegan Paul, 1964.

———. *Two Essays on Analytical Psychology*. CW7. Translated by R. F. C. Hull. London: Routledge & Kegan Paul, 1966.

———. *Symbols of Transformation*. CW5. Translated by R. F. C. Hull. London: Routledge & Kegan Paul, 1967.

———. *Alchemical Studies*. CW13. Translated by R. F. C. Hull. London: Routledge & Kegan Paul, 1967.

———. *Psychological Types*. CW6. Translated by R. F. C. Hull. London: Routledge & Kegan Paul, 1971.

———. *The Symbolic Life*. CW18. Translated by R. F. C. Hull. London: Routledge & Kegan Paul, 1977.

———. *Septem Sermones ad Mortuos: Written by Basilides in*

Alexandria, the City Where East and West Meet. Translated by Stephan A. Hoeller. In Stephan A. Hoeller, *The Gnostic Jung and the Seven Sermons to the Dead*, pp. 44–58. Wheaton, IL: Theosophical Publishing House, 1982.

———. 'Psychology and Alchemy'. In *Collected Works*. Translated by R. F. C. Hull. Princeton, NJ: Princeton University Press, 1983.

———. *The Red Book: Liber Novus*. Edited by Sonu Shamdasani. Translated by Mark Kyburz, John Peck, and Sonu Shamdasani. New York: W. W. Norton, 2009.

Kaes, Anton. *Kino-Debatte: Texte zum Verhältnis von Literatur und Film*. Munich: dtv, 1978.

Kahn, Didier. 'Alchimie et littérature à Paris en des temps de trouble: Le Discours d'Autheur incertain sur la pierre des philosophes (1590)'. *Réforme, Humanisme, Renaissance* 41, no. 1 (1995): pp. 75–122.

Kandinsky, Wassily. *Concerning the Spiritual in Art*. Translated by M. T. H. Sadler. New York: Dover Publications, 1977.

Karr, Don. 'Notes on the Study of Later Kabbalah in English: The Safed Period and Lurianic Kabbalah'. In Don Karr, *Collected Articles on the Kabbalah*, Vol. 2, pp. 23–31. Ithaca, NY: KoM, No. 6, 1985.

Kassell, Lauren. *Medicine & Magic in Elizabethan London: Simon Forman, Astrologer, Alchemist, & Physician*. Oxford: Oxford University Press, 2005.

Kelley, David H. and Milone, Eugene F. *Exploring Ancient Skies: An Encyclopedic Survey of Archaeoastronomy*. New York: Springer Science + Business Media, Inc., 2005.

Kelly, Ivan. 'Why Astrology Doesn't Work'. *Psychological Reports* 82 (1998): pp. 527–46.

Khalid. *Liber Trium Verborum Kalid Regis*. In *Theatrum chemicum*, Vol. 5, pp. 186–90. 1622.

Khunrath, Heinrich. *De Igne Magorum Philosophorumque secreto externo et visibili*. Strasburg, 1608.

Kieckhefer, Richard. *Magic in the Middle Ages*. 1989; reprinted by Cambridge: Cambridge University Press, 2003.

Kiesewetter, Carl. *Faust in der Geschichte und Tradition*. Leipzig, 1893. Reprinted New York: Georg Olms Verlag, 1978.

Knapp, Bettina L. *A Jungian Approach to Literature*. Carbondale, IL: Southern Illinois University Press, 1984.

Knorr von Rosenroth, Christian. *Kabbala denudata, seu, Doctrina Hebraeorum transcendentalis et metaphysica atque theologica: opus antiquissimae philosophiae barbaricae...in quo, ante ipsam translationem libri...cui nomen Sohar tam veteris quam recentis, ejusque tikkunim...praemittitur apparatus [pars 1–4]*. 3 volumes. Sulzbach/Frankfurt: Abraham Lichtenthal, 1677–1684.

Koch-Westenholz, Ulla. *Mesopotamian Astrology: An Introduction to Babylonian and Assyrian Celestial Divination*. Copenhagen: Museum Tusculanum Press, 1995.

Krupp, E.C. 'Sky Tales And Why We Tell Them'. In Helaine Selin, ed., *Astronomy Across Cultures: the History of Non-Western Astronomy*. Dordrecht: Kluwer Academic Publishers, 2000.

Kugelman, Robert. 'Review of *The Red Book*'. *Journal of the History of the Behavioral Sciences* 47, no. 1 (2011): pp. 101–4.

Kusukawa, Sachiko. 'The Uses of Pictures in the Formation of Learned Knowledge: The Cases of Leonhard Fuchs and Andreas Vesalius'. In Sachiko Kusukawa and Ian Maclean, eds., *Transmitting Knowledge: Words, Images, and Instruments in Early Modern Europe*, pp. 73–96. Oxford: Oxford University Press, 2006.

———. and Ian Maclean, eds. *Transmitting Knowledge: Words, Images, and Instruments in Early Modern Europe*. Oxford: Oxford University Press, 2006.

Kuyt, Annelies. 'Hasidut Ashkenaz on the Angel of Dreams: A Heavenly Messenger Reflecting or Exchanging Man's Thoughts'. In Rachel Elior and Peter Schäfer, eds., *Creation and Re-Creation in Jewish Thought: Festschrift in Honor of Joseph Dan on the Occasion of his Seventieth Birthday*, pp. 147–63. Tübingen: Mohr Siebeck, 2005.

Langerman, Y. Tzvi. 'Some Astrological Themes in the Thought of Abraham ibn Ezra'. In Isadore Twersky and Jay M. Harris, eds., *Rabbi Abraham ibn Ezra: Studies in the Writings of a Twelfth-Century Jewish Polymath*, pp. 28–85. Cambridge, MA: Harvard University Press, 1993.

Leader, E. 'The Somerset Zodiac: Myths and Legends'. In Mary Williams, ed., *Glastonbury: A Study in Patterns*, p. 8. London: R.I.L.K.O., 1969.

Leitch, Y., ed. *Signs and Secrets of the Glastonbury Zodiac: An Anthology from the Maltwood Moot*. Glastonbury: Avalonian Aeon Publications, 2013.

Leo, Alan. *How to Judge a Nativity*. 1903. Reprinted by London: Modern Astrology, 1922.

———. *Esoteric Astrology*. London: N.Fowler, 1925.

Leslie, C. R. *Memoirs of the Life of John Constable, Esq, R.A.* London: Longman, Brown, Green, and Longmans, 1845.

Lévy-Bruhl, Lucien. *How Natives Think*. Princeton: Princeton University Press, 1985.

Libavius, Andreas. *Rerum Chymicarum Epistolica Forma ad Philosophos et Medicos quosdam in Germania, Liber Tertius*. Frankfurt, 1599.

———. *Examen Philosophiae Novae, quae veteri abrogandae opponitur*. Frankfurt, 1615.

Lilly, William. *Christian Astrology*. London, 1647.

Linden, Stanton J., ed., *The Alchemy Reader: From Hermes Trismegistus to Isaac Newton*. Cambridge: Cambridge University Press, 2003.

Ling, Martin. *Symbol and Archetype: A Study of the Meaning of Existence*. Louisville: Fons Vitae, 2005.

Link-Salinger, Ruth, ed. *Torah and Wisdom-Studies in Jewish Philosophy, Kabbalah, and Halacha: Essays in Honor of Arthur Hyman*. New York: Shengold Publishers, 1992.

Lipiński, Edward. *Resheph: A Syro-Canaanite Deity*. Orientalia

Lovaniensia Analecta 181, Studia Phoenicia XIX. Leuven: Uitgeverij Peeters en Departement Oosterse Studies, 2009.

Magid, Shaul. 'From Theosophy to Midrash: Lurianic Exegesis and the Garden of Eden'. *Association for Jewish Studies Review* 22, no. 1 (1997): pp. 37–75.

Magness, J. 'Heaven on Earth: Helios and the Zodiac Cycle in Ancient Palestinian Synagogues'. *DOP* 59 (2005): pp. 1–52.

Maier, Emanuel. *The Psychology of C. G. Jung in the Works of Hermann Hesse*. unpublished PhD dissertation, New York University, 1953.

Maidenbaum, Aryeh and Stephen A. Martin, eds. *Lingering Shadows: Jungians, Freudians, and Anti-Semitism*. Boston: Shambhala, 1991.

———, ed. *Jung and the Shadow of Anti-Semitism: Collected Essays*. Berwick, ME: Nicolas-Hays, 2002.

Main, Roderick, ed., *Jung on Synchronicity and the Paranormal*. Princeton: Princeton University Press, 1997.

Maltwood, K. E. *Air View Supplement to a Guide to Glastonbury's Temple of the Stars*. London: John M. Watkins, 1937.

———. *The Enchantments of Britain*. Victoria: Victoria Publishing Company, 1944.

———. Draft Article, 'King Solomon's Table'. undated. Maltwood Archive, McPherson Library, University of Victoria, BC, Canada.

———. Undated Paper, 'Blavatsky says....'. Maltwood Archive, McPherson Library, University of Victoria, BC, Canada.

Mancuso, Piergabriele. *Shabbatai Donnolo's Sefer Ḥakhmoni: Introduction, Critical Text, and Annotated Translation*. Leiden: Brill, 2010.

Manget, Jean-Jacques, ed. *Bibliotheca chemica curiosa*, 2 volumes. Geneva, 1701–1702.

Marcus, Ivan G. *Piety and Society: The Jewish Pietists of Medieval Germany*. Leiden: E. J. Brill, 1981.

———. 'History, Story, and Collective Memory: Narrativity in Early Ashkenazic Culture'. *Prooftexts* 10 (1990): pp. 365–88.

———. 'The Historical Meaning of Hasidei Ashkenaz: Fact, Fiction or Cultural Self-Image?' In Peter Schäfer and Joseph Dan, eds., *Gershom Scholem's Major Trends in Jewish Mysticism 50 Years After: Proceedings of the Sixth International Conference on the History of Jewish Mysticism*, pp. 103–14. Tübingen: J. C. B. Mohr, 1993.

———. 'Prayer Gestures in German Hasidism'. In Karl Erich Grözinger and Joseph Dan, eds., *Mysticism, Magic and Kabbalah in Ashkenazi Judaism: International Symposium Held in Frankfurt a. M. 1991*, pp. 44–59. Berlin: Walter de Gruyter, 1995.

Marcus, Joseph. 'Studies in the *Chronicle of Ahimaaz*'. *Proceedings of the American Academy of Jewish Research* 5 (1933/34): pp. 85–93.

Margolin, Jean-Claude and Sylvain Matton, eds. *Alchimie et philosophie à la Renaissance. Actes du colloque international de Tours (4–7 Déc. 1991)*. Paris: Vrin, 1993.

Mattheeuws, Christel. 'Towards an Anthropology in Life: The Astrological Architecture of Zanadroandrena Land in West

Bezanozano, Madagascar'. PhD Thesis, University of Aberdeen, 2008.

Matton, Sylvain. 'Marsile Ficin et l'alchimie. Sa position, son influence'. In Jean-Claude Margolin and Sylvain Matton, eds., *Alchimie et philosophie à la Renaissance. Actes du colloque international de Tours (4-7 Déc. 1991)*, pp. 123–92. Paris: Vrin, 1993.

Maurach, Gregor. 'Daniel von Morley, "Philosophia"' *Mittellateinisches Jahrbuch* 14 (1979): pp. 204–55.

Mayo, Jeff. *Astrology: A Key to Personality*. London: Penguin, 1995.

——— and Ramsdale, Christine. *Teach Yourself Astrology*. London: Hodder Headline, 1996.

Mazzucchelli, Aldo. 'Celestial Weathercock, Diagrams & Metaphorical Web: Some Semiotic Considerations on Western Astrology'. *European Journal for Semiotic Studies* 12, no. 4 (2000), reproduced as 'Astrology, Hermeneutics and Metaphorical Web'. http://cura.free.fr/xv/12mazz-e.html, accessed 15 November 2012.

McEvoy, James. 'The Chronology of Robert Grosseteste's Writings on Nature and Natural Philosophy'. *Speculum* 58, no. 3 (1983): pp. 614–55.

Mead, G. R. S., ed. and trans., *Thrice-Greatest Hermes*. London: Theosophical Publishing House, 1906.

———, ed. and trans., *The Mysteries of Mithra*. London: Theosophical Publishing House, 1907.

———, ed. and trans., *Pistis Sophia*. London: J. M. Watkins, 1921.

Merkur, Dan. *Gnosis: An Esoteric Tradition of Mystical Visions and Unions*. Albany, NY: SUNY Press, 1993.

———. 'Stages of Ascension in Hermetic Rebirth'. *Esoterica* 1 (1999): pp. 79–96.

Merleau-Ponty, Maurice. *The Phenomenology of Perception*. London: Routledge, Kegan and Paul, 1962.

Mettinger, Tryggve. *The Riddle of Resurrection: 'Dying and Rising Gods' in the Ancient Near East*. Coniectanea Biblica Old Testament Series 50. Stockholm: Almqvist & Wiksell International, 2001.

Michelspacher, Stephan. *Cabala, Spiegel der Kunst und Natur: in Alchymia*. Augspurg, 1616.

Möller, Kai. *Paul Wegener. Sein Leben und seine Rollen. Ein Buch von ihm und über ihn. Eingerichtet von Kai Möller*. Rheinbeck bei Hamburg: Rohwolt, 1954.

Moran, Bruce T. *Andreas Libavius and the Transformation of Alchemy: Separating Chemical Cultures with Polemical Fire*. Sagamore Beach, MA: Science History Publications, 2007.

Moskowitz, Anita Fiderer. *Italian Gothic Sculpture c.1250–c.1400*. Cambridge: Cambridge University Press, 2001.

———. *Pisano Pulpits*. London: Harvey Miller Publishers, 2005.

Multhauf, Robert P. *The Origins of Chemistry*. London: Oldbourne, 1966.

Murnau, F. W. *Faust, eine deutsche Volksage* (1926). Film.

Mylius, Johann Daniel. *Opus Medico-Chymicum*. Frankfurt, 1618.
———. *Anatomia auri, sive Tyrocinium medico-chymicum*. Frankfurt, 1628.
Nasmyth, James. *James Nasmyth, Engineer: An Autobiography*. New York, 1883.
Naydler, Jeremy, ed. *Goethe on Science: A Selection of Goethe's Writings*. Edinburgh: Floris Books, 1996.
Neu, Erich. *Das Hurritische Epos der Freilassung I: Untersuchungen zu einem hurritisch-hethitischen Textensemble aus Ḫattuša*. Studien zu den Boğazköy-Texten 32. Wiesbaden: Harrasowitz, 1996.
Neumann, Erich. *The Origins and History of Consciousness*. Princeton, NJ: Princeton University Press, 1973.
Neusner, Jacob. 'Rabbi and Magus in Third-Century Sasanian Babylonia'. *History of Religions* 6, no. 2 (1966): pp. 169–78.
Newman, William R. *The Summa Perfectionis of Pseudo-Geber: A Critical Edition, Translation & Study*. Leiden: Brill, 1991.
———. *Gehennical Fire: The Lives of George Starkey, An American Alchemist in the Scientific Revolution*. Chicago: University of Chicago Press, 2003.
———, and Anthony Grafton. 'Introduction: the Problematic Status of Astrology and Alchemy in Premodern Europe'. In William R. Newman and Anthony Grafton, eds., *Secrets of Nature: Astrology and Alchemy in Early Modern Europe*, pp. 24–25. Cambridge, MA: MIT Press, 2001.
———, eds. *Secrets of Nature: Astrology and Alchemy in Early Modern Europe*. Cambridge, MA: MIT Press, 2001.
———, and Lawrence M., *Alchemy Tried in the Fire: Starkey, Boyle, and the Fate of Helmontian Chymistry*. Chicago: University of Chicago Press, 2002.
Noll, Richard. *The Jung Cult: Origins of a Charismatic Movement*. Princeton, NJ: Princeton University Press, 1994.
———. *The Aryan Christ*. New York: Random House, 1997.
Norman, Diana. 'Change and Continuity: Art and Religion after the Black Death'. In Diana Norman, ed., *Siena, Florence and Padua: Art, Society and Religion 1280–1400, Volume I: Interpretive Essays*, pp. 177–195. New Haven, CT: Yale University Press in association with the Open University, 1995.
Norton, Thomas. *The Ordinall of Alchimy*. In Elias Ashmole, ed., *Theatrum Chemicum Britannicum*, pp. 1–106. London, 1652.
Obrist, Barbara. *Les Débuts de l'Imagerie Alchimique (XIVe-XVe siècles)*. Paris: Éditions le Sycomore, 1982.
———. 'Visualization in Medieval Alchemy'. *HYLE: International Journal for Philosophy of Chemistry* 9, no. 2 (2003): pp. 131–70.
Oestmann, Günther, H. Darrel Rutkin, and Kocku von Stuckrad, eds., *Horoscopes and Public Spheres: Essays on the History of Astrology*. Berlin: Walter de Gruyter, 2005.
Origen. *Contra Celsum*. Edited and translated by Henry Chadwick. Cambridge: Cambridge University Press, 1953.

Ortner, Sherry. 'On key symbols'. *American Anthropologist* 75, no. 6 (1973): pp 1338–46.

Osler, Margaret J. 'Book Note on *Campanus of Novara and medieval planetary theory. Theorica planetarum*'. *Journal of the History of Philosophy* 14, no. 4 (October 1976): p. 498.

Ovid. *Metamorphoses*. Translated by Samuel Garth, John Dryden, *et al.*, at <http://classics.mit.edu/Ovid/metam.mb.txt>

Owens, Lance S. 'The Hermeneutics of Vision: C. G. Jung and *Liber Novus*'. *The Gnostic* 3 (2010): pp. 23–46.

Oyvind Dahl, Oyvind. 'When the Future Comes from Behind: Malagasy and other Time Concepts and some Consequences for Communication'. *International Journal for Intercultural Relations* 19, no. 2 (1995): pp. 197–209.

Pagano, Cathy. 'Symbolic Language Is Our Mother Tongue'. 23 October 2008. http://www.astrology.com/symbolic-language-our-mother-tongue/2-d-d-48089

Pagel, Walter. *Paracelsus: An Introduction to Philosophical Medicine in the Era of the Renaissance*. Basel: S. Karger, 1958.

———. *The Smiling Spleen: Paracelsianism in Storm and Stress*. Basel: S. Karger, 1984.

Paracelsus. *Aurora Thesaurusque Philosophorum*. Basel, 1577.

———. *De Tinctura Physicorum*. In *Aureoli Philippi Theophrasti Bombasts von Hohenheim Paracelsi, Opera*. Edited by Johann Huser. 2 volumes. Vol. 1, pp. 922–25. Strasburg, 1603.

———. *Aureoli Philippi Theophrasti Bombasts von Hohenheim Paracelsi, Opera*. Edited by Johann Huser. 2 volumes. Strasburg, 1603.

———. *Operum Medico-Chimicorum sive Paradoxorum, Tomus Genuinus Undecimus*. Frankfurt, 1605.

———. *Of The Chymical Transmutation of Metals & Minerals*. Translated by R. Turner. London, 1657.

———. *His Aurora, & Treasure of the Philosophers*. London, 1659.

———. *The Hermetic and Alchemical Writings of Paracelsus*. Translated by A. E. Waite. 1894.

———. *Paracelsus (Theophrastus Bombastus von Hohenheim, 1493-1541): Essential Theoretical Writings*. Edited and translated by Andrew Weeks. Leiden: Brill, 2008.

Pardee. Dennis. *Ritual and Cult at Ugarit*, SBL Writings of the Ancient World 10. Edited by Theodore J. Lewis. Atlanta, GA: Scholars Press, 2002.

Partington, James R. *A History of Chemistry*. 4 volumes. London: Macmillan & Co., 1961–1970.

Partridge, John. *Defectio Geniturarum: Being an Essay toward the Reviving and Proving the True Old Principles of Astrology*. London, 1697.

Pehnt, Wolfgang. 'Wille zum Ausdruck. Zu Leben und Werk Hans Poelzigs'. In Wolfgang Pehnt and Matthias Schirren, eds., *Hans Poelzig. Architekt Lehrer Künstler. 1869 bis 1936*, pp. 10–51. Munich: DVA, 2007.

Penrose, Roger. *The Emperor's New Mind: Concerning Computers, Minds and the Laws of Physics*. London: Vintage, 1991.

Pereira, Michela. 'Heavens on Earth. From the Tabula Smaragdina to the Alchemical Fifth Essence'. *Early Science and Medicine* 5, no. 2, *Alchemy and Hermeticism* (2000): pp. 131–44.

Pernety, Antoine-Joseph. *Dictionnaire mytho-hermétique*. Paris, 1758.

Peterson, Joseph, ed., *The Sixth and Seventh Books of Moses*. Lake Worth, FL: Ibis Press, 2008.

Pines, Shlomo. 'On the Term *Ruḥaniyyot* and its Origin and on Judah Halevi's Doctrine'. *Tarbiz* 57 (1988): pp. 511–40.

Plait, Philip. *Bad Astronomy: Misconceptions and Misuses Revealed, from Astrology to the Moon Landing 'Hoax'*. New York: Wiley, 2002.

Plato, *Timaeus*. Translated by R. G. Bury. Cambridge MA: Harvard University Press, 1931.

Platonis Quartorum cum commento Hebuhabes Hamed. In *Theatrum Chemicum*, Vol. 5, pp. 101–85. 1622.

Plessner, M. 'Hermes Trismegistus and Arab Science'. *Studia Islamica* 2 (1954): pp. 45–59.

Plotinus. 'On Whether the Stars are Causes'. In *Enneads*, Vol., 2. Translated by A. H. Armstrong. Cambridge MA: Harvard University Press, 1929.

———. 'The Second Ennead, Third Tractate—Are the Stars Causes?', Translated by Stephen MacKenna and B. S. Page http://classics.mit.edu/Plotinus/enneads.2.second.html, accessed 17 March 2013.

Poelzig, Hans. 'Festspielhaus in Salzburg'. *Das Kunstblatt* (1921). Reprinted in Nendeln/Liechtenstein: Kraus Reprint, 1978, pp. 77–88.

Poliakov, Léon. *The Aryan Myth: A History of Racist and Nationalist Ideas in Europe*. Translated by Edmund Howard. New York: Basic Books, 1971.

Pope-Hennessy, John. *Italian Gothic Sculpture*. Vol. 1, An Introduction to Italian Sculpture. London: Phaidon Press Limited, 2000.

Popović, M. *Reading the Human Body: Physiognomics and Astrology in the Dead Sea Scrolls and Hellenistic-Early Judaism*. STDJ 67. Leiden: Brill, 2007.

Prawer S. S. *Between Two Worlds: The Jewish Presence in German and Austrian Film, 1910–1933*. New York: Berghan, 2005.

Priesner, Claus and Karin Figala, eds., *Alchemie: Lexikon einer hermetischen Wissenschaft*. Munich: Verlag C. H. Beck, 1998.

Principe, Lawrence M. *The Aspiring Adept: Robert Boyle and his Alchemical Quest*. Princeton, NJ: Princeton University Press, 1998.

———, ed., *Chymists and Chymistry: Studies in the History of Alchemy and Early Modern Chemistry*. Sagamore Beach: Science History Publications, 2007.

Progoff, Ira. *The Symbolic and the Real: A New Psychological Approach to the Fuller Experience of Personal Existence*. New York: McGraw-

Hill, 1973.
———. *At a Journal Workshop: Writing to Access the Power of the Unconscious and Evoke Creative Ability*. New York: Penguin-Tarcher, 1992.
Ptolemy. *Centum Sententiae, quod Centiloquium dicunt*. In *Claudii Ptolemaei Omnia, quae extant, Opera*, pp. 500–4. Basel, 1541.
———. 'Centiloquium'. In John Partridge, *Mikropanastron, or an Astrological Vade Mecum, briefly Teaching the whole Art of Astrology—viz., Questions, Nativities, with all its parts, and the whole Doctrine of Elections never so comprised nor compiled before, &c.* William Bromwich: London, 1679.
Quispel, Gilles. *Gnosis als Weltreligion: Die Bedeutung der Gnosis in der Antike*. Zürich: Origo Verlag, 1951.
———. 'C. G. Jung und die Gnosis'. *Eranos-Jahrbuch* 37 (1968): pp. 277–98.
Raff, Jeffrey. *Jung and the Alchemical Imagination*. York Beach, ME: Nicolas-Hays, 2000.
Raguseius, Georgius. *Epistolarum mathematicarum seu De Divinatione, Libri Duo*. Paris, 1623.
Rahtz, P. and L. Watts. *Glastonbury: Myth and Archaeology*. Stroud: Tempus, 2003.
Raine, Kathleen. *Yeats, the Tarot, and the Golden Dawn*. Dublin: Dolmen Press, 1972.
Rayner, Sheila, ed. *Dane Rudhyar: Interviewed by Sheila Finch Rayner, Clare G. Rayner and Rob Newell*. Long Beach, CA: California State University Library, 1977.
Read, John. *Prelude to Chemistry: an Outline of Alchemy, its Literature and Relationships*. London: G. Bell & Sons, 1936; reprinted by Cambridge, MA: The MIT Press, 1966.
———. *From Alchemy to Chemistry*. London: G. Bell & Sons, 1957; reprinted by NY: Dover Publications, 1995.
Reed, Annette Yoshiko. 'Abraham as Chaldean Scientist and Father of the Jews: Josephus *ANT*. 1.154–168, and the Greco-Roman Discourse About Astronomy/Astrology'. *Journal for the Study of Judaism* 35, no. 2 (2004): pp. 119–58.
———. 'Was There Science in Ancient Judaism? Historical and Cross-Cultural Reflections on "Religion" and "Science"'. *Studies in Religion* 36, nos. 3-4 (2007): pp. 461–95.
Reichmann, Hans-Peter. *Hans Poelzig. Bauten für den Film*. Exhibition catalogue. Kinematograph 12. Frankfurt am Main: Deutsches Filmmuseum, 1997.
Reiner, Erica. *Astral Magic in Babylonia*. Philadelphia: American Philosophical Society, 1995.
Rewald, Sabine. *Caspar David Friedrich: Moonwatchers*. New York: Metropolitan Museum of Art, 2002.
Riess, J. B. *Political Ideals in Medieval Italian Art: The Frescoes in the Palazzo Dei Priori, Perugia (1297)*. Ann Arbor: UMI Research Press, 1981.
Riffard, Pierre. *L'esoterisme*. Paris: Laffont, 1990.

Ripley, George. *Liber Duodecim Portarum*. In Jean Jacques Manget, *Bibliotheca Chemica Curiosa*. 2 volumes Vol. 2, pp. 275–85. Geneva, 1702.

Robb, D. M. 'Niccolò: A North Italian Sculptor of the Twelfth Century'. *Art Bulletin* 12, no. 4 (1930): pp. 374–420.

Robb, John E. 'The Archaeology of Symbols'. In *Annual Review of Anthropology* 27 (1998): pp. 329–46.

Roberts, J. J. M. 'Erra: Scorched Earth'. *Journal of Cuneiform Studies* 24, no. 1/2 (1971): pp. 11–16.

Robinson, James M., ed., *The Nag Hammadi Library in English*. Leiden: Brill, 1988.

Rochberg, Francesca. *The Heavenly Writing: Divination, Horoscopy, and Astronomy in Mesopotamian Culture*. Cambridge: Cambridge University Press, 2004.

Röder, Sabine. 'Traumspiel mit Kulissen'. In Sabine Röder and Gerhard Storck. *Der dramatische Raum. Hans Poelzig. Malerei Theater Film*, pp. 8–22. Krefeld: Kat. Museum Haus Lange, Haus Esters, 1986.

Roitman, Adolfo D. 'This People are Descendants of Chaldeans' (Judith 5:6): Its Literary Form and Historical Setting'. *Journal of Biblical Literature* 113, no. 2 (1994): pp. 245–63.

Roob, Alexander. *The Hermetic Museum: Alchemy & Mysticism*. Cologne: Taschen, 2003.

Rosenfeld, Beate. *Die Golemsage und ihre Verwertung in der deutschen Literatur*. Sprache und Kultur der germanischen und romanischen Völker. B. Germanistische Reihe, Vol. 5. Breslau: H. Priebatsch, 1934.

Rossi, Raffaele , Attilio Bartoli Angeli, and Roberta Sottani. *Perugia*. Vol. 1. Milan: Elio Sellino Editore, 1993.

Roth-Scholtz, Friedrich, ed. *Deutsches Theatrum Chemicum*. 3 volumes. 1728–1733.

Rudhyar, Dane. *The Astrology of Personality*. 1936. Reprinted by Garden City, NY: Doubleday, 1970.

———. *New Mansions for New Men*. New York: Lucis Publishing Company, 1938.

———. Dane, *The Lunation Cycle: A Key to the Understanding of Personality*. Santa Fe: Aurora Press, 1967.

———. Dane, *Occult Preparations for a New Age*. Wheaton, IL: The Theosophical Publishing House, 1975.

———. *The Planetarisation of Consciousness*. New York: Aurora Press, 1977.

———. *Person Centered Astrology*. New York: Aurora Press, 1980.

———. 'Symbols and the Cyclic Character of Human Existence', http://mindfire.ca/The%20Sabian%20Symbols/An%20Astrological%20Mandala%20-%20Symbols%20&%20the%20Cyclic%20Character%20of%20Existence.htm, accessed 4 Nov. 2012.

Ruland, Martin. *Lapidis Philosophici Vera Conficiendi Ratio*. Frankfurt, 1606.

Ruperti, Alexander. *Cycles of Becoming*. Vancouver, WA: CRCS Publications, 1978.

———. *Meaning of Humanistic Astrology*. 2002. http://www.stand.cz/astrologie/czech/texty/rez-ru-a/rez-ru-a.htm, consulted 25 Jan. 2003.

Sagerman, Robert. 'Ambivalence toward Christianity in the Kabbalah of Abraham Abulafia'. PhD dissertation, New York University, 2008.

Sahlins, Marshall. 'Two or Three Things that I Know about Culture'. *The Journal of the Royal Anthropological Institute* 5, no. 3 (Sept. 1999): pp. 400–3.

Samuels, Andrew. *Jung and the Post-Jungians*. London: Routledge, 1985.

Sandivogius, Micheel [*sic*]. *A New Light of Alchymie*. London, 1650.

Savedow, Steve, trans., *Sepher Rezial Hemelach: The Book of the Angel Rezial*. York Beach, ME: WeiserBooks, 2001.

Sawyer, J. F. A. and F. R. Stephenson. 'Literary and Astronomical Evidence for a Total Eclipse of the Sun in Ancient Ugarit on 3 May 1375 BC'. *Bulletin of the School of Oriental and African Studies, University of London* 33, no. 3 (1970): pp. 467–89.

Schäfer, Peter and Joseph Dan, eds. *Gershom Scholem's Major Trends in Jewish Mysticism 50 Years After: Proceedings of the Sixth International Conference on the History of Jewish Mysticism*. Tübingen: J. C. B. Mohr, 1993.

Schaeffer, Claude. 'The Last Days of Ugarit: Drought, Famine, Earthquakes and, Ultimately, Fire Ended Civilisation at Ugarit'. *Biblical Archaeology Review* 9 (1983): pp. 74–75.

Schaffer, Simon. 'On Astronomical Drawing'. In Caroline A. Jones and Peter Galison, *Picturing Science, Producing Art*. New York and London: Routledge, 1998.

Schmidt, Francis. 'Ancient Jewish Astrology: An Attempt to Interpret 4QCRYPTIC (4Q186)'. In Michael E. Stone and Esther G. Chazon, eds., *Biblical Perspectives: Early Use and Interpretation of the Bible in Light of the Dead Sea Scrolls*, pp. 189–205. Leiden: Brill, 1998.

Schmitz, Oscar A. H. *Der Geist der Astrologie*. Munich: Georg Müller, 1922.

Schönemann, Heidi. *Paul Wegener: frühe Moderne im Film*. Stuttgart: Edition Axel Menges, 2003.

Scholem, Gershom. 'Eine kabbalistische Erklärung der Prophetie als Selbstbegegnung'. *Monatsschrift für Geschichte und Wissenschaft des Judentums* 74 (1930).

———. *Major Trends in Jewish Mysticism*. New York: Schocken Books, 1961.

———. *On the Kabbalah and Its Symbolism*. Translated by Ralph Mannheim. New York, NY: Schocken Books, 1965.

———. *Kabbalah*. New York: Keter Publishing House, 1974.

———. *Origins of the Kabbalah*. Edited by R. J. Zwi Werblowsky. Translated by Allan Arkush. Princeton, NJ: Princeton University

Press, 1987.

———. *On the Mystical Shape of the Godhead: Basic Concepts in the Kabbalah*. Translated by Joachim Neugroschel. Edited by Jonathan Chipman. New York: Schocken Books, 1991.

———. 'The Idea of the Golem'. In Gershom Scholem, *On Kabbalah and its Symbolism*, pp. 158–204. Translated by R. Manheim. New York: Schocken, 1996.

Schopenhauer, Arthur. *The World as Will and Representation*. 2 vols. Translated by E. F. J. Payne. New York: Dover, 1958.

Schwartz, Dov. *Studies on Astral Magic in Medieval Jewish Thought*. Translated by David Louvish and Batya Stein. Leiden: Brill, 2005.

Secret, François. 'Sur quelques traductions du Sefer Raziel'. *Revue des Études Juives* 128 (1969): pp. 223–45.

Segal, Robert A., ed., *The Gnostic Jung*. Princeton, NJ: Princeton University Press, 1992.

———. 'Jung as psychologist of religion and Jung as philosopher of religion'. *Journal of Analytical Psychology* 55 (2010): pp. 361–84.

Seiling, Max. *Goethe als Okkultist*. Berlin: Joh. Baum Verlag, 1919. Reprinted by Leipzig: Bohmeier Verlag, 2008.

Sela, Shlomo. *Astrology and Biblical Exegesis in Abraham Ibn Ezra's Thought*. Rama-Gan: Bar-Ilan University, 1999. [Hebrew].

Sells, Benjamin, ed., *Working with Images: The Theoretical Base of Archetypal Psychology*. Woodstock, CT: Spring Publications, 2000.

Serrano, Miguel. *C. G. Jung and Hermann Hesse: A Record of Two Friendships*. Einsiedeln, CH: Daimon Verlag, 1998.

Shamdasani, Sonu. 'Who is Jung's Philemon? Unpublished Letter to Alice'. *Jung History* 2, no. 2 (2011), at <https://www.philemonfoundation.org/resources/jung_history/volume_2_issue_2>

———. *C. G. Jung: A Biography in Books*. New York: W. W. Norton, 2012.

Sharf, Andrew. *The Universe of Shabbetai Donnolo*. New York: Ktav, 1976.

Shaw, Gregory. *Theurgy and the Soul: The Neoplatonism of Iamblichus*. University Park, PA: Penn State Press, 1971.

Shumaker, Wayne. *The Occult Sciences of the Renaissance: A Study in Intellectual Patterns*. 1972; reprinted by Berkeley: University of California Press, 1979.

Sibley, Ebenezer. *A new and complete illustration of the celestial science of Astrology*. London, 1822.

Smith, George. 'The Chaldean Account of the Deluge'. *Transactions of the Society of Biblical Archaeology* 1–2 (1872): pp. 213–34.

Smith, Mark. 'Baal'. In Simon B. Parker, ed., *Ugaritic Narrative Poetry*, SBL Writings from the Ancient World 9, pp. 81–176. Atlanta, GA: Scholars Press, 1997.

Smith, Morton. *The Secret Gospel: The Discovery and Interpretation of the Secret Gospel According to Mark*. London: Victor Gollancz, 1974.

Soloveitchik, Haym. 'Three Themes in the *Sefer Ḥasidim*'. *AJS Review* 1 (1976): pp. 311–58.

Somers, Barbara. *The Fires of Alchemy: A Transpersonal Viewpoint*. Bourne: Archive Publishing, 2004.

Spencer, Neil. 'Just Whose 'Consciousness' Are We Talking About, Anyway?'. Short Cuts. *The Guardian*. 19 August 2003.

Stafford, Barbara Maria. *Artful Science*. Cambridge, MA: MIT Press, 1994.

Stapleton, H. E., G. L. Lewis, and F. Sherwood Taylor. 'The sayings of Hermes quoted in the *Ma Al-Waraqi* of Ibn Umail'. *Ambix* 3 (1949): pp. 69–90.

Stern, Sacha, *Calendar and Community: A History of the Jewish Calendar Second Century* BCE*–Tenth Century* CE. Oxford: Oxford University Press, 2001.

Stieglitz, Robert R. 'The Hebrew Names of the Seven Planets.' *Journal of Near Eastern Studies* 40, no. 2 (1981): pp. 135–37.

Stratton, Kimberly. 'Imagining Power: Magic, Miracle, and the Social Context of Rabbinic Self-Representation'. *Journal of the American Academy of Religion* 73, no. 2 (2005): pp. 361–93.

Stratton-Kent, Jake. *The True Grimoire*. Scarlet Imprint, 2009.

Strauß, Heinz Artur. *Astrologie: Grundsätzliche Betrachtungen*. München: Kurt Wolff, 1927.

Struck, Peter. *The Birth of the Symbol*. Princeton: Princeton University Press, 2004.

Sugg, Richard P., ed., *Jungian Literary Criticism*. Evanston, IL: Northwestern University Press, 1992.

Swartz, Michael D. 'Divination and its Discontents: Finding and Questioning Meaning in Ancient and Medieval Judaism'. In Scott Noegel, Joel Walker, and Brannon Wheeler, eds., *Prayer, Magic, and the Stars in the Ancient and Late Antique World*, pp. 155–66. University Park, PA: The Pennsylvania State University, 2003.

Swimme, Brian and Thomas Berry. *The Universe Story*. San Francisco: Harper, 1992.

Szanto, Gregory. *The Marriage of Heaven and Earth: the Philosophy of Astrology*. London: Penguin-Arkana, 1985.

Tarnas, Richard. *Cosmos and Psyche*. New York: Penguin Group, 2006.

Taylor, Thomas. *Collected Writings on the Gods and the World*. Frome: Prometheus Trust, 1994.

Telle, Joachim. 'Die Magia naturalis Wolfgang Hildebrands'. *Sudhoffs Archiv* 60, no. 2 (1976): pp. 105–22.

———. 'Astrologie et alchimie au XVIe siècle: à propos des poèmes didactiques astro-alchimiques de Christoph von Hirschenberg et de Basile Valentin'. *Chrysopoeia* 3, no. 2 (April/June 1989): pp. 163–92.

———. 'Astrologie und Alchemie im 16. Jahrhundert. Zu den astroalchemischen Lehrdichtungen von Christoph von Hirschenberg und Basilius Valentinus'. In August Buck, ed., *Die okkulten Wissenschaften in der Renaissance*, pp. 227–53. Wiesbaden: Harrassowitz, 1992.

Tester, Jim. *A History of Western Astrology*. 1987; reprinted by Woodbridge: The Boydell Press, 1999.

Theatrum Chemicum, praecipuos selectorum auctorum tractatus de chemiae et lapidis philosophici antiquitate, veritate, iure, praesentia et operationibus. 6 volumes. Ursel/Strasbourg, 1602–1661.

The Freud-Jung Letters. Edited by William McGuire. Translated by Ralph Manheim and R. F. C. Hull. London: Hogarth Press/Routledge & Kegan Paul, 1977.

Thurneysser zum Thurn, Leonhart. *Megale Chymia, seu Magna Alchymia*. Berlin, 1583.

Trachtenberg, Joshua. *Jewish Magic and Superstition: A Study in Folk Religion*. Philadelphia: University of Pennsylvania Press, 2004.

Trismosin, Salomon. *Splendor Solis*. Translated by Joscelyn Godwin. Introduction and Commentary by Adam McLean. Grand Rapids, MI: Phanes Press, 1991.

Turba Philosophorum. In *Theatrum Chemicum*, Vol. 5, pp. 1–51. 1622.

Turner, Jane, ed. *The Dictionary of Art*. London, Macmillan, 1996.

Twersky, Isadore and Jay M. Harris, eds. *Rabbi Abraham ibn Ezra: Studies in the Writings of a Twelfth-Century Jewish Polymath*. Cambridge, MA: Harvard University Press, 1993.

Valentin, Basile. *Révélation des Mystères des Teintures Essentielles des Sept Metaux, & de leurs Vertus Medicinales*. Paris, 1678.

Valentine, Basil. *Zwölf Schlüssel, in Fratris Basilii Valentini Benedictiner Ordens Chymische Schriften alle/ so viel derer verhanden*. Hamburg, 1700.

Valentinus, Basilius. *Letztes Testament*. Strassburg, 1651.

———. *His Last Will and Testament*. London, 1657.

Vallensis, Roberto. *Gloria Mundi, sonsten Paradeiss-Taffel*. In Friederich Roth-Scholtz, ed., *Deutsches Theatrum Chemicum*. 3 volumes. Vol. 3, pp. 372–536. 1728–1733.

van den Broek, Roelof and Wouter J. Hanegraaff, eds., *Gnosis and Hermeticism: From Antiquity to Modern Times*. Albany, NY: SUNY Press, 1998.

van Soldt, W.H. ''*atn prln*, "'attā/ēnu the Diviner"'. *Ugarit-Forschungen* 21 (1989): pp. 365–68.

VanderKam, James C. *Enoch and the Growth of An Apocalyptic Tradition*. Washington, DC: The Catholic Biblical Association of America, 1984.

———. 'Prophecy and Apocalyptics in the Ancient Near East'. In Jack M. Sasson, ed., *Civilizations of the Ancient Near East: Volumes Three & Four*, pp. 2083–94. Peabody, MA: Hendrickson Publishers, 1995.

———. *From Revelation to Canon: Studies in the Hebrew Bible and Second Temple Literature*. Leiden: Brill, 2000.

———. *An Introduction to Early Judaism*. Grand Rapids/Cambridge: Wm. B. Eerdmans Publishing Co., 2001.

Verene, Donald Philip. *The Origins of the Philosophy of Symbolic Forms: Kant, Hegel, and Cassirer*. Evanston, IL: Northwestern University Press, 2011.

Villani, Giovanni. *Cronica Di Giovanni Villani*. Florence: Per Il Magheri, 1823.

von Franz, Marie-Louise. *Alchemical Active Imagination*. Irving, TX: Spring Publications, 1979.

von Grunebaum, G. E. and Roger Caillois, eds. *The Dreams and Human Societies*. Berkeley: University of California Press, 1966.

von Siebenstern, Christian Friedrich Sendimir. *Chymischer Mondenschein*. Frankfurt, 1760.

von Stuckrad, Kocku. 'Jewish and Christian Astrology in Late Antiquity: A New Approach'. *Numen* 47 (2000): pp. 1–40.

———. 'From Astronomy to Naturphilosophie, From Matter to Spirit: Astrology in German Romanticism'. Paper presented at 'Astrologies', the eighth annual Sophia Centre Conference, Bath, 24–25 July 2010.

———. *Locations of Knowledge in Medieval and Early Modern Europe: Esoteric Discourse and Western Identities*. Leiden: Brill, 2010.

Waite, A. E. *The Key to the Tarot: Being Fragments of a Secret Tradition under the Veil of Divination*. London: William Rider & Son, 1909.

Wasserstrom, Stephen. *Religion after Religion: Gershom Scholem, Mircea Eliade, and Henry Corbin at Eranos*. Princeton, NJ: Princeton University Press, 1999.

Webb, James. *The Occult Establishment*. La Salle, IL: Open Court Publishing Co. 1976.

Webster, James Carson. *The Labors of the Months*, Northwestern University Studies in the Humanities, no. 4. New York: AMS Press, 1938.

Wegener, Paul. *Der Golem* (1914). Film.

———. *Der Golem und die Tänzerin* (1917). Film.

———. *Der Golem, wie er in die Welt kam* (1920). Film.

———. Zensurkarte Prüf-Nr. 29156 (Berlin, 1931), 3, in Golem file: 4359. Stiftung Deutsche Kinemathek, Berlin. Schriftgutarchiv.

———. 'Die künstlerlischen Möglichkeiten des Films'. In Rudolph S. Joseph, *Paul Wegener. Der Regisseur und Schauspieler*, pp. 6–8. Munich: Photo- u. Filmmuseum im Münchner Stadtmuseum, 1967.

———, and Henrik Galeen. 'Golem I Drehbuch'. 1914. Stiftung Deutsche Kinemathek, Berlin. NachlassArchiv. Golem file: 4359. Folder 4.4-80/18,1.

Werblowsky, R. J. Zvi. 'Mystical and Magical Contemplation: The Kabbalists in Sixteenth-Century Safed'. *History of Religions* 1, no. 1 (1961): pp. 9–36.

Weston, P. *Mysterium Artorius: Arthurian Grail Glastonbury Studies, An Introduction and Invocation*. Glastonbury: Avalon Aeon, 2007.

White, Gavin. *Star-Lore: An Illustrated Guide to the Star-lore and Constellations of Ancient Babylonia*. London: Solaria Publications, 2008.

White, John. *Art and Architecture in Italy 1250–1400*, Pelican History of Art. New Haven: Yale University Press, 1993.

———. 'The Reconstruction of Nicola Pisano's Perugia Fountain'. *Journal of the Warburg and Courtauld Institutes* 33 (1970): pp. 70–83.

Wieruszowski, Helene. 'Art and the Commune in the Time of Dante'. *Speculum* 19, no. 1 (1944): pp. 14–33.

Wikander, Ola. 'Solen och pestguden i Ugarit'. *Aorta* 27 (2011): pp. 17–21.

Winnicott, D. W. 'Review of C. G. Jung, *Memories, Dreams, Reflections*'. *International Journal of Psycho-analysis* 45 (1964): pp. 450–55.

Wise, Michael, Abegg Jr., Martin, Cook, Edward, eds. *The Dead Sea Scrolls: A New Translation*. New York: HarperCollins, 1996.

Wixom, William D. 'A Glimpse at the Fountains of the Middle Ages'. *Cleveland Studies in the History of Art* 8 (2003): pp. 6–23.

Wolfson, Elliot R. 'The Theosophy of Shabbetai Donnolo, with Special Emphasis on the Doctrine of *Sefirot* in his *Sefer Ḥakhmoni*'. *Jewish History* 6 (1992): pp. 281–316.

———. 'The Mystical Significance of Torah Study in German Pietism'. *Jewish Quarterly Review* 84 (1993): pp. 43–77.

———. *Through a Speculum That Shines: Vision and Imagination in Medieval Jewish Mysticism*. Princeton, NJ: Princeton University Press, 1994.

———. *Through a Speculum That Shines: Vision and Imagination in Medieval Jewish Mysticism*. Princeton, NJ: Princeton University Press, 1997.

———. *Abraham Abulafia—Kabbalist and Prophet: Hermeneutics, Theosophy, and Theurgy*. Los Angeles: Cherub Press, 2000.

———. 'Text, Context, Pretext: A Review Essay of Yehuda Liebes's "Ars Poetica in Sefer Yetsirah"'. *Studia Philonica Annual* 16 (2004): pp. 218–28.

———. *Venturing Beyond: Law and Morality in Kabbalistic Mysticism*. Oxford: Oxford University Press, 2006.

———. 'Imago Templi and the Meeting of the Two Seas'. *RES: Journal of Anthropology and Aesthetics* 51 (2007): pp. 121–35.

———. 'Review of Moshe Halbertal, *Concealment and Revelation: Esotericism in Jewish Thought and its Philosophical Implications*'. *Journal of Religion in Europe* 2 (2009): pp. 314–18.

———. *A Dream Interpreted Within a Dream: Oneiropoiesis and the Prism of Imagination*. New York: Zone Books, 2011.

Wright, J. Edward. *The Early History of Heaven*. Oxford: Oxford University Press, 2000.

Wyatt, Nicholas. *Religious Texts from Ugarit*, 2nd ed., The Biblical Seminar 53. New York: Sheffield Press, 2002.

Yeats, William Butler. *The Autobiography of William Butler Yeats*. New York: Macmillan, 1953.

Yoshida, Hiromi. *Joyce and Jung: The 'Four Stages of Eroticism' in* A Portrait of the Artist as a Young Man. New York: Peter Lang,

2007.

Xella, Paolo. 'The Omen Texts'. In Wilfred G.E. Watson and Nicolas Wyatt, eds., *Handbook of Ugaritic Studies*, pp. 353–58. Leiden: Brill, 1999.

Zarka, Philippe. 'Astrology and Astronomy'. In David Valls-Gabaud and Alec Boksenberg, eds. *The Role of Astronomy in Society and Culture*, pp. 420–25. International Astronomical Union Symposium 260, UNESCO, Paris, 19–23 January 2008. Cambridge: Cambridge University Press, 2008.

<http://faculty.indwes.edu/bcupp/solarsys/Names.htm>

<http://wildhunt.org/blog/tag/liber-novus>.

<http://www.celtnet.org.uk/gods_c/cocidius.html>

<http://www.levity.com/alchemy/emerald.html>.

<http:planetarynames.wr.usgs.gov>

<http://www.essex.ac.uk/centres/psycho/>

<http://marygreer.wordpress.com/2008/04/18/carl-jung-on-the-major-arcana/>

<http://www.tairis.co.uk/index.php?option=com_content&view=article&id=125:the-dagda-part-1&catid=45:gods&Itemid=8>

INDEX

www.ingramcontent.com/pod-product-compliance
Lightning Source LLC
Chambersburg PA
CBHW072047020426
42334CB00017B/1418
* 9 7 8 1 9 0 7 7 6 7 0 3 6 *